Nuclear IEEE Standards

and American National Standards on Nuclear Instrumentation Published by IEEE

Volume 2

Library of Congress Catalog Number 78-70587

Published by
The Institute of Electrical and Electronics Engineers, Inc

Distributed in cooperation with
Wiley-Interscience, a division of John Wiley & Sons, Inc

How To Use This Book.

The standards contained in this volume are listed in a tab-key on page 3. To find the text of the standard look for the black marking on the right-hand edge of the pages when the book is closed. This marking will be found in the same position as the tab appearing on page 1.

Chronological Presentation of the Standards in this Book

1965	1969	1970	1971	1972	1973	1974	1975	1976	1977	1978
N42.5										
N42.6										
	300									
		309								
			N13.4							
			N42.4							
			279							
			325							
				398	420	308				
						323				
						334				
						382				
						383				
						494				
						N13.10				
							344			
							352			
							380			
							450			
							484			
							498			
								301		
								317		
								415		
								577		
									336	
									338	
									379	
									381	
									384	
									387	
									497	
									566	
									603	
									645	
									N322	
										485
										634
										680
										C37.98
										N323
										N42.13
										N42.14

For current terminology please refer to the most recently approved standards.

ANSI/IEEE STD 383-1974
(ANSI N41.10-1975)

An American National Standard

IEEE Standard for Type Test of Class 1E Electric Cables, Field Splices, and Connections for Nuclear Power Generating Stations

Sponsor

**Nuclear Power Engineering Committee of the
IEEE Power Engineering Society**

Approved April 30, 1975
American National Standards Institute

© Copyright 1974 by

The Institute of Electrical and Electronics Engineers, Inc.

Foreword

(This foreword is not a part of IEEE Std 383-1974, IEEE Standard for Type Test of Class IE Electric Cables, Field Splices, and Connections for Nuclear Power Generating Stations.)

The Institute of Electrical and Electronics Engineers has generated this document to provide guidance for developing a program to type test cables, field splices, and connections and obtain specific type test data. It supplements IEEE Std 323-1974 Standard for Qualifying Class IE Equipment for Nuclear Power Generating Stations, which describes basic requirements for equipment qualification.

Each applicant to the Atomic Energy Commission for a license to operate a nuclear power generating station has the responsibility to assure himself and others that this standard, if used in whole or part, is pertinent to his application and that the integrated performance of his station is adequate.

It is the integrated performance of the structures, fluid systems, the electrical systems, the instrumentation systems of the station, and, in particular, the plant protection system, that limits the consequences of accidents. Seismic effects on installed cable systems are not within the scope of this document.

Section 2 of this guide is an example of type tests. It is the purpose of this guide to deal with cable and connections; however, at the time of issue, detailed examples of tests for connections were not available.

The performance criteria for Class IE service have been expanding in scope during the preparation of this document, and the state of the technology has been continually advancing.

This standard will be revised from time to time to incorporate the latest information available. Topics presently under consideration for future inclusion are: (1) aging correlation procedure, (2) connections, and (3) the corrosive effects from burning cables.

Comments on this document supported by data will be reviewed for later issues.

This standard was prepared by working group 12-32 of the Insulated Conductors' Committee of the IEEE Power Engineering Society. Members of this group were:

Alfred Garshick, *Chairman*

Russ Budrow	John Ferencik	John G. Quin
George S. Buettner	Harry Hilberg	R. E. Sharp
Carmen M. Chiappetta	Alan S. Hintze	Joseph L. Steiner
John Conley	Frank E. La Fetra	H. K. Stolt
John T. Corbett	T. H. Ling	J. R. Tuzinski
Edward Donegan	Reinhold Luther	Chuck vonDamm
Irvine N. Dwyer	E. E. McIlveen	William G. White
	Cutter D. Palmer	

At the time it approved this standard, Subcommittee 2 (Qualification) of the Nuclear Power Engineering Committee of the IEEE Power Engineering Society had the following membership:

A. J. Simmons, *Chairman*   **L. D. Test,** *Vice Chairman*

A. Kaplan, *Secretary*

J. F. Bates	D. J. Meraner	D. G. Woodward
J. T. Bauer	F. W. Chandler	G. T. Dowd, Jr
L. J. Blasiak	H. E. McConnell	W. J. Denkowski
J. T. Keiper	C. E. Corley	R. F. Edwards
F. Campbell	W. Dalos	C. F. Miller
W. J. Foley	E. P. Donegan	A. S. Hintze
J. B. Gardner	W. D. Loftus	W. G. Stiffler
T. H. Ling	W. H. Steigelmann	B. Gregory
T. R. Beans	S. Carfagno	A. Garshick
G. W. Hammond	W. A. Szelistowski	E. E. McIlveen
	F. Trunzo	

At the time it approved this standard, The Nuclear Power Engineering Committee had the following membership:

Contents

An American National Standard

IEEE Standard for Type Test of Class IE Electric Cables, Field Splices, and Connections for Nuclear Power Generating Stations

1. General Provisions

1.1 Scope

1.1.1 This standard provides direction for establishing type tests which may be used in qualifying Class IE electric cables, field splices, and other connections for service in nuclear power generating stations. General guidelines for qualifications are given in IEEE Std 323-1974, Standard for Qualifying Class IE Electric Equipment for Nuclear Power Generating Stations. Categories of cables covered are those used for power control and instrumentation services.

1.1.2 Though intended primarily to pertain to cable for field installation, this guide may also be used for the qualification of internal wiring of manufactured devices.

1.1.3 This guide does not cover cables for service within the reactor vessel.

1.2 Definitions [1]

cable type. A cable type for purposes of qualification testing shall be representative of those cables having the same materials, similar construction, and service rating, as manufactured by a given manufacturer.

Class IE. The safety classification of the electric equipment and systems that are essential to emergency reactor shutdown, containment isolation, reactor core cooling and containment, and reactor heat removal or otherwise are essential in preventing significant release of radioactive material to the environment.

connection. A cable terminal, splice, or hostile environment boundary seal at the interface of cable and equipment.

containment. That portion of the engineered safety features designed to act as the principal barrier, after the reactor system pressure boundary, to prevent the release, even under conditions of a reactor accident, of unacceptable quantities of radioactive material beyond a controlled zone.

design basis events. Postulated abnormal events used in the design to establish the performance requirements of the structures, systems, and components (IEEE Std 323-1974).

field splice. A permanent joining and reinsulating of conductors in the field to meet the service conditions required.

installed life. The interval from installation to removal, during which the equipment or component thereof may be subject to design service conditions and system demands (IEEE Std 323-1974).

NOTE: Equipment may have an installed life of 40 years with certain components of the equipment changed periodically; thus, the installed life of the components would be less than 40 years.

qualified life. The period of time for which satisfactory performance can be demonstrated for a specific set of service conditions. (IEEE Std 323-1974).

type tests. Tests made on one or more units to verify adequacy of design (IEEE Std 380-1972).

1.3 Type Tests As Qualification Method.
As described in IEEE Std 323-1974, type tests are the preferred method to demonstrate or assist in demonstrating that electric equipment is capable of meeting performance requirements under service conditions which include normal and design basis event environments. To perform type tests for cable, field splices, and connections requires: (1) description (identification) of cable, (2) description of significant aspects of the environment, and (3) description of cable performance required. These,

---

[1] Other definitions related to this document may be found in IEEE Std 100-1972, (ANSI C42.100-1972), Dictionary of Electrical and Electronics Terms, IEEE Std 323-1974, and IEEE Std 380-1972, Definitions of Terms Used in IEEE Nuclear Power Generating Stations Standards.

then, with engineering knowledge and experience in insulating materials and systems form a basis for designing type tests to demonstrate the capabilities. Qualification of one cable may permit extrapolation of results to qualify other cables of the same type, with consideration being given to cable dimensions and probable modes of failure.

A sample field splice or connection or both must be type tested with the cable to demonstrate its electrical, mechanical, and chemical compatibility in the environments.

1.3.1 *Cable Description.* This description or specification should include as a minimum:

1.3.1.1 Conductor — material identification, size, stranding, coating.

1.3.1.2 Insulation — material identification, thickness, method of application.

1.3.1.3 Assembly (multiconductor cables only) — number and arrangement of conductors, fillers, binders.

1.3.1.4 Shielding — tapes, extrusions, braids, or others.

1.3.1.5 Covering — jacket or metallic armor or both, material identification, thickness, method of application.

1.3.1.6 Characteristics — voltage and temperature rating (normal and emergency). For instrumentation cables — capacitance, attenuation, characteristic impedance, microphonics, insulation resistance, as applicable.

1.3.1.7 Identification — manufacturer's trade name, catalog number.

1.3.2 *Field Splice or Connection Description or Both.* This description or specification should include as a minimum:

1.3.2.1 Whether factory or field assembled to cable.

1.3.2.2 Conductor connection — type, material identification, and method of assembly.

1.3.2.3 Items from Sections 1.3.1.2 through 1.3.1.7.

1.3.3 *Description of Significant Environmental Conditions.* Both normal operating and design basis event conditions, as well as their sequence and duration, are relevant for type testing. Separate requirements for post design basis event conditions may be required in recognition of momentary or accumulative changes in material properties due to aging, radiation, heat, and steam exposure. Environmental factors, the limits of which may be significant to the cable's operation are as follows:

1.3.3.1 *Atmosphere.* Maximum and average ambient or normal operation condition and design basis event condition or profile for the following:

(1) Gas composition and velocity
(2) Moisture content
(3) Temperature
(4) Pressure

1.3.3.2 *Radiation.*

(1) Normal dose rate and type
(2) Total normal installed life dosage
(3) Design basis event dose rate. Maximum dose rate and approximate profile
(4) Total design basis event dosage
(5) Total for the installed life plus design basis event

1.3.3.3 *Chemicals*

(1) Type of chemicals and concentration
(2) Spray or immersion rate and time
(3) Temperature of exposure

1.3.3.4 *Mechanical.* Normal operating condition and design basis event condition for the following:

(1) Bending or flexing
(2) Vibration
(3) Tension
(4) Sidewall pressure

1.3.3.5 *Fire*

1.3.4 *Operating Requirements*

1.3.4.1 *Meeting Service Conditions.* The cable, as installed, should be suitable for operation at maximum ambient temperature, radiation, and atmospheric conditions and normal electrical and physical stresses for its installed life, as specified. Evidence of this suitability may be based on compliance with appropriate published industry standards, past documented operating experience, component tests, or a combination of these.

The total station may be subdivided into zones with substantially different ambient conditions, and if segregation of cables to certain areas is assured, a cable need only be suitable for meeting service conditions in those zones in which it is located.

1.3.4.2 *Design Basis Event Conditions for Qualifying Cables*

1.3.4.2.1 *Design Basis Event — Loss-of-Coolant Accident (LOCA) (for cables in containment only).* The cable, field splices, and connections should throughout their normal lives be capable of operating through postulated environmental conditions resulting from a LOCA. Conditions of loading and signal levels shall be assumed to be those most unfavorable for cable operation which may be anticipated under such circumstances.

1.3.4.2.2 *Design Basis Event — Fire.* The cable should not propagate fire under conditions of installation.

1.3.4.2.3 *Other Design Basis Events.* These should also be considered in case they represent different types or more severe hazards to cable operation.

1.3.5 *Type Test Conditions and Sequences*

1.3.5.1 *General.* Type tests are used primarily to indicate that the cables, field splices, and connections can perform under the conditions of a design basis event. Because the design basis events may occur at any time in the station life, the thermal and radiation aging required in type tests to simulate these conditions may at the same time indicate the ability of cable types to operate under the normal service conditions within the station.

1.3.5.2 *Aging.* The effect of normal operating conditions with time may either add to or reduce the ability of cable, field splices, and connections to withstand the extreme environments and loads imposed during and following a design basis event. Thus, the type testing for design basis event conditions shall involve both aged and nonaged samples. *Aging* pertains to temperature, radiation, and atmospheric effects applied in sequence or simultaneously in an accelerated manner.

The basis for establishing time and temperature conditions for aging of samples to simulate their qualified life may be that of Arrhenius plotting (IEEE Std 1-1969, General Principles for Temperature Limits in the Rating of Electric Equipment, IEEE Std 98-1972, Guide for the Preparation of Test Procedures for the Thermal Evaluation and Establishment of, Temperature Indices of Solid Electrical Insulating Materials, IEEE Std 99-1970, Guide for the Preparation of Test Procedures for the Thermal Evaluation of Insulation Systems for Electric Equipment, and IEEE Std 101-1972, Guide for Statistical Analysis of Thermal Life Test Data) or other method of proven validity and applicability for the materials in question.

1.3.5.3 *Test Design Basis Event.* Type tests for design basis event conditions should consist of subjecting nonaged and aged cables, field splices, and connections to a sequence of environmental extremes which simulate the most severe postulated conditions of a design basis event and specified conditions of installation. Type tests shall demonstrate margin by application of multiple transients, increased level, or other justifiable means. Satisfactory performance of the cable will be evaluated by electrical and physical measurements appropriate to the type of cable during or following the environmental cycle or both.

The values of pressure, temperature, radiation, chemical concentrations, humidity, and time in Section 2 do not represent acceptable limitations for all nuclear power stations. The user of this guide should assure that the values used in the required type tests represent acceptable limits for the service conditions in which the cable or connections will be installed.

1.4 Documentation

1.4.1 *General.* Type test data used to demonstrate the qualification of cables should be organized in an auditable form. The documentation should include:

1.4.1.1 Description or specification of cable.

1.4.1.2 Description or specification of field splice or connection.

1.4.1.3 Identification of the specific environmental features.

1.4.1.4 Identification of the specific performance requirements to be demonstrated.

1.4.1.5 The test program outline.

1.4.1.6 The test results.

1.4.1.7 Approving signature and date.

1.4.2 *Test Program Outline.* For cable and connections, this outline shall include:

1.4.2.1 The physical arrangement of the cable and test equipment description.

1.4.2.2 Time program and sequence of all environmental factors.

1.4.2.3 The type and location of all environmental and cable monitoring sensors for each variable.

1.4.2.4 The voltages or currents programmed in conjunction with Section 1.4.2.1 above.

1.4.2.5 The electrical, thermal, or mechanical tests to be performed during environmental exposure.

1.4.2.6 Testing or examinations subsequent to environmental cycle.

1.4.3 *Test Results.* Test results should demonstrate that:

1.4.3.1 The intended environmental sequences were achieved.

1.4.3.2 The cable or field splice (or connection) or both was capable of performing its intended function.

1.4.4 *Test Evaluation.* An evaluation of data should be made to demonstrate the adequacy of cable performance as outlined in Section 1.4.1.4.

1.5 Modifications. When modification in the materials or design of cables or in the conditions of installation or in the postulated environments are made, prior type tests shall be reviewed to determine the effect on the cable qualification. This evaluation shall indicate whether or not new type tests are required. The analysis of data and evaluation that demonstrates the effect of the modification on the equipment performance shall be added to the qualification documentation.

2. Examples of Type Tests

2.1 Introduction. Type tests described in this document are examples of methods which may be used to qualify electrical cables, field splices and connections for use in nuclear power generating stations. Tests of the cable or connection assembly, as applicable, should then supplement the cable tests in order to qualify the connections and other aspects unique to planned usage.

The values of pressure, temperature, radiation, chemical concentrations, humidity, and time used do not represent acceptable limits for all nuclear power generating stations. The user of this guide should assure that the values used in the required type tests represent acceptable limits for the service conditions in which the cable or connections, or both will be installed.

Results of prior tests that are being used as the bases for the present tests should be referenced in the documentation.

2.2 Type Test Samples. The samples tested should contain the conductor, insulation, fillers, jacket, binder tape, overall jacket, shielding, and field splices which are representative of the cable category being qualified. Table 1

lists sizes which have been considered representative of these categories. The sample lengths should be sufficient to permit reliable test readings and evaluation consistent with good testing practice.

2.3 Testing to Qualify for Normal Operation

2.3.1 *Temperature and Moisture Resistance.* Evidence of qualification for normal operation may be demonstrated by providing certified evidence that the cable has been manufactured and tested and passed in accordance with the provisions of one or more of the following industry standards or criteria.

ANSI C83.21-1972 Requirements for Solid Dielectric Transmission Lines

ANSI C96.1-1964 (R1969) Temperature Measurement Thermocouples

ANSI C1-1971 National Electrical Code, NFPA 70-1971, Sections on Types RHH, RHW, and XHHW [2]

IPCEA S-19-81 Rubber-Insulated Cable

IPCEA S-66-524 Cross-Linked-Polyethylene-Insulated Cable

IPCEA S-68-516 Interim Standards for Ethylene-Propylene-Rubber-Insulated Wire and Cable. Number 1, Cables Rated 0-35 000 V. Number 2, Cables Rated 2000 V, Integral Insulation and Jacket.

AEIC 5-71 Specifications for Polyethylene and Cross-Linked-Polyethylene-Insulated, Shielded Power Cables rated 5000 – 35 000 V

AEIC 6-73 Specifications for Ethylene-Propylene-Rubber-Insulated Shielded Power Cables Rated 5 – 46 kV

2.3.2 *Long-Term Physical Aging Properties.* Aging data should be submitted to establish long-term performance of the insulation. Data may be evaluated using the Arrhenius technique. A minimum of 3 data points, including 136 °C and two or more others at least 10 °C apart in temperature, should be used.

2.3.3 *Thermal and Radiation Exposure.* The following test sequence may be used to demonstrate that the cable will be operational after exposure to simulated thermal and radiation aging.

---

[2] Cable types RHH, RHW, and XHHW, as specified in the National Electrical Code should meet the requirements established by the applicable standards of Underwriters' Laboratories, Inc or other recognized agencies.

Table 1
Represensentative Cables for Type Tests

Type	Test	Section	Size
Up to 2000 V multiconductor control cable or Shielded multiconductor signal cable (see list below for individual component) or Single conductor power cable	temperature and moisture resistance	2.3.1	1/C — 14 or 12 AWG
	thermal and radiation exposure	2.3.3	1/C or M/C — 14 or 12 AWG
	design basis event simulation	2.4	1/C or M/C — 14 or 12 AWG 1/C — 6, 4 or 2 AWG
	vertical flame test singles from cable assembly	2.5.6	1/C — 14 or 12 AWG
	vertical tray flame test	2.5.4	7/C — 16, 14 or 12 AWG
Shielded pairs, triple or quad from multiconductor signal cable	temperature and moisture resistance	2.3.1	1 pair shielded 16 AWG or actual cable
	thermal and radiation exposure	2.3.3	
	design basis event simulation	2.4	
	vertical flame test	2.5.6	
Coaxial, triaxial or special instrument cable	temperature and moisture resistance	2.3.1	actual size
	thermal and radiation exposure	2.3.3	
	design basis event simulation	2.4	
	vertical flame test singles from cable assembly	2.5.6	
Single pair thermocouple extension cable	temperature and moisture resistance	2.3.1	2/C — 20 AWG or actual size if smaller
	thermal and radiation exposure	2.3.3	
	design basis event simulation	2.4	
	vertical tray flame test	2.5.4	
	vertical flame test singles from cable assembly	2.5.6	
2001—15 000 V power cable 1/C triplexed and multiconductor	vertical tray flame test	2.5.4	6 AWG (2-5kV) 2/O or 4/O or 4/O (2-15kV)

2.3.3.1 Form suitable lengths of insulated conductor which conform to the applicable standards into test coils so that the effective section of each coil under test will be not less than 10 ft.

2.3.3.2 Subject the coils to circulating air oven aging at a temperature and time developed by plotting data using the Arrhenius technique or other method of proven validity to simulated installed life.

2.3.3.3 The specimens with conditioning as covered in Section 2.3.3.2 should be sub-jected in air to gamma radiation from a source such as ^{60}Co to a dosage of 5×10^7 rd at a rate not greater than 1×10^6 rd per hour.

2.3.3.4 After the radiation exposure of Section 2.3.3.3 the specimen should be straightened and recoiled with an inside diameter of approximately 20 times the cable overall diameter and immersed in tap water at room temperature. While still immersed, these specimens should pass a voltage withstand test for 5 minutes at a potential of 80 V/mil ac or 240 V/mil dc.

2.4 Testing for Operation During Design Basis Event

2.4.1 *General.* This section is predicated upon a loss of coolant accident (LOCA) but not necessarily limited thereto.

Prepare two sets of specimens in accordance with the following.

2.4.1.1 One set to be unaged.

2.4.1.2 The other set to be heat aged specimens in accordance with Sections 2.3.3.1 and 2.3.3.2.

NOTE: The requirements of Sections 2.3.3.3 and 2.3.3.4 may be omitted if Section 2.4 is followed as a guide since the requirements of Section 2.4 exceed those of Sections 2.3.3.3 and 2.3.3.4.

2.4.2 *Radiation Exposure — Total.* Exposure specimens to the maximum total cumulative radiation dosage expected over the installed life (see Section 2.3.3.3) plus one LOCA exposure to radiation for the particular installation involved as covered in IEEE Std 323-1974 Appendix A or B. The rate of exposure shall not be greater than 1×10^6 rd per hour. This restriction is removed when simulation of the LOCA profile requires a greater dose rate.

2.4.3 *LOCA Simulation.* Test irradiated specimens in a pressure vessel so constructed that the specimens can be operated under rated voltage and load while simultaneously exposed to the pressure, temperature, humidity and chemical spray of a LOCA event. Chamber designs should have provisions for monitoring and varying temperature and steam pressure, for recycling chemical spray, and for electrically loading the specimens as specified herein.

2.4.3.1 After conditioned specimens are installed inside the pressure vessel they should be energized at rated voltage and loaded with rated service current while under the average normal operating condition. The energized specimens should be exposed to one cycle of the environmental extremes according to the schedule postulated for the particular installation, see IEEE Std 323-1974.

2.4.3.2 The cable should function electrically throughout its exposure to the environmental extremes within the specified electrical parameters.

2.4.4 *Post LOCA Simulation Test.* Upon completion of the LOCA simulation, the specimens should be straightened and recoiled around a metal mandrel with a diameter of approximately 40 times the overall cable diameter and immersed in tap water at room temperature. While still immersed, these specimens should again pass the same voltage withstand test performed under Section 2.3.3.4.

NOTE: The post LOCA simulation test demonstrates an adequate margin of safety by requiring mechanical durability (mandrel bend) following the environmental simulation and is more severe than exposure to two cycles of the environment.

2.5 Flame Tests

2.5.1 *General.* This section describes the method for type testing of grouped cables via the vertical tray flame test to determine their relative ability to resist fire.

2.5.2 *Criteria*

2.5.2.1 The fire test should demonstrate that the cable does not propagate fire even if its outer covering and insulation have been destroyed in the area of flame impingement.

2.5.2.2 The fire test should approximate installed conditions and should provide consistent results.

2.5.3 *Test Specimens*

2.5.3.1 The tests proposed are for power, control, and instrumentation cables.

2.5.3.2 Sizes recommended for type tests may be as listed in Table 1 but not necessarily limited thereto.

2.5.4 *Fire Test Facility and Procedure*

2.5.4.1 Test should be conducted in a naturally ventilated room or enclosure free from excessive drafts and spurious air currents.

2.5.4.2 The vertical tray configuration is recommended as the best arrangement to establish whether or not a cable could propagate a fire. The tray should be a vertical, metal, ladder type, 3 in deep, 12 in wide, and 8 ft long. The tray may be bolted at the bottom to a length of horizontal tray for support.

2.5.4.3 Test sample arrangement — multiple lengths of cable should be arranged in a single layer filling at least the center six inch portion of the tray with a separation of approximately 1/2 the cable diameter between each cable. The test should be conducted 3 times to demonstrate reproducibility using different samples of cable.

2.5.4.4 Flame source, when specified, the procedure detailed below shall be followed:

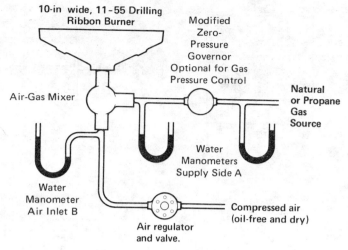

NOTE: All pressures measured under dynamic conditions.

A   Schematic Drawing

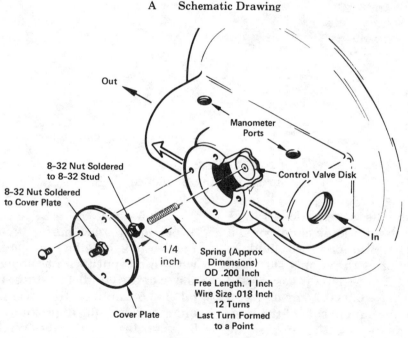

B   Detail Drawing of Zero
Pressure Governor Modification

Fig 1
Flame Source

2.5.4.4.1 The ribbon gas burner[3] shall be mounted horizontally such that the flame impinges on the specimen midway between the tray rungs, and so that the burner face is 3 in behind and approximately 2 ft above the bottom of the vertical tray. Because of its uniform heat content natural grade propane is preferred to commercial gas.

2.5.4.4.2 The flame temperature should be approximately 1500°F when measured by a thermocouple located in the flame close to, but not touching the surface of the test specimens (about 1/8 in spacing).

---

[3] An American Gas Furnace Co 10 in, 11-55 drilling, ribbon type, catalog no 10X 11-55 with an air-gas Venturi mixer, catalog no 14-18 (2 lbf/in² max gauge pressure) is the only presently available model that has been found satisfactory for purposes of these tests.

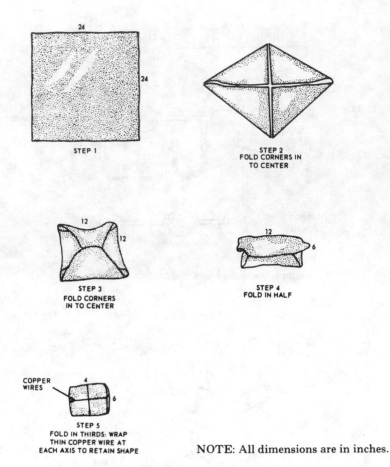

STEP 1

STEP 2
FOLD CORNERS IN
TO CENTER

STEP 3
FOLD CORNERS
IN TO CENTER

STEP 4
FOLD IN HALF

COPPER WIRES

STEP 5
FOLD IN THIRDS: WRAP
THIN COPPER WIRE AT
EACH AXIS TO RETAIN SHAPE

NOTE: All dimensions are in inches.

Fig 2
Burlap Folding Sequence

2.5.4.4.3 For the schematic arrangement see Fig 1. Under dynamic conditions, if propane gas is used the pressure shall be −2.6±0.3 cm of water at the supply side A to the Venturi mixer. If commercial gas is used the pressure shall be −0.9±0.1 cm of water when measured at the supply side of the Venturi mixer. For propane gas, the air pressure should be 4.3±0.5 cm of water. For commercial gas it shall be 5.6±0.5 cm of water, measured at the air inlet B to the mixer. In practice the flame length will be approximately 15 in when measured along its path.

2.5.4.4.4 Gas-burner procedure — ignite the burner and allow it to burn for 20 minutes. Record temperatures at point of impingement throughout the duration of the test, length of time flame continues to burn after gas burner is shut off, jacket char distance, and distance insulation is damaged.

2.5.4.5 Alternative flame source, oil or burlap — when specified, the procedure detailed below shall be followed.

2.5.4.5.1 Use a 24 in square piece of 9 oz per square yard burlap, folded as shown in Fig 2 into a bundle 4 in × 4 in × 6 in. Wrap with fine copper wire as shown, to retain the shape of the bundle. Immerse in a container of oil[4] for 5 minutes. Remove, hang free in air, allow to drain for approximately 15 minutes. The burlap ignitor is weighed before immersion and after draining, and the fuel pickup should be 160±5 g. The repeatability of this test is derived from constant fuel pickup in ignitors of constant size and weight. Temperature should be monitored at point of maximum flame impingement upon the test cables.

2.5.4.5.2 After draining, the ignitor should be placed in front of and approximately 2 ft above the bottom of the tray with the 4 in × 6 in face of the ignitor held in place against the cables by a suitable metal wire or band.

<hr>

[4]Such as Mobilect 33.

2.5.4.5.3 Ignite the oil soaked burlap. The applied flame should be allowed to burn itself out naturally.

2.5.5 *Evaluation.* Cables which propagate the flame and burn the total height of the tray above the flame source fail the test. Cables which self-extinguish when the flame source is removed or burn out pass the test. Cables which continue to burn after the flame source is shut off or burns out should be allowed to burn in order to determine the extent.

2.5.6 *Instrument Cable and Single Conductors from Multiconductor Assembly.* A specimen of each type of instrument cable or the individually insulated or insulated and jacketed conductors removed from each multiconductor control cable which is type tested should pass a flame resistance test in accordance with ASTM D2220-68, Vinyl Chloride Plastic Insulation for Wire and Cable, Section 5 (IPCEA Standard S-19-81, Section 6.19.6), except the weight may be omitted if the specimen is securely clamped.

2.6 Documentation of Type Testing. Following the procedures outlined in this guide, provide data necessary to document satisfactory compliance. Certification of prior test results will be provided when required.

Section
2.3.1   Temperature and Moisture
2.3.2   Long-Term Physical Aging Properties
2.3.3   Thermal and Radiation Exposure
2.4   Testing for Operation During Design Basis Event (LOCA)
2.5.1   Flame Test on Grouped Cables in Vertical Tray
2.5.6   Flame Test on Single Conductor

3. References

IEEE Std 1-1969, General Principles for Temperature Limits in the Rating of Electric Equipment

IEEE Std 98-1972, Guide for the Preparation of Test Procedures for the Thermal Evaluation, and Establishment of Temperature Indices of Solid Electrical Insulating Materials

IEEE Std 99-1970, Guide for the Preparation of Test Procedures for the Thermal Evaluation of Insulation Systems for Electric Equipment

IEEE Std 100-1972 (ANSI C42.100-1972), Dictionary of Electrical and Electronics Terms

IEEE Std 101-1972, Guide for the Statistical Analysis of Thermal Life Test Data

IEEE Std 279-1971 (ANSI N42.7-1972), Criteria for Protection Systems for Nuclear Power Generating Stations

IEEE Std 308-1974, IEEE Standard Criteria for Class IE Power Systems for Nuclear Power Generating Stations

IEEE Std 317-1971, Electrical Penetration Assemblies in Containment Structures for Nuclear Power Generating Stations

IEEE Std 323-1974, Standard for Qualifying Class IE Electric Equipment for Nuclear Power Generating Stations

IEEE Std 334-1971, Type Tests of Continuous Duty Class I Motors Installed Inside the Containment of Nuclear Power Generating Stations

IEEE Std 336-1972 (ANSI N45.2.4-1972), Installation, Inspection, and Testing Requirements for Instrumentation and Electric Equipment During the Construction of Nuclear Power Generating Stations

IEEE Std 380-1972, Definitions of Terms Used in IEEE Nuclear Power Generating Stations Standards

ASTM D2220-68, Vinyl Chloride Plastic Insulation for Wire and Cable

IEEE Standard
Criteria for Independence of
Class 1E Equipment and Circuits

Sponsor

Nuclear Power Engineering Committee
of the
IEEE Power Engineering Society

Foreword

(This foreword is not a part of IEEE Std 384-1977, Criteria for Independence of Class 1E Equipment and Circuits.)

This document provides criteria and requirements for establishing and maintaining the independence of Class 1E equipment and circuits and auxiliary supporting features by physical separation and electrical isolation.

This 1977 revision of IEEE Standard 384-1974, Trial-Use Standard Criteria for Separation of Class 1E Equipment and Circuits, reflects an expansion in scope responsive to various expressed needs to address the following areas of concern:

(1) Electrical independence of Class 1E circuits, equipment, and systems, which includes the concepts of shielding and electrical isolation, in addition to the physical separation.

(2) Separation and other requirements for auxiliary supporting features, including the separation and isolation characteristics of interfacing mechanical and structural plant systems and features required to permit the electrical and physical independence of Class 1E systems to be achieved.

(3) Consideration of the Nuclear Regulatory Commission (NRC) regulatory positions stated in Regulatory Guide 1.75, Revision 1.

(4) Isolation devices; types and application criteria.

(5) Fire caused by ignition sources external to the raceway in addition to internally generated fires. Consideration has been given to NRC Publication NUREG-0050, Recommendations Related to Brown's Ferry Fire, as well as to the Nuclear Engineering Liability Property Insurance Association (NELPIA) recommendations. This aspect has been correlated with work being done on ANSI N18.10, Draft Standard on Generic Requirements for Nuclear Power Plant Fire Protection, IEEE Std 383-1974, Type Test of Class 1E Electric Cables, Field Splices, and Connections for Nuclear Power Generating Stations (ANSI N41.10-1975), and various cable fire tests sponsored by industry, Underwriters' Laboratories, and the NRC.

The distances that are given for separation between trays required to be separated in areas of limited hazard potential are based on current available data from actual cable fire situations and are considered to provide an adequate degree of separation. However, other events such as industrial sabotage, airplane crash, or other special considerations which could require increased separation requirements are beyond the scope of this document.

The Working Group has followed the development of IEEE Std 420-1973, Trial Use Guide for Class 1E Control Switchboards for Nuclear Power Generating Stations, developed by the Nuclear Power Plant Control and Protection Working Group of the Power Generation Committee. The separation requirements herein are considered not to be in conflict with IEEE Std 420-1973 which defines a design that, if followed, would be expected to result in an acceptable installation.

Two areas were identified during final Nuclear Power Engineering Committee (NPEC) approval that will be pursued in future revisions to this document:

(1) Proximity criteria relating to both associated cables and loads comprising the associated circuits.

(2) More definitive criteria relating to energy levels of circuits, specifically low energy instrumentation and control circuits.

The Institute of Electrical and Electronics Engineers (IEEE) has developed these criteria to provide guidance in the determination of the independence requirements related to the Class 1E systems of the nuclear facility. Each applicant for construction permit or an operating license for a nuclear facility in the United States is required to develop these items to comply with the NRC's Code of Federal Regulations, Part 50.

Adherence to these criteria may not suffice for assuring the public health and safety, because it is the integrated performance of the structures, the fluid systems, the instrumentation and the electric systems of the station that establishes the consequences of accidents. Failure to meet these requirements may be an indication of system inadequacy. Each applicant has the responsibility to assure himself and others that this integrated performance is adequate.

The Working Group feels that the criteria herein represent an industry and government consensus for ascertaining the adequacy of the independence of Class 1E systems.

The IEEE will maintain this standard current with the state of the technology. Comments are invited on this standard as well as suggestions for additional material that should be included in future revisions. These should be addressed to:

Secretary
IEEE Standards Board
The Institute of Electrical and
 Electronics Engineers, Inc
345 East 47th Street
New York, NY 10017

The revision to this standard was prepared by Working Group SC1.4 of Subcommittee 1 under the Nuclear Power Engineering Committee (NPEC). The members of Working Group SC1.4 were:

B. M. Rice, *Chairman*

J. H. Boehms	R. Isernhagen
J. V. Bonucchi	T. S. Killen
J. A. Graham	M. R. Lane
A. H. Imagawa	D. R. Lasher

J. W. Wanless

At the time it approved this standard, The Nuclear Power Engineering Committee had the following membership:

T. J. Martin, *Chairman*   **A. J. Simmons,** *Vice Chairman*

L. M. Johsnon, *Secretary*   **R. E. Allen,** *Coordinator*

J. F. Bates	T. J. McGrath
J. T. Bauer	G. M. McHugh
F. D. Baxter	W. C. McKay
J. T. Beard	W. P. Nowicki
R. G. Benham	W. E. O'Neal
J. T. Boettger	E. S. Patterson
D. F. Brosnan	J. R. Penland
F. W. Chandler	D. G. Pitcher
C. M. Chiappetta	H. V. Redgate
N. C. Farr	B. M. Rice
J. M. Gallagher	J. C. Russ
J. B. Gardner	W. F. Sailer
R. I. Hayford	J. H. Smith
T. A. Ippolito	A. J. Spurgin
I. M. Jacobs	L. Stanley
A. Kaplan	W. Steigelmann
R. F. Karlicek	H. K. Stolt
A. Laird	D. F. Sullivan
J. I. Martone	P. Szabados

Contents

IEEE Standard
Criteria for Independence of
Class 1E Equipment and Circuits

1. Scope

The scope of this document is the independence requirements of the circuits and equipment comprising or associated with Class 1E systems. It sets forth criteria for the independence that can be achieved by physical separation and electrical isolation of circuits and equipment which are redundant but does not address the determination of what is to be considered redundant.

2. Purpose

The purpose of this document is to establish the criteria for implementation of the independence requirements of IEEE Std 279-1971, Criteria for Protection Systems for Nuclear Power Generating Stations (ANSI N42. 7-1972), and IEEE Std 308-1974, Criteria for Class 1E Power Systems for Nuclear Power Generating Stations.

3. Definitions

acceptable. Demonstrated to be adequate by the safety analysis of the station.

associated circuits. Non-Class 1E circuits that share power supplies, signal sources, enclosures, or raceways with Class 1E circuits or are not physically separated or electrically isolated from Class 1E circuits by acceptable separation distance, barriers, or isolation devices.

NOTE: Circuits include the interconnecting cabling and the connected loads as defined in IEEE Std 100-1972, Dictionary of Electrical and Electronics Terms (ANSI C42.100-1972).

auxiliary supporting features. Systems or components which provide services (such as cooling, lubrication, and energy supply) which are required for the safety system to accomplish its protective functions.

NOTE: Examples of auxiliary supporting features are ventilation systems for Class 1E switchgear, cooling water systems for Class 1E motors, and fuel oil supply systems for emergency diesel generators.

barrier. A device or structure interposed between Class 1E equipment or circuits and a potential source of damage to limit damage to Class 1E systems to an acceptable level.

Class 1E. The safety classification of the electric equipment and systems that are essential to emergency reactor shutdown, containment and reactor heat removal, or are otherwise essential in preventing significant release of radioactive material to the environment.

design basis events. Postulated events specified by the safety analysis of the station used in the design to establish the acceptable performance requirements of the structures and systems.

division. The designation applied to a given system or set of components that enables the establishment and maintenance of physical, electrical, and functional independence from other redundant sets of components.

NOTE: The terms division, train, channel, separation group, when used in this context, are interchangeable.

enclosure. An identifiable housing such as a cubicle, compartment, terminal box, panel, or enclosed raceway used for electrical equipment or cables.

flame retardant. Capable of limiting the propagation of a fire beyond the area of influence of the energy source that initiated the fire.

hazard. A specified result of a design basis event that could cause unacceptable damage to systems or components important to safety.

isolation device. A device in a circuit which prevents malfunctions in one section of a circuit from causing unacceptable influences in other sections of the circuit or other circuits.

raceway. Any channel that is designed and used expressly for supporting or enclosing wires, cable, or busbars. Raceways consist primarily of, but are not restricted to, cable trays and conduits.

redundant equipment or system. An equipment or system that duplicates the essential function of another equipment or system to the extent that either may perform the required function regardless of the state of operation or failure of the other.

safety class structures. Structures designed to protect Class 1E equipment against the effects of the design basis events.

NOTE: For the purposes of this document, separate safety class structures can be separate rooms in the same building. The rooms may share a common wall.

separation distance. Space without interposing structures, equipment, or materials that could aid in the propagation of fire or that could disable Class 1E systems or equipment.

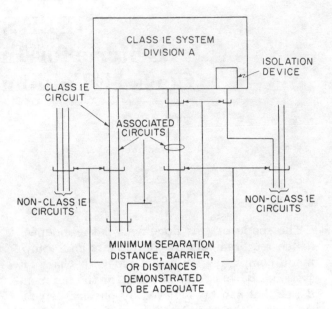

Fig 1
Examples of
Acceptable Circuit Arrangements

4. General Independence Criteria

4.1 Required Independence. Physical separation and electrical isolation shall be provided to maintain the physical and electrical independence of a sufficient number of circuits and equipment so that the protective functions required during and following any design basis event can be accomplished.

4.2 Methods of Achieving Independence. The physical separation of circuits and equipment shall be achieved by the use of safety class structures, distance, or barriers or any combination thereof. Electrical isolation shall be achieved by the use of physical separation, isolation devices, or shielding and wiring techniques.

Fig 1 shows examples of acceptable circuit arrangements.

4.3 Equipment and Circuits Requiring Independence. Equipment and circuits requiring independence shall be determined and delineated

early in the plant design and shall be identified on documents and drawings in a distinctive manner.

4.4 Compatibility with Auxiliary Supporting Features. The independence of Class 1E circuits and equipment shall not be compromised by the functional failure of auxiliary supporting features. For example, an auxiliary supporting feature (such as Class 1E switchgear room ventilation) shall be assigned to the same division as the Class 1E system it is supporting in order to prevent loss of mechanical function in one division from causing loss of electrical function in another division.

4.5 Associated Circuits

4.5.1 *General Criteria.* Associated circuits shall comply with one of the following requirements:

(1) They shall be uniquely identified as such or as Class 1E and shall remain with, or be physically separated the same as, those Class 1E circuits with which they are associated.

(2) They shall be in accordance with (1) above from the Class 1E equipment to and including an isolation device. Beyond the isolation device such a circuit is not subject to the requirements of this document provided it does not again become associated with a Class 1E system.

(3) They shall be analyzed or tested to demonstrate that Class 1E circuits are not degraded below an acceptable level.

NOTE: Preferred power supply circuits from the transmission network and those similar power supply circuits from the unit generator that become associated circuits solely by their connection to the Class 1E distribution system input terminals are exempt from the requirements for associated circuits.

4.5.2 Associated Circuits.
Associated circuits shall be subject to the requirements placed on Class 1E circuits to the extent necessary to assure that the Class 1E circuits are not degraded below an acceptable level. Associated circuit cables shall meet the requirements of IEEE Std 383-1974, Type Test of Class 1E Electric Cables, Field Splices, and Connections for Nuclear Power Generating Stations (ANSI 41.10-1975).

4.6 Non-Class 1E Circuits

4.6.1 General Criteria. The isolation of non-Class 1E circuits from Class 1E circuits or associated circuits shall be achieved by complying with the following requirements:

(1) Non-Class 1E circuits shall be physically separated from Class 1E circuits and associated circuits by the minimum separation requirements specified in 5.1.3, 5.1.4, or 5.6 [except as permitted in 4.6.1 (4)] or they become associated circuits.

(2) Non-Class 1E circuits shall be electrically isolated from Class 1E circuits and associated circuits by the use of isolation devices, shielding and wiring techniques, physical separation, or an appropriate combination thereof as specified in Section 6 [except as permitted in 4.6.1 (4)] or the non-Class 1E circuits become associated circuits.

(3) The effects of lesser separation or the absence of electrical isolation between the non-Class 1E circuits and the Class 1E circuits or associated circuits shall be analyzed to demonstrate that Class 1E circuits are not degraded below an acceptable level or they become associated circuits.

(4) Low energy non-Class 1E instrumentation and control circuits are not required to be physically separated or electrically isolated from associated circuits provided (a) the non-Class 1E circuits are not routed with associated cables of a redundant division, and (b) they are analyzed to demonstrate that Class 1E circuits are not degraded below an acceptable level. As part of the analysis consideration shall be given to potential energy and identification of the circuits involved.

4.7 Effects of Hazards

4.7.1 General. Hazards which could affect Class 1E systems include the effects of fire, failure, or misoperation of mechanical and structural components.

4.7.2 Mechanical Systems. Class 1E circuits shall be routed or protected such that failure of related mechanical equipment of one division cannot disable Class 1E circuits or equipment essential to the performance of the protective function. The effects of failure or misoperation of a mechanical system on its own division shall be considered when the Class 1E circuits or equipment are required to mitigate consequences of such failure or misoperation. The effects of pipe whip, jet impingement, radiation, pressurization, and elevated temperature or humidity on redundant electrical systems caused by failure, misoperation, or operation of mechanical systems shall be considered. The potential generation of missiles resulting from failure of rotating equipment or high energy systems shall be considered.

4.7.3 Non-Category I Structures and Equipment. Independence and redundance of Class 1E systems shall be maintained during and subsequent to failure of non-Category I structures and equipment due to design basis events.

NOTE: The effects of failure of non-Category I structures and equipment on a single division need not be addressed unless that division system is required to mitigate the consequences of such a failure or misoperation.

4.7.4 Fire Protection Systems. In areas where redundant division equipment and circuits must be placed within the area of influence of a fixed fire protection system, the design of the equipment and circuits and the fire protection system shall be coordinated such that the independence of the Class 1E system will not be compromised.

4.7.5 Fire.

4.7.5.1 Hazards Analysis. A fire hazard analysis based on nuclear safety considerations and criteria given in 5.1.7 shall be performed and documented in the initial design phase of the nuclear power plant to ensure the independence requirements of this document are met.

The analysis shall be reviewed and updated

whenever changes are made in design that may affect the results of the previous analysis.

4.7.5.2 *Criterion.* A fire in one Class 1E cable division shall not cause a loss of functions in its redundant cable division.

4.7.5.3 *External Source.* The independence of redundant systems shall not be compromised by externally generated fires. This shall be achieved by physical separation, barriers, or combinations thereof providing that:

(1) Space between redundant wiring divisions does not contain interposing structures, equipment, or materials that could aid in the propagation of the fire.

(2) Barriers when used between divisions shall have a fire rating commensurate with the fire hazard being protected against. Otherwise a 3 hour barrier shall be used.

5. Specific Separation Criteria

5.1 Cables and Raceways
5.1.1 General
5.1.1.1 *Classification of Areas.* The areas through which Class 1E and associated circuit cables are routed and within which equipment served by those Class 1E circuits is located shall be reviewed for the existence of potential hazards such as high energy piping, missiles, flammable material and ignition sources, and flooding. These areas shall be classified as follows:

(1) Cable spreading areas (See 5.1.3)
(2) General plant areas (See 5.1.4)
(3) Hazard areas (See 5.1.5, 5.1.6, 5.1.7)

In each class of area, a degree of physical separation commensurate with the damage potential of the hazard shall be provided such that the independence of redundant Class 1E systems is maintained at an acceptable level. The physical separation of Class 1E circuits and equipment shall make effective use of features inherent in the plant design such as the use of separate rooms. Opposite sides of rooms or areas may be used provided that there is adequate heat removal capability.

Independence of redundant Class 1E systems must be maintained at an acceptable level in hazard areas by cable routing restrictions or by a combination of cable routing restrictions and special physical separation.

5.1.1.2 *Minimum Separation Distances.* The minimum separation distances specified in 5.1.3 and 5.1.4 are based on open ventilated cable trays of either the ladder or the trough type as defined in NEMA VE 1-1971, Cable Tray Systems. Where these distances are used to provide adequate physical separation:

(1) Cables involved shall meet the requirements of IEEE Std 383-1974.

(2) Cable trays and raceways shall be noncombustible per ASTM E-136. Cable trays shall not be filled above the side rails.

(3) Hazards shall be limited to failures or faults internal to the electrical equipment or cables.

If lesser separation distances are used they shall be established as in 5.1.1.3.

5.1.1.3 *Separation By Analysis.* In those areas where the damage potential is limited to failures or faults internal to the electrical equipment or circuits, the minimum separation distance can be established by analysis of the proposed cable installation. This analysis shall be based on tests performed to determine the flame retardant characteristics of the proposed cable installation considering features such as cable insulation and jacket materials, cable tray fill, and cable tray arrangements.

5.1.1.4 *Raceway Fire Stops.* Raceway fire stops through fire barriers shall have a fire resistance rating commensurate with the fire hazards being protected against. Consideration shall be given to the effect on ampacity of cables passing through fire stops.

5.1.2 *Identification.* Exposed Class 1E raceways shall be marked in a distinct permanent manner at intervals not to exceed 15 ft (4.5 m) and at points of entry to and exit from enclosed areas. Class 1E raceways shall be marked prior to the installation of their cables. Cables installed in these raceways shall be marked in a manner of sufficient durability and at intervals of approximately 5 ft (1.5 m) to facilitate initial verification that the installation is in conformance with the separation criteria. These cable markings shall be applied prior to or during installation.

Class 1E cables shall be identified by a permanent marker at each end in accordance with the design drawings or cable schedule.

The method of identification used to meet the above requirements shall readily distinguish between redundant Class 1E systems, between Class 1E and non-Class 1E systems, and between non-Class 1E systems associated with different redundant Class 1E systems without need for frequent consultation of reference material.

5.1.3 *Cable Spreading Area.* The cable spreading area is the space or spaces adjacent to the main control room where instrumentation and control cables converge prior to entering the control, termination, or instrument panels. An area designated as a cable spreading area shall meet the following criteria:

(1) The area shall not contain high energy equipment such as switchgear, transformers, rotating equipment, or potential sources of missiles or pipe whip.

(2) The area shall not be used for storing flammable materials.

(3) Circuits in the area should be limited to control and instrument functions and those power supply circuits and facilities serving the control room and instrument systems and equipment located within the area.

(4) Power supply feeders to instrument and control room distribution panels in these areas shall be installed in enclosed raceways that qualify as barriers.

(5) Power circuits that must traverse this area shall be routed in embedded conduit or in a separate enclosure designed as a safety class structure (for example, a concrete duct bank or other suitable enclosure) which in effect removes them from the area defined as the cable spreading area.

(6) The area shall be bounded by and separated from any other adjacent area by a fire barrier having a fire resistance rating commensurate with the fire hazard that can exist. In lieu of this, a 3 hour barrier shall be used.

(7) The area shall be bounded by and separated from any adjacent pipe hazard area or missile hazard area by a barrier capable of withstanding the design basis hazard at that location.

(8) Administrative control of operations and maintenance activities shall control and limit introduction of hazards into the area.

(9) The minimum separation distance between redundant Class 1E cable trays shall be determined by 5.1.1.3 or, where conditions of 5.1.1.2 are met, shall be 1 ft (0.3 m) between trays separated horizontally and 3 ft (0.9 m) between trays separated vertically.

NOTE: Horizontal separation is measured from the side rail of one tray to the side rail of the adjacent

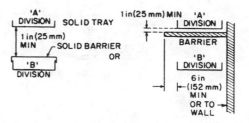

Fig 2
Example of Acceptable Arrangement
Without Consideration of External Hazards
Where Vertical Separation Distance
Cannot be Maintained

tray. Vertical separation is measured from the bottom of the top tray to the top of the side rail of the bottom tray (see also 5.1.4).

(10) Where termination arrangements preclude maintaining the minimum separation distance, the redundant circuits shall be run in enclosed raceways that qualify as barriers or other barriers shall be provided between redundant circuits. The minimum distance between these redundant enclosed raceways and between barriers and raceways shall be 1 in (25 mm). Figs 2 through 5 illustrate examples of acceptable arrangements of barriers and enclosed raceways where the minimum separation distance cannot be maintained.

5.1.4 *General Plant Areas.* The general plant areas are those plant areas from which potential hazards such as missiles, external fires, and pipe whip are excluded, the minimum separation distance between redundant cable trays shall be determined by 5.1.1.3 or, where the conditions of 5.1.1.2 are met, shall be 3 ft (0.9 m) between trays separated horizontally and 5 ft (1.5 m) between trays separated vertically.

If, in addition, high energy electric equipment such as switchgear, transformers, and rotating equipment is excluded and power cables are installed in enclosed raceways that

Fig 3
Example of Acceptable Arrangement
Without Consideration of External Hazards
Where Horizontal Separation Distance
Cannot be Maintained

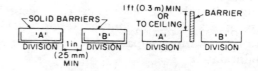

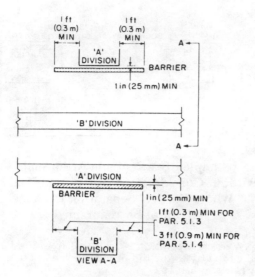

**Fig 4
Example of Acceptable Arrangements
Without Consideration of External Hazards
for Redundant Cable Tray Crossings
Where Vertical Separation Distance
Cannot be Maintained**

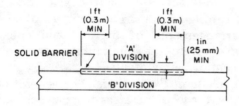

**Fig 5
Example of Acceptable Arrangement Without
Consideration of External Hazards for
Redundant Cable Tray Crossings Where
Vertical Separation Distance
Cannot be Maintained**

qualify as barriers, or there are no power cables, the minimum separation distance may be as specified in 5.1.3 for cable spreading areas.

Where the plant arrangements preclude maintaining the minimum separation distance, the redundant circuits shall be run in enclosed raceways that qualify as barriers or other barriers shall be provided between redundant circuits. The minimum distance between these redundant enclosed raceways and between barriers and raceways shall be 1 in (2.5 cm). Figs 2 through 5 illustrate examples of acceptable arrangements of barriers and enclosed raceways where the minimum separation distance cannot be maintained.

5.1.5 *Pipe Failure Hazard Areas.*

5.1.5.1 *Area Designation.* An area shall be designated a pipe failure hazard area if it contains piping normally operating at high or moderate energies.

For moderate energy piping, pipewhip and jet impingement need not be considered; however, the wetting and environmental effects must be.

5.1.5.2 *Separation Methods.* Separation of nonhazard areas from pipe failure hazard areas shall be accomplished by the use of barriers, restraints, separation distance, or appropriate combination thereof.

5.1.5.3 *Routing Requirements.* The routing of Class 1E or associated cables or raceways in pipe failure hazard areas shall conform to the following requirements unless it can be demonstrated that a pipe failure cannot prevent the Class 1E circuits from performing their protective function:

(1) Where the piping involved is not assignable to a single division, and the pipe failure requires no protective action, Class 1E or associated cables or raceways routed through the area shall be limited to a single division.

(2) Where the pipe failure requires protective action, Class 1E or associated cables or raceways shall not be routed through the area except those which must terminate at devices or loads within the area.

NOTE: Special provisions may be required for these circuits to meet the single failure criteria.

(3) Where the piping involved is assignable to a single division, and the pipe failure requires no protective action, Class 1E or associated cables or raceways routed through the area shall be limited to the same division as the piping.

5.1.6 *Missile Hazard Areas.*

5.1.6.1 *Area Designation.* An area shall be designated a missile hazard area if it contains any missile source having sufficient kinetic energy under design basis event conditions which could damage redundant Class 1E circuits considered as if they were routed in the area and separated as stated in 5.1.4.

Examples of missile sources include:

(1) Rotating equipment

(2) Crane loads

(3) Tensioned cables

(4) Non-Category I structures

(5) Unrestrained pressurized vessels

5.1.6.2 *Separation Methods.* Separation of nonhazard areas from missile hazard areas shall be accomplished by the use of barriers, orientation, separation distance, or appropriate combination thereof.

5.1.6.3 *Routing Requirements.* The routing of Class 1E or associated cables through the same area shall conform to the following requirements:

(1) Where the missile source involved is not assignable to a single division, and the effect of the missile does not require protective action, Class 1E or associated cables or raceways through the area shall be limited to a single division.

(2) Where the effect of the missile source involved requires protective action, Class 1E or associated cables or raceways shall not be routed through the area except those which must terminate at devices or loads within the area.

NOTE: Special provisions may be required for these circuits to meet the single failure criteria.

(3) Where the missile source involved is assignable to a single division and protective action is not required, Class 1E or associated cables or raceways routed through the area shall be limited to the same division as the missile source.

5.1.7 *Fire Hazard Areas.*

5.1.7.1 *Area Designation.* An area shall be designated a fire hazard area if it contains any of the following hazards:

(1) Open flames and welding operations

(2) Liquids which are classified as flammable or combustible per NFPA 321-1973

(3) Solids[1] exhibiting a flame spread classification of 26 or higher per ASTM E84-1975

An area need not be designated a fire hazard area if administrative control provides suppression measures for temporary ignition source use or the introduction of the above hazards is temporary or limited to an acceptable quantity.

5.1.7.2 *Separation Method.* Separation of a nonhazard area from a fire hazard shall be accomplished by the use of fire barriers or separation distance, or both, as follows:

$$SD + 17B \geqslant 50$$

where SD is the separation distance between hazard and non-hazard areas in feet, and B is

---

[1]Excluding cables.

the fire barrier fire resistance rating per ASTM-E-119 in hours (derived from NFPA 80A-1975. See 4.2.4).

Lesser separation may be utilized if test or analysis taking into account the potential duration and intensity of the fire source demonstrates that the effects of lesser separation do not degrade Class 1E circuits below an acceptable level.

5.1.7.3 *Routing Requirements.* The routing of Class 1E or associated cables or raceways in fire hazard areas shall conform to the following requirements:

(1) Where the fire hazard involved is not assignable to a single division, and the fire requires no protective action, Class 1E or associated cables or raceways routed through the area shall be limited to a single division.

(2) Where the effect of the fire hazard requires protective action, Class 1E or associated cables or raceways shall not be routed through the area except those which must terminate at devices or loads within the area.

NOTE: Special provisions may be required for these circuits to meet the single failure criteria.

(3) Where the fire hazard involved is assignable to a single division and protective action is not required, Class 1E or associated cables or raceways routed through the area shall be limited to the same division as the fire hazard.

5.2 Standby Power Supply

5.2.1 *Standby Generating Units.* Redundant Class 1E standby generating units shall be placed in separate safety class structures.

5.2.2 *Auxiliaries and Local Controls.* The auxiliaries and local controls for redundant standby generating units shall be located in the same safety class structure as the unit they serve or be physically separated in accordance with the requirements of Section 4.

5.3 DC System

5.3.1 *Batteries.* Redundant Class 1E batteries shall be placed in separate safety class structures.

5.3.2 *Battery Chargers.* Battery chargers for redundant Class 1E batteries shall be physically separated in accordance with the requirements of Section 4.

5.4 Distribution System

5.4.1 *Switchgear.* Redundant Class 1E distribution switchgear groups shall be physically separated in accordance with the requirements of Section 4.

5.4.2 *Motor Control Centers.* Redundant Class 1E motor control centers shall be physically separated in accordance with the requirements of Section 4.

5.4.3 *Distribution Panels.* Redundant Class 1E distribution panels shall be physically separated in accordance with the requirements of Section 4.

5.5 Containment Electrical Penetrations. Redundant Class 1E containment electrical penetrations shall be physically separated in accordance with the requirements of Section 4. Compliance with Section 4 will generally require that redundant penetrations be widely dispersed around the circumference of the containment. The minimum physical separation for redundant penetrations shall meet the requirements for cables and raceways given in 5.1.4.

Non-Class 1E circuits routed in penetrations containing Class 1E circuits shall be treated as associated circuits in accordance with the requirements of 4.5.

5.6 Control Switchboards

5.6.1 *Location and Arrangement.* Main control switchboards shall be located in a control room within a safety class structure. The control room shall protect the control switchboards from and shall not contain high energy switchgear, transformers, rotating equipment, or potential sources of missiles or pipe whip.

Local control switchboards shall be located so that hazards such as fires, missiles, vibration, pipe whip, and water sprays shall not cause failures that affect redundant Class 1E functions.

Separation of redundant Class 1E equipment and circuits may be achieved by locating them on separate control switchboards physically separated in accordance with the requirements of Section 4. Where operational considerations dictate that redundant Class 1E or Class 1E and non-Class 1E equipment be located on a single control switchboard or cabinet the requirements of 5.6.2, 5.6.3, and 5.6.6 shall apply.

5.6.2 *Internal Separation.* The minimum separation distance between redundant Class 1E equipment and wiring internal to the control switchboards can be established by analysis of the proposed installation. This analysis shall be based on tests performed to determine the flame retardant characteristics of the wiring, wiring materials, equipment, and other materials internal to the control switchboard. Where the control switchboard materials are flame retardant and analysis is not performed, the minimum separation distance shall be 6 in (15.24 cm). In the event the above separation distances are not maintained, barriers shall be installed between redundant Class 1E equipment and wiring.

5.6.3 *Internal Wiring Identification.* Class 1E wire bundles or cables internal to the control boards shall be identified in a distinct permanent manner at a sufficient number of points to readily distinguish between redundant Class 1E wiring and between Class 1E and non-Class 1E wiring.

For a cabinet or compartment containing only Class 1E wiring of a single division no distinctive identification is required.

5.6.4 *Common Terminations.* Where redundant Class 1E wiring is terminated on a common device, the provisions of 5.6.2 shall be met.

5.6.5 *Non-Class 1E Wiring.* Non-Class 1E wiring not separated from Class 1E or associated wiring by the minimum separation distance (determined in 5.6.2) or by a barrier shall become associated circuits and shall be subject to the requirements of 4.5.

5.6.6 *Cable Entrance.* Redundant Class 1E cable entering the control board enclosure shall meet the requirements of 5.1.3.

5.7 Instrumentation Cabinets. Redundant Class 1E instruments shall be located in separate cabinets or compartments of a cabinet complying with the requirements of 5.6.

Where the redundant Class 1E instruments are located in separate compartments of a single cabinet, attention must be given to routing of external cables to the instruments to assure that cable separation is retained.

In locating Class 1E instrument cabinets, attention shall be given to the effects of all pertinent design basis events.

5.8 Sensors and Sensor to Process Connections. Redundant Class 1E sensors and their connections to the process system shall be independent and sufficiently separated that functional capability of the protection system will be maintained despite any single design basis event or result therefrom. If a protective function is required for a design basis event the remaining equipment not damaged by the event must meet the single failure criterion. Consideration shall be given to secondary effects of design basis events such as pipe whip, steam release, radiation, missiles, or flooding.

Large components such as the reactor vessel can be considered a suitable barrier if the sensor to process connecting lines are brought out at widely divergent points and routed so as to keep the component between the redundant lines. Redundant pressure taps located on opposite sides of a large pipe may be considered to be separated by the pipe, but the lines leaving the taps must be protected against damage from a credible common cause unless other redundant or diverse instrumentation not vulnerable to damage from the same cause is provided. The power supply divisional assignment shall be consistent with that of the sensors to assure the maintenance of sensor independence.

5.9 Actuated Equipment. Locations of Class 1E actuated equipment, such as pump drive motors and valve operating motors, are normally dictated by the location of the driven equipment. The resultant locations of this equipment must be reviewed to ensure that separation of redundant Class 1E actuated equipment is acceptable.

6. Specific Electrical Isolation Criteria

6.1 Power Circuits

6.1.1 *General.* Electrical isolation of power circuits shall be achieved by Class 1E isolation devices applied to interconnections of the following kinds of circuits (see Fig 6):

(1) Non-Class 1E and Class 1E circuits

(2) Associated circuits and non-Class 1E circuits

**Fig 6
Examples of Isolation Device
Application in Power Circuits**

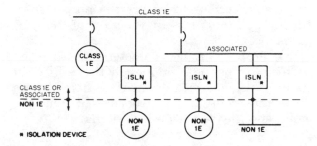

* ISOLATION DEVICE

6.1.2 *Isolation Devices.* A device is considered to be a power circuit isolation device if it is applied such that the maximum credible voltage or current transient applied to one side of the device will not degrade below an acceptable level the operation of the circuit on the other side of that device. The following devices are considered to be acceptable power circuit isolation devices.

6.1.2.1 *Circuit Breaker Tripped by Fault Currents.* A circuit breaker automatically tripped by fault current qualifies as an isolation device provided the following coordination criteria are met:

(1) The breaker time-overcurrent trip characteristic for all circuit faults will cause the breaker to interrupt the fault current prior to initiation of a trip of any upstream breaker. Periodic testing shall demonstrate that the overall coordination scheme remains within the limits specified in the design criteria. This testing may be performed as a series of overlapping tests.

(2) The power source shall supply the necessary fault current for sufficient time to ensure the proper coordination without loss of function of Class 1E loads.

NOTE: For example, diesel generator excitation systems should be capable of providing the required transient current during faults.

6.1.2.2 *Circuit Breaker Tripped by Accident Signals.* A circuit breaker not meeting the requirements of 6.1.2.1 qualifies as an isolation device if it is automatically tripped by an accident signal generated within the same division as that to which the device is applied provided that the time delay involved in generating the accident signal and tripping the breaker does not cause unacceptable degradation of the Class 1E power system.

6.1.2.3 *Input Current Limiters.* Devices which will limit the input current to an acceptable value under faulted conditions of the output qualify as isolation devices. Periodic testing shall verify that the current limiting characteristic has not been compromised or lost.

NOTE: Devices in this category may include inverters, regulating transformers, and battery chargers with current limiting characteristics.

6.1.2.4 *Current Transformers.* Current transformers can be used as isolation devices.

6.2 Instrumentation and Control Circuits

6.2.1 General

6.2.1.1 *Electrical Isolation.* Electrical isolation methods shall be used as required in instrumentation and control circuits to maintain the independence of redundant circuits and equipment such that protective functions required during and following any design basis event can be accomplished. This electrical isolation of instrumentation and control circuits shall be achieved through the use of Class 1E isolation devices applied to interconnections of (a) Class 1E and non-Class 1E circuits, (b) associated circuits and non-Class 1E circuits. (See Fig 7). Shielding and wiring techniques may also be necessary to achieve and maintain the independence of redundant circuits and equipment.

6.2.1.2 *Types of Electrical Interference.* The types of electrical interference that instrumentation and control circuits must be protected against are:

(1) Electrostatic (capacitive) coupling
(2) Electromagnetic induction
(3) Common-mode interference
(4) Radio frequency interference

6.2.1.3 *Methods for Reducing Electrical Interference.* Electrical interference shall be reduced by the use of the following methods as applicable and where necessary:

(1) Filtering circuits
(2) Grounding
(3) Shielding
(4) Physical separation
(5) Wiring techniques

6.2.2 Isolation Devices

6.2.2.1 *General.* A device may be considered to be an electrical isolation device for instrumentation and control circuits if it is applied such that (a) the maximum credible voltage or current transient applied to the device output will not degrade below an acceptable level the operation of the circuit connected to the device input; and (b) shorts, grounds, or open circuits occurring in the output will not degrade below an acceptable level the circuit connected to the device input.

The highest voltage to which the isolation device output wiring is exposed shall determine the minimum voltage level that the device shall withstand between input and output terminals and from these terminals to ground. Transient voltages that may appear in the output circuit must also be considered.

The separation of the wiring at the input and output terminals of the isolation device may be less than 6 in as required in 5.6.2 provided it is not less than the distance between input and output terminals.

Minimum separation requirements do not apply for wiring and components within the isolation device if testing verifies its acceptability; however, separation shall be provided wherever practicable.

The capability of the device to peform its isolation function shall be demonstrated by test.

6.2.2.2 *Acceptable Isolation Devices.* When the requirements of 6.2.2.1 are met, the following devices are considered to be acceptable isolation devices for instrumentation and control circuits:

(1) Amplifiers
(2) Control switches (contact to contact on different stages except between divisions)
(3) Current transformers
(4) Fiber optic couplers
(5) Fuses
(6) Photo-optical couplers
(7) Relays (coil to contact and except between divisions contact to contact)
(8) Transducers

NOTE: In using contact-to-contact isolation, consideration shall be given to the effect on independence that may occur from welding of contacts.

6.2.2.3 *Fuses.* A fuse may be used as an isolation device except between redundant divisions if the following criteria are met:

(1) Each fuse shall be factory tested to verify overcurrent protection as designed.

(2) Fuses shall provide the design overcurrent protection capability for the life of the fuse.

(3) The fuse time-overcurrent trip characteristic for all circuit faults shall cause the fuse to open prior to the initiation of an opening of any upstream interrupting device.

(4) The power source shall supply the necessary fault current to ensure the proper coordination without loss of function of Class 1E loads.

(5) Proper fusing characteristics shall be verified by periodic nondestructive tests such as by resistance measurement to demonstrate that the coordination remains within the limits specified in the design criteria.

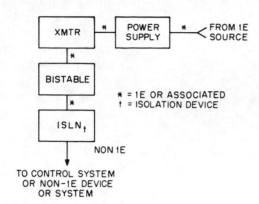

(A) Protection and Controls

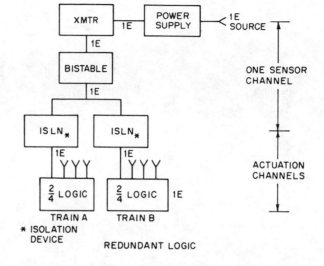

(B) Redundant Logic

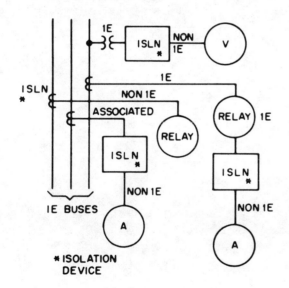

(C) Power and Control or Instrumentation

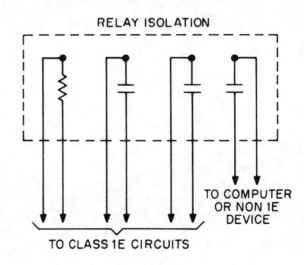

(D) Relay Isolation

Fig 7
Examples of Isolation Device Application
in Control and Instrumentation Circuits

IEEE Standard Criteria for Diesel-Generator Units Applied as Standby Power Supplies for Nuclear Power Generating Stations

Sponsor

**Nuclear Power Engineering Committee of the
IEEE Power Engineering Society**

Foreword

(This foreword is not a part of IEEE Std 387-1977, Criteria for Diesel-Generator Units Applied as Standby Power Supplies for Nuclear Power Generating Stations.)

This document is supplementary to IEEE Std 308-1974, Standard Criteria for Class 1E Power Systems for Nuclear Power Generating Stations, and specifically amplifies paragraph 5.2.4, "Standby Power Supplies," of that document with respect to the application of diesel-generator units.

The IEEE has developed this document to provide the principal design criteria, design features, qualification considerations, and testing requirements for individual diesel-generator units including auxiliary equipment and controls within the scope of this document used in the standby power supply of a nuclear facility, which comply with the Nuclear Regulatory Commission's code of Federal Regulations (10 CFR 50). This document presents specific procedures and criteria applicable to qualifying the diesel-generator unit and supplements the criteria described in IEEE Std 323-1974, "Standard for Qualifying Class 1E Equipment for Nuclear Power Generating Stations."

Operating experience is generally available on diesel-generator units similar to those covered by this document to support the position that when rated in accordance with this document the equipment should provide many years of continuous operating life. However, it should be noted that the application of these standby diesel-generator units for nuclear service is such that the actual operating time under loaded conditions may only be equivalent to one year continuous service in the 40 year life expectancy of the plant.

The principal concerns are the ability of the diesel-generator unit to operate at design loads whenever necessary during the life of the plant, and the ongoing maintenance procedures which should be followed to maintain this equipment in a "ready state." These concerns may be met by following the manufacturer's recommendations and performing periodic tests in accordance with this document and site testing programs.

Components that may deteriorate primarily with age, such as seals, hoses, gaskets, etc, can be replaced long before failure of these components is expected. Such components should be identified and documented, and the required action included as part of the maintenance program.

Engine auxiliaries, including electric motors, may be selected on a conservative basis, with recommendations for replacement periodically during the life of the plant. Replacement interval should be based on conservative judgement of component life supported by operating experience as described in IEEE Std 232-1974. The Generator excitation system, and other electrical equipment should be periodically inspected and tested, including insulation megger or leakage current tests, or both, where practical according to manufacturer's recommendations, and this information should be recorded and compared with previously recorded results to determine if there is any sign of degradation.

It is the intent of the IEEE to add a Section 5.6.4 entitled "Protection" at some future date. In the interim, users of this document are referred to IEEE Std 308-1974 for general requirements covering the area of protection.

Additional work is in progress to expand on the following sections of this document:

(a) Section 5.1.2 (3) and Section 6.4.2 — Light Load or No Load Operation.

(b) Section 5.5 — Design and Application Consideration. (Work is in progress to include interface considerations with associated systems outside the scope of this document.)

(c) Section 6.2 — Factory Production Tests.

(d) Section 6.7 — Preventive Maintenance, Inspection, and Testing.

NOTE: ANS59 is preparing a series of standards applicable to the fuel oil, combustion air, starting, coolant, and lube oil systems of the diesel-generator units, and these standards will be referenced, as appropriate, in future revisions of this document.

Adherence to these criteria may not suffice for assuring the public health and safety because it is the integrated performance of the structures, the fluid systems, the instrumentation systems, and the electrical systems of the station that establishes the consequences of accidents. Each applicant has the responsibility to assure himself and others that this integrated performance is adequate.

Working Group 4.2C had the following membership at the time it prepared this standard:

L.C. Madison, *Chairman*

S. Chakraborty	T. Luke
D.H. Clark	C. Manning
H.W. Falter	G.L. Miller
A.R. Fleischer	C.H. Moeller
L.H. Flisher	F. Rosa
H.M. Hardy	R.L. Spetka
F.H. Lamoureaux	G.J. Taras
M.R. Lane	H.F. Thieme
B.R. Little	D.J. Williams

D.W. Wilson

At the time it approved this standard, the Nuclear Power Engineering Committee had the following membership:

T. J. Martin, *Chairman*　　　　　　　　　　　　　　　**A. J. Simmons,** *Vice Chairman*

L. M. Johnson, *Secretary*

R. E. Allen	R. I. Hayford	J. R. Penland
J. F. Bates	T. A. Ippolito	D. G. Pitcher
J. T. Bauer	I. M. Jacobs	H. V. Redgate
F. D. Baxter	A. Kaplan	B. M. Rice
J. T. Beard	R. F. Karlicek	J. C. Russ
R. G. Benham	A. Laird	W. F. Sailer
J. T. Boettger	J. I. Martone	J. H. Smith
K. J. Brockwell	T. J. McGrath	A. J. Spurgin
D. F. Brosnan	G. M. McHugh	L. Stanley
F. W. Chandler	W. C. McKay	W. Steigelmann
C. M. Chiappetta	W. P. Nowicki	H. K. Stolt
N. C. Farr	W. E. O'Neal	D. F. Sullivan
J. M. Gallagher	E. S. Patterson	P. Szabados
J. B. Gardner		H. A. Thomas

Contents

IEEE Standard Criteria for Diesel-Generator Units Applied as Standby Power Supplies for Nuclear Power Generating Stations

1. Scope

1.1 General. This document applies to the aplication of diesel-generator units as individual units of the standby power supplies in stationary nuclear power generating stations.

1.2 Inclusions. The following are within the scope of this document:

(1) The diesel engine, including:
  (a) the flywheel
  (b) the combustion air system, starting at the engine air intake connection including the affects of any remote air intake filter or silencer, or both
  (c) the starting system
  (d) the starting system energy sources
  (e) the fuel oil system starting at the filters and strainers ahead of the engine fuel oil reservoir
  (f) the lubricating oil system
  (g) the cooling system, starting at the point where the cooling medium is introduced to the diesel-generator unit
  (h) the exhaust system to the downstream side of the exhaust silencer, but excluding piping from the engine exhaust connection to the inlet of the silencer and silencer tail pipe
  (i) the governor system.
(2) The generator, including:
  (a) the main leads stopping at the generator terminals
  (b) the excitation and voltage regulation systems.
(3) The control, protection, and surveillance systems associated with the diesel engine, the generator, and their auxiliary equipment and systems cited above.
(4) The ac and dc distribution systems associated with the diesel engine, the generator, and their auxiliary equipment and systems cited above, exclusive of the auxiliary power system beyond the generator terminals.
(5) Those elements of the unit necessary for maintaining the diesel-generator in a warm standby condition and essential to the safety function.

1.3 Exclusions. The following are outside of the scope of this document:

(1) The diesel-generator unit enclosure and foundations.
(2) The external service equipment and systems which are a part of or which are housed in the diesel-generator unit enclosure, other than those tabulated in 1.2, such as equipment for providing and conveying combustion air, ventilating air, etc, to the vicinity of the diesel-generator unit.
(3) The auxiliary power system beyond the generator terminals of the diesel-generator unit, including:
  (a) the conductors for conveying power from the generator
  (b) the diesel-generator unit main disconnecting and protective device
  (c) the generator circuit instrument transformers, whether furnished with the diesel-generator unit or not
  (d) the generator protective relays.

(4) The control, surveillance, and protection systems for:
  (a) initiating the "Start Diesel Signal"
  (b) loading the diesel-generator unit
  (c) protecting the loads energized by the diesel-generator unit
  (d) disconnecting the loads energized by the diesel-generator unit
  (e) prevention of common-mode failure between the preferred power supply and the standby power supply.
(5) Determination of the characteristics of the service environment.

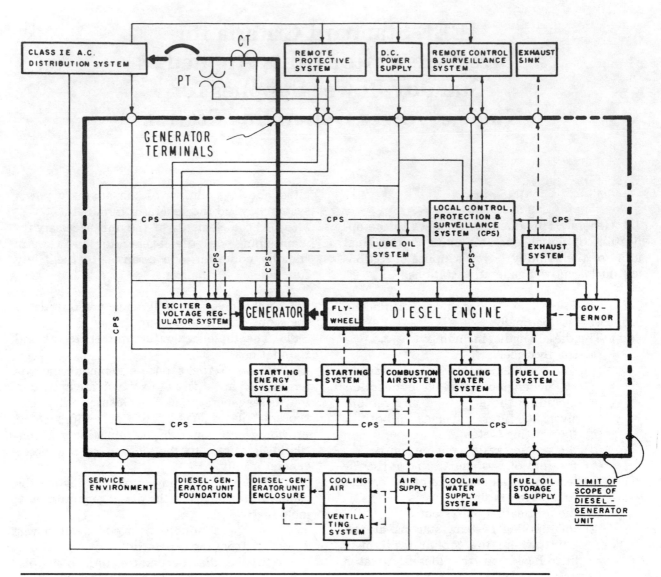

LEGEND:

——— ; ▬▬▬▬   ELECTRIC CIRCUITS (CONTROL, AUXILIARY POWER, ETC; MAIN POWER)

— — — ; ▬ ▬ ▬   NON-ELECTRICAL CHANNELS (OIL, AIR, EXHAUST, ETC; DRIVE)

▬O▬▬O▬   LIMIT OF SCOPE OF DIESEL-GENERATOR UNIT WITH INTERFACE (▬O▬)

CPS   Control Protection and Surveillance Systems

Fig. 1
Scope Diagram

[Abbreviations herein are in accordance with ANSI Y1.1-1972, "Abbreviations for Use on Drawings and in Text." Graphic Symbols and designations are in accordance with IEEE Std 315-1975 (ANSI Y32.2-1975), "Graphic Symbols for Electrical and Electronics Diagrams" (CSA Z99-1975).]

1.4 Scope Diagram.

1.4 Scope Diagram. The scope diagram presented in Fig. 1 illustrates the delineation of the scope that is stated above.

2. Purpose

The purpose of this document is to provide the principal design criteria, the design features, the qualification considerations, and the testing requirements for the individual diesel-generator units which enable them to meet their functional requirements as a part of the standby power supply under the conditions produced by the design basis events catalogued in the Plant Safety Analysis.

3. Definitions

3.1 acceptable. Demonstrated to be adequate by the safety analysis of the plant.

3.2 common failure mode. A mechanism by which a single design basis event can cause redundant equipment to be inoperable.

3.3 design basis events. Postulated events used in the design to establish the performance requirements of the structures and systems.

3.4 design load. That combination of electric loads, having the most severe power demand characteristic, which is provided with electric energy from a diesel-generator unit for the operation of engineered safety features and other systems required during and following shutdown of the reactor.

3.5 diesel-generator unit. The assembly or aggregate of assemblies of one or more single or multiple diesel-engine generators, associated auxiliary systems and control, surveillance, and protection systems that make up an individual unit of a diesel-generator standby power supply.

3.6 preferred power supply. That power supply that is preferred to furnish electric energy under accident or post-accident conditions.

3.7 rating of diesel-generator unit.

3.7.1 *continuous rating.* The electric power output capability that the diesel-generator unit can maintain in the service environment for 8760 h of operation per (common) year with only scheduled outages for maintenance.

3.7.2 *short time rating.* The electric power output capability that the diesel-generator unit can maintain in the service environment for 2 h in any 24-h period, without exceeding the manufacturer's design limits and without reducing the maintenance interval established for the continuous rating.

NOTE: Operation at this rating does not limit the use of the diesel-generator unit at its continuous rating.

3.8 redundant equipment or system. An equipment or system that duplicates the essential function of another equipment or system to the extent that either may perform the required function regardless of the state of operation or failure of the other.

3.9 service environment. The aggregate of conditions surrounding the diesel-generator unit in the diesel generator unit enclosure, while serving the design load during normal, accident, and post-accident operation.

3.10 standby power supply. The power supply that is selected to furnish electric energy when the preferred power supply is not available.

3.11 start diesel signal. That input signal to the diesel-generator unit start logic which initiates a diesel-generator unit start sequence.

3.12 surveillance. The determination of the state or condition of a system or subsystem.

3.13 qualified diesel generator unit. A diesel-generator unit that meets the qualification requirements of this document.

4. Reference Standards

4.1 Standards. The equipment and accessories of the diesel-generator unit shall conform to the applicable portions of the following standards and the latest revisions thereof, as of the date of approval of this document.[1]

[1] ANSI C50.5-1955, Rotating Exciters for Synchronous Machines.

[2] ANSI C50.10-1977, General Requirements for Synchronous Machines.

[3] ANSI C50.12-1965, Requirements for Salient Pole Synchronous Generators and Condensers.

[4] API Std 650, Welded Steel Tanks for Oil Storage.

[5] DEMA, Standard Practices for Low and Medium Speed Stationary Diesel and Gas Engines.

[6] IEEE Std 115-1965, Test Procedures for Synchronous Machines.

[7] IEEE Std 308-1974, Standard Criteria for Class 1E Power Systems for Nuclear Power Generating Stations.

[8] IEEE Std 323-1974, Standard for Qualifying Class 1E Equipment for Nuclear Power Generating Stations.

NOTE: The requirements for qualification stated in IEEE Std 387-1977, Criteria for Diesel-Generator Units Applied as Standby Power Supplies for Nuclear Power Generating Stations, are based on an interpretation of IEEE Std 323-1974, as applicable to these diesel-generator units.

[9] NEMA MG-1-1972, Motors and Generators.

[10] NFPA No 37, Standard for the Installation and Use of Stationary Combustion Engines and Gas Turbines.

[11] TEMA, Standards of Tubular Exchanger Manufacturers' Association.

[12] IEEE Std 344-1975, Recommended Practices for Seismic Qualification of Class 1E Equipment for Nuclear Power Generating Stations.

---

[1]*Legend for Standards Organization:*

ANSI	—	American National Standards Institute
API	—	American Petroleum Institute
DEMA	—	Diesel Engine Manufacturers' Association
IEEE	—	Institute of Electrical and Electronic's Engineers
NEMA	—	National Electrical Manufacturers' Association
NFPA	—	National Fire Protection Association
TEMA	—	Tubular Exchanger Manufacturers' Association

4.2 Conflicts. Where conflicts occur between this standard and any reference standards listed in Subsection 4.1, the provisions set forth herein shall govern.

5. Principal Design Criteria

5.1 Capability.

5.1.1 *General.* When in service, each diesel-generator unit shall have the capability of performing as a redundant unit of a standby power supply, in accordance with the requirements stated in IEEE Std 308-1974 [7].

5.1.2 *Mechanical and Electrical Capabilities.* The diesel-generator unit shall also have each of the following specific capabilities:

(1) *Service Environment.* Operation in its service environment during and after any design basis event, without support from the preferred power supply.

(2) *Starting and Loading.* Starting, accelerating, and being loaded with the design load, within an acceptable time

(a) from the normal standby condition

(b) with no cooling available, for a time equivalent to that required to bring the cooling equipment into service with energy from the diesel-generator unit

(c) on a restart with an initial engine temperature equal to the continuous rating full-load engine temperature.

(3) *Light Load or No Load Operation.* Accepting design load following operation at light load or no load for an acceptable time.

(4) *Design Load Profile.* Carrying the design load for an acceptable duration of time.

(5) *Quality of Power.* Maintaining voltage and frequency at the generator terminals within limits that will not degrade the performance of any of the loads comprising the design load below their minimum requirements, including the duration of transients caused by load application or load removal.

5.2 Ratings.

5.2.1 *General.* The diesel-generator unit shall have continuous and short-time ratings which shall reflect the output capabilities of the diesel-generator unit as constrained by the capability requirements of 5.1 and the application rules of 5.2.2 and 5.2.3.

5.2.2 *General Application Rules* (See 3.7).

Rule 1. Inspections and scheduled maintenance shall be performed periodically using the manufacturer's recommendations and procedures or operating experience, or both.

Rule 2. Unscheduled maintenance shall be performed in accordance with need as indicated by the periodic inspections as suggested by the manufacturer's recommendations or as based on operating experience, or both.

5.2.3 *Operation Application Rule* (See 3.7.1 and 3.7.2). The diesel-generator units may be utilized to the limit of their power capabilities as defined by the continuous and short time ratings.

5.3 Interactions. Mechanical and electric system interactions between a particular diesel-generator unit and other units of the standby power supply, the nuclear plant, the conventional plant, and the Class 1E electric system shall be coordinated in such a way that the diesel-generator units design function, and capability requirements of 5.1, may be realized for any design basis event, except failure of that diesel-generator unit.

5.4 Qualification. The design of the diesel-generator unit for application as part of the standby power supply and requiring the capabilities listed in 5.1 shall be qualified in accordance with IEEE Std 323-1974 [8] and 6.3, Type Qualification Testing, based on the following considerations:

5.4.1 The effect of aging components on the capability to perform in accordance with Subsection 5.1 may be established by previous operating experience and a program for periodic preventive maintenance, inspection, testing, and parts replacement in accordance with 6.6 and 6.7, to be conducted throughout the operating life of the plant.

5.4.2 Major changes to a qualified engine such as differences in the number of cylinders, changes in stroke or bore, brake mean effective pressure, speed, or diesel-generator arrangement in unique or different configuration, shall be requalified in accordance with 6.3.

5.4.3 Modifications to a qualified diesel-gen-

erator unit such as governor, generator, overall system $W_R{}^2$, excitation characteristics, and other accessories/auxiliaries that may change the capability or performance of a previously qualified engine-generator unit shall be qualified by analysis or further testing, or both.

5.4.4 Minor changes to a previously qualified engine-generator unit, such as component parts replacement, shall be qualified as follows:

(a) When replacement does not alter the original design, qualification shall be by analysis or testing, or both, in accordance with 6.6.

(b) When replacement alters the original design, qualification shall be by analysis or 6.3.1, Load Capability Qualification, and 6.3.3, Margin Qualification, or all.

(c) When replacement degrades the engine starting or load acceptance capability, qualification shall be in accordance with all requirements stated under 6.3.

5.5 Design and Application Considerations. Design and application considerations shall include but not necessarily be limited to the considerations listed in Table 1.

5.6 Design Features.

5.6.1 *Mechanical and Electrical Design Features*.

5.6.1.1 *Vibration.* Harmful vibration stresses shall not occur during acceleration or deceleration.

5.6.1.2 *Torsional Vibration.* Harmful torsional vibration stresses shall not occur within a range from 10 percent above to 10 percent below rated idle speed and from 5 percent above to 5 percent below rated synchronous speed.

5.6.1.3 *Overspeed.* Moving parts shall be designed to withstand, without damage, that level of overspeed that is caused by the following:

(1) Full short-time load rejection; plus

(2) Margin to allow the overspeed device to be set sufficiently high to guarantee that the unit will not trip on full short-time load rejection.

(3) As a minimum, the generator, exciter, and flywheel shall be designed to withstand an overspeed of 25 percent without damage.

5.6.1.4 *Governor Operation.* If the diesel engine is equipped to operate in either the isochronous or the droop mode, provisions shall be included to automatically place the engine governor in an acceptable mode of operation when the diesel-generator unit is required to operate automatically.

5.6.1.5 *Voltage Regulator Operation.* If the voltage regulator is equipped to operate in either the paralleled or nonparalleled mode, provisions shall be included to automatically place the voltage regulator in an acceptable mode of operation when the diesel-generator unit is required to operate automatically.

5.6.2 *Control.*

5.6.2.1 *Control Modes.* The diesel-generator unit shall be provided with control systems permitting automatic and manual control.

5.6.2.2 *Automatic Control.* Upon receipt of a start-diesel signal the automatic control system shall provide automatic startup and automatic adjustment of speed and voltage to a ready-to-load condition.

(1) A start-diesel signal shall override all other operating modes and return control of

Table 1
Design and Application Considerations

Consideration	Of Intereset to	
	User/ Designer	Manu-fac-turer
1. Common failure mode between units of the standby power supply	x	
2. Single failure criterion as applied to the standby power supply	x	
3. Matching of diesel engine, alternator, excitation system, and voltage regulator		x
4. Energy for operation of the control, surveillance, and protection systems	x	x
5. Control, surveillance, and protection systems	x	x
6. Lubrication system and equipment		x
7. Selection of air, water, or other means of cooling	x	
8. Supply of cooling medium	x	
9. Cooling system and equipment	x	x
10. Selection of electric, pneumatic, or other means of starting	x	
11. Supply of starting energy	x	x
12. Starting system and equipment		x
13. Supply of combustion air	x	
14. Combustion air system and equipment	x	x
15. Supply of fuel	x	x
16. Fuel supply system and equipment	x	x
17. Removal of products of combustion	x	x
18. Equipment design life	x	x
19. Service environment	x	x
20. Seismic design	x	x
21. Design load	x	x
22. Time available between receipt of start diesel signal and initiation of load sequence	x	
23. Description of loading sequence with time durations of application of individual loads	x	
24. Maximum time available between receipt of start diesel signal and acceptance of design load	x	
25. Accommodation of loading sequence and time duration for application of individual loads		x
26. Load performance characteristics	x	x
27. Continuous rating	x	x
28. Short time rating	x	x
29. Light load and no load operation	x	x
30. Diesel-generator unit performance characteristics		x
31. Electric fault conditions	x	x
32. Electric transients	x	x
33. Insulation and temperature rating of electric equipment insulation systems for operating and quiescent conditions		x
34. Creepage and clearance distances for electric equipment contacts		x
35. Electrically induced thermal effects		x
36. Mechanically induced thermal effects		x
37. Thermal shock		x
38. Mechanical shock		x
39. Operating cycles that may cuse thermally induced stresses	x	x
40. Physical configuration and mechanical support of attached auxiliaries, accessories, hardware, piping, wire and cable and raceways		x
41. Handling during manufacture, shipping, storage, and installation	x	x
42. Fire protection system	x	x

the diesel-generator unit to the automatic control system.

(2) A start-diesel signal shall not override any manual nonoperating modes such as those for repair and maintenance.

5.6.2.3 *Control Points.* Provisions shall be made for control from the control room and external to the control room.

5.6.3 *Surveillance.*

5.6.3.1 *Surveillance Systems.* The diesel-generator unit shall be provided with surveillance systems permitting romote and local surveillance and to indicate the occurence of abnormal, pretrip, or trip conditions.

5.6.3.2 *Modes Surveyed.* As a minimum the following conditions of operation shall be surveyed:

(1) unit not running
(2) unit running — not loaded
(3) unit running — loaded
(4) unit out of service.

5.6.3.3 *Surveillance Instrumentation.* The following systems shall have sufficient mechanical and electric instrumentation to survey the variables required for successful operation and to generate the abnormal, pretrip, and trip signals required for alarm of such conditions:

(1) starting system
(2) lubricating system
(3) fuel system
(4) primary cooling system
(5) secondary cooling system
(6) combustion air system
(7) exhaust system
(8) generator
(9) excitation system
(10) voltage regulation system
(11) governor system
(12) auxiliary electric system.

6. Requirements for Testing and Analyses

6.1 General.

6.1.1 *Implementation.* The requirements of Section 6 shall be implemented in accordance with a written test plan which shall be consistent with the reference standards listed in Section 4 of this document.

6.1.2 *Break-in Run.* Break-in runs shall be performed on each new diesel-generator unit for the length of time required to pass through the initial failure period of the unit. The break-in run on the diesel may be performed before the diesel is assembled to the generator.

Length of time shall be based upon previous operating and test experience of the manufacturer.

6.1.3 *Service Environment.* Results of tests shall be corrected to the condition of the service environment including site exhaust muffler and air intake air filter-silencer systems.

6.1.4 *Documentation.* Tests shall be completely documented, including records of failures, their repair, and retesting. *Type Qualification* test data shall contain the following:

(1) The equipment performance specifications.

(2) Identification of the specific feature(s) to be demonstrated by the test.

(3) Test plan.

(4) Report of test results. The report shall include:

(a) objective
(b) equipment tested
(c) description of test facility (test setup), instrumentation used including calibration records reference, and test environment
(d) test procedures
(e) test data and accuracy (results)
(f) summary, conclusions, and recommendations
(g) supporting data
(h) approval signature and date.

6.1.5 *Analyses.* Although testing is preferred, analyses may supplement or be substituted for tests, where testing is not practical, to demonstrate conformance to the criteria stated in Section 5 of this document.

6.1.6 If type qualification tests are performed at the engine manufacturer's or assembler's facilities, and not at the site, the exhaust muffler and intake air filter-silencer normally used for shop tests may be substituted in place of the equipment to be provided for a specific site, since it is not practical to duplicate the air intake and exhaust equipment and piping which will exist at the site, or future sites for which the diesel-generator unit is being qualified.

6.2 Factory Production Tests. The following minimum production tests shall be performed by the equipment manufacturers for each unit:

(1) *Diesel-Engine* — In accordance with manufacturer's standard test procedure.

(2) *Generator* — In accordance with the latest NEMA Publication MG-1-22.50.

(3) *Excitation, Control, and other Accessories/Auxiliaries* — In accordance with manufacturer's production test procedure.

6.3 Type Qualification Testing Procedures and Methods. Diesel-generators of types not previously qualified as a standby power source for nuclear power generating stations shall be subject to a *type qualification* testing program consisting of *load capability qualification, start and load acceptance qualification*, and *margin qualification*. It is preferred that these qualification tests be performed at the engine manufacturer's or assember's factory; however, they may be conducted at the site if certified calibrated instrumentation is provided to measure and record the same functions and characteristics normally measured under factory testing conditions. Qualification tests may be performed on one or more units, although qualification of one unit will qualify like units of that *type* for equal or less severe service. If *start* and *load acceptance qualification* tests (see 6.3.2) are paerformed using more than one identical unit, then each of these units must be tested for *load capability qualification* (see 6.3.1) and *margin qualification* (see 6.3.3).

Type Qualification tests on the complete diesel-generator unit included in scope diagram figure No 1 shall be performed in addition to seismic analysis or seismic testing by the equipment manufacturers in accordance with IEEE Std 344-1975 [12].

Type Qualification tests shall be performed following successful completion of the diesel break-in run, and the Factory Production Tests.

Following the successful completion of these *type qualification* tests, the equipment shall be inspected in accordance with the manufacturer's standard procedure, and inspection results shall be documented.

6.3.1 *Load Capability Qualification.* This test is to demonstrate the capability of the diesel-generator set to carry the following rated loads at rated power factor for the period of time indicated, and to successfully reject rated load in accordance with 6.4.5. One successful completion of the test sequence shall satisfy this particular *type qualification* requirement.

(1) Load equal to the continuous rating for the time required to reach engine temperature equilibrium, plus 22 h. The engine temperature equilibrium is defined as jacket water and lube oil temperatures within ±10° F (5½°C) of normal operating temperatures as established by the engine manufacturer.

(2) Immediately following the load in 6.3.1 (1), the rated short-time load shall be applied for a period of 2 h.

(3) The continuous rating load rejection test shall be performed. The load rejection test will be acceptable if the increase in speed of the diesel does not exceed 75 percent of the difference between nominal speed and the overspeed trip set point, or 15 percent above nominal, whichever is lower.

(4) Light load equal to the design basis light loads for the required duration.

6.3.2 *Start and Load Acceptance Qualification.* A series of tests shall be conducted to establish the capability of the diesel-generator unit to start and accept load within the period of time to satisfy the plant design requirement. An acceptable *start* and *load acceptance* test is defined as follows; however, other methods with proper justification may be found equivalent for the level of reliability to be demonstrated:

A total of 300 valid start and loading tests shall be performed with no more than 3 failures allowed. If the 300 tests are spread over more than one unit, each unit shall be started and loaded at least 100 times. Failure of the unit or units to successfully complete this series of tests, as prescribed, will require a review of the system design adequacy, the cause of the failures to be corrected, and the tests continued until 300 valid tests are achieved without exceeding the 3 failures allowed.

The start and load tests shall be conducted as follows:

(1) Engine cranking shall begin upon receipt of the start signal, and the diesel-generator set shall accelerate to specified frequency and voltage within the required time interval.

(2) Immediately following (1), the diesel-generator set shall accept a single step load equal to or greater than 50 percent of the generator nameplate continuous kW rating. Load may be totally resistive, or a combination of

resistive and inductive loads. Voltage and frequency shall stabilize to within specified limits within the required time interval.

(3)   At least 270 of these tests shall be performed with the diesel-generator set initially at "warm standby," based on jacket water and lube-oil temperatures at or below values recommended by the engine manufacturer. After load is applied, the diesel-generator set shall continue to operate until jacket water and lube-oil temperatures are within ±10°F (5½°C) of the normal engine operating temperatures for the corresponding load.

(4)   At least 30 tests shall be performed with the engine initially at normal operating temperature equilibrium defined as jacket water and lube-oil temperature within ±10°F (5½°C) of normal operating temperatures as established by the engine manufacturer for the correseponding load.

(5)   If these tests are performed on more than one unit, the number of starts on each unit at "warm standby" and "normal operating temperature" shall be in proportion to the start tests stated under (3) and (4) above.

If the cause for failure to start or accept load in accordance with the preceding sequence falls under any of the categories listed below, that particular test may be disregarded, and the test sequence resumed without penalty following identification of the cause for the unsuccessful attempt:

(a)   Unsuccessful start attempts which can definitely be attributed to operator error, including setting of alignment control switches, rheostats, potentiometers, or other adjustments that may have been changed inadvertently prior to that particular start test.

(b)   A starting or loading or both tests performed for verification of a scheduled maintenance procedure required during this series of tests. This maintenance procedure shall be defined prior to conducting the *start* and *load acceptance qualification* tests and will then become a part of the normal maintenance schedule after installation.

(c)   Tests performed in the process of troubleshooting (tests performed to verify correction of the problem may be counted as valid tests).

(d)   Successful start attempts which were terminated intentionally without loading.

(e)   Failure of any of the temporary service systems such as dc power source, output circuit breaker, load, interconnecting piping and wiring, and any other temporary setup which will not be part of the permanent installation.

6.3.3   *Margin Qualification.* Tests shall be conducted to demonstrate the diesel-generator set capability to start and carry loads that are greater than the most severe step load change within the plant design loading sequence. These tests may be combined with the *load capability* or *start* and *load acceptance* qualification tests. At least two margin tests shall be performed using either the same or different load arrangement. A margin test load at least 10 percent greater than the most severe single step load within the design load sequence is considered sufficient for the margin test. The frequency and voltage excursions recorded may exceed those values specified for the plant design load. The criteria for *margin qualification* are as follows:

(1)   Demonstrate the ability of the generator and excitation system to accept the most severe electrical load (usually the low power factor, high inrush, starting current to a pump motor) without experiencing instability resulting in generator voltage collapse, or significant evidence of the inability of the voltage to recover.

(2)   Demonstrate that there is sufficient engine torque available to prevent engine stall, and to permit the engine speed to recover, when experiencing the most severe load requirement.

6.4 Site Test Categories.

6.4.1   *Starting Test.*   Starting tests shall demonstrate the capability to attain and stabilize frequency and voltage within the acceptable limits and time.

6.4.2   *Load Acceptance Test.*   Load acceptance tests shall demonstrate the capability to accept the individual loads that make up the design load, in the desired sequence and time duration, and to maintain the voltage and frequency within the acceptable limits.

NOTE: If the diesel-generator unit has a light load or no load operation capability, the load acceptance test sequence shall include considerations of the potential effects on load acceptance following such operation.

6.4.3   *Rated Load Test.*   Rated load tests shall demonstrate the capability of carrying the following loads for the indicated times without exceeding the **manufacturer's** design limits:

(1) A load equal to the continuous rating for a time required to reach a temperature equilibrium plus 1 h.

(2) A load equal to the short time rating for 2 h.

6.4.4 *Design Load Test*. Design load tests shall demonstrate the capability of carrying the design load for a time required to reach a temperature equilibrium plus 1 h, without exceeding the manufacturer's design limits.

6.4.5 *Load Rejection Test*. Load rejection tests shall demonstrate the capability of rejecting the maximum rated load without exceeding speeds or voltages which will cause tripping, mechanical damage, or harmful overstresses.

6.4.6 *Electrical Test*. Electrical tests shall demonstrate that the electrical properties of the generator, excitation system, voltage regulation system, engine governor system, and the control and surveillance systems are acceptable for the intended application.

6.4.7 *Subsystem Test*. Tests shall demonstrate the capability of the control, surveillance, and protection systems to function in accordance with the requirements of the intended application.

6.5 Site Acceptance Testing. After final assembly and preliminary startup testing, each diesel-generator unit shall be tested at the site to demonstrate that the capability of the unit to perform its intended function is acceptable.

6.5.1 *Test Loads*. Loads to be applied, carried, and rejected during site testing shall be the design load auxiliaries located at the station. Equivalent loads may be used if these auxiliaries cannot be operated for testing.

6.5.2 *Test Conduct*. Test loads shall be applied in the sequence and timing specified in the Plant Safety Analysis and shall be carried at least until a steady-state operating temperature is reached.

6.5.3 *Tests*. The tests to be given to the diesel-generator unit shall be as follows:

(1) starting tests
(2) load acceptance tests
(3) rated load tests
(4) design load tests
(5) load rejection tests
(6) electrical tests
(7) subsystem tests.

6.6 Periodic Testing. After being placed in service, the diesel-generator unit shall be tested periodically to demonstrate that the continued capability and availability of the unit to perform its intended function is acceptable.

6.6.1 *Availability Test*. The diesel-generator unit shall be started and loaded at intervals of no longer than 1 month to the capacity recommended by the manufacturer, operating for a period necessary to normalize all operating temperatures in order to demonstrate its continued availability for operation.

6.6.2 *Operational Test*. The diesel-generator unit shall be given one cycle of each of the following tests, at acceptable intervals, to demonstrate its continued capability of performing its required function:

(1) starting test
(2) load acceptance test
(3) design load test
(4) load rejection test
(5) subsystem tests.

6.7 Preventive Maintenance, Inspection, and Testing. Separate preventive maintenance, inspection, and testing programs shall be established for the engine-generator and all supporting systems based on manufacturer's recommendations, including time interval for parts replacement. These procedures shall be supplemented based upon operating experience. Procedures related to maintaining qualification of the unit in accordance with 5.4 shall be made mandatory provisions of this program.

An American National Standard

IEEE Standard Test Procedures for Photomultipliers for Scintillation Counting and Glossary for Scintillation Counting Field

Sponsor

**Nuclear Instruments and Detectors Committee of the
IEEE Nuclear Science Group**

**Secretariat for American National Standards Committee N42
Institute of Electrical and Electronics Engineers**

**Approved April 28, 1972
American National Standards Institute**

Foreword

(This Foreword is not a part of IEEE Std 398-1972, Test Procedures for Photomultipliers for Scintillation Counting and Glossary for Scintillation Counting Field, ANSI N42.9-1972.)

Photomultipliers are extensively used in scintillation counting for the detection and analysis of ionizing radiation. The utilization of these detectors in a variety of technical disciplines have made standard test procedures desirable so that measurements may have the same meaning to all manufacturers and users.

This standard is not intended to imply that all tests and procedures described herein are mandatory for every application, but only that such tests as are carried out on photomultipliers for scintillation and Cerenkov counting should be performed in accordance with the procedures given in this document.

This publication was prepared by the Nuclear Instruments and Detectors Committee of the IEEE Nuclear Science Group:

G. L. Miller, *Chairman*   Louis Costrell, *Secretary*

R. K. Abele	F. S. Goulding	W. G. Spear
J. L. Blankenship	T. R. Kohler	J. H. Trainor
W. L. Brown	H. R. Krall	S. Wagner
R. L. Butenhoff	W. W. Managan	F. J. Walter
J. A. Coleman	H. M. Mann	H. R. Wasson
D. C. Cook	D. E. Persyk	A. L. Whetstone

Project leaders for the development of this IEEE Standard were:

D. E. Persyk   H. R. Krall

At the time it approved this standard, the American National Standards Committee N42 on Radiation Instrumentation had the following personnel:

Louis Costrell, *Chairman*

David C. Cook, *Recording Secretary*   Sava I. Sherr, *Executive Secretary*

Organization Represented	Name of Representative
American Chemical Society	Louis P. Remsberg, Jr
American Conference of Governmental Industrial Hygienists	Jesse Lieberman
American Industrial Hygiene Association	W. H. Ray
American Nuclear Society	W. C. Lipinski
	Thomas Mulcahey (*Alt*)
American Society of Mechanical Engineers	R. C. Austin
American Society of Safety Engineers	*Representation Vacant*
American Society for Testing and Materials	John L. Kuranz
	Jack Bystrom (*Alt*)
Atomic Industrial Forum	*Representation Vacant*
Electric Light and Power Group	G. S. Keeley
	G. A. Olson (*Alt*)
Health Physics Society	J. B. Horner Kuper
	Robert L. Butenhoff (*Alt*)
Institute of Electrical and Electronics Engineers	Louis Costrell
	Lester Kornblith, Jr
	P. J. Spurgin
	J. Forster (*Alt*)
Instrument Society of America	M. T. Slind
	J. E. Kaveckis (*Alt*)
Manufacturing Chemists Association	Mont G. Mason
National Electrical Manufacturers Association	Theodore Hamburger
Oak Ridge National Laboratory	Frank W. Manning
Scientific Apparatus Makers Association	Robert Breen
Underwriters' Laboratories	Leonard Horn
U. S. Atomic Energy Commission, Division of Biology and Medicine	Hodge R. Wasson

Contents

An American National Standard

IEEE Standard Test Procedures for Photomultipliers for Scintillation Counting and Glossary for Scintillation Counting Field

1. General

The photomultiplier is an essential component in scintillation and Cerenkov counting. In these applications there are special requirements with regard to pulse-height characteristics, spurious pulses, and timing.

1.1 The Scintillation Counter. The scintillation counter is a radiation detector that consists of three major components: a scintillating medium that produces a flash of light when ionizing radiation interacts with it; one or more photomultipliers, optically coupled to the scintillator, which converts the light flash to an amplified electrical impulse; and associated electronic instrumentation which powers the photomultiplier and processes the output signal.

1.2 The Cerenkov Counter. The Cerenkov counter is a radiation detector that consists of three major parts: a medium in which light is produced by the Cerenkov effect; one or more photomultipliers, optically coupled to the Cerenkov medium; and the associated electronic instrumentation which powers the photomultiplier and processes the output signal.

2. Photomultiplier Characteristics

2.1 General Characteristics. The tests herein described for photomultipliers to be used in scintillation counters are supplementary to those tests described in IEEE Std 158-1962, Methods of Testing Electron Tubes, which covers the following basic characteristics commonly requiring specification for photomultipliers (numbers in brackets refer to section numbers of IEEE Std 158-1962, Part 5):

Radiant sensitivity [2.4]
Uniformity of sensitivity [2.7]
Current amplification [3.2]
Current–voltage relationship [4.2]
Dynamic performance, pulse performance [5.2]
Electrode dark current [6]
Noise [7]
Peak-output-current limitations [9]

2.2 Additional Characteristics Specific to Scintillation and Cerenkov Counters. Additional specifications and tests required for photomultipliers used in scintillation and Cerenkov counting in connection with the preceding basic characteristics are the following:

(1) Pulse-height characteristics
(2) Spurious-pulse characteristics
(3) Pulse-timing characteristics

3. Testing of Photomultiplier Characteristics

3.1 Pulse-Height Characteristics. The following sections deal with pulse characteristics of photomultipliers used in counting applications. Throughout this section it is assumed that the pulse-height analyzer is linear and that channel numbers are with respect to the extrapolated pulse-height analyzer zero.

3.1.1 *Pulse Height.* A photomultiplier consists of a photocathode which produces photoelectrons and an electron multiplier structure which provides gain by secondary electron multiplication. For a given number N of photoelectrons emitted from the photocathode, a number $G \times N$ electrons are observed at the anode. The factor G is the multiplier gain, which depends upon the interstage potential differences applied to the electrodes.

The counting rates and resolving times shall be such that pulse pileup is sufficiently low so as not to significantly affect the accuracy of the results.

When the input consists of photon packets sufficiently well spaced in time, the photomultiplier can resolve them as separate events, and the output signal consists of separate charge pulses. These pulses can be amplified with a charge-sensitive preamplifier, followed by a main amplifier, and sorted with a pulse-height analyzer. The term pulse height is commonly used to designate the charge associated with a PMT (photomultiplier tube) output pulse.

Pulse height is measured with a charge-sensitive preamplifier and a pulse-height analyzer. The system can be calibrated in terms of coulombs per channel with a precision pulser. The slope and zero intercept shall be determined.

3.1.2 *Pulse-Height Resolution.* A photomultiplier produces a charge output proportional to the number of photons incident on the photocathode. Because of the statistical variations inherent in the conversion of photons to photoelectrons, together with the statistical nature of the secondary emission process, the output-signal charge varies from one pulse to the next even for equal numbers of incident photons. The resulting distribution in pulse height limits the photon resolution of the device and hence limits the energy resolution of a scintillator–photomultiplier combination. For this reason a figure of merit called PHR (pulse-height resolution) is introduced to characterize the device's ability to discriminate between slightly different input-signal amplitudes.

PHR is the fractional full width at half maximum of the pulse-height-distribution curve (FWHM/$A1$) of the peak of interest, where $A1$ is the pulse height corresponding to the maximum of the distribution curve. It is customary to state PHR in units of percent. The following discussion outlines several kinds of PHR measurements that are used to characterize photomultipliers.

In general there are five distinct PHR measurements that serve to define the photon-and-electron resolution of photomultipliers and photomultiplier–scintillator combinations.

(1) ^{137}*Cs PHR for a Scintillation-Crystal–PMT Combination.* This PHR is principally a function of the photocathode quantum efficiency and spatial uniformity, as well as the resolution of the scintillation crystal.

(2) *Light-Emitting Diode PHR.* This PHR is obtained with an LED (light-emitting diode) calibrated to provide a ^{137}Cs-equivalent signal (or stated number of photoelectrons per pulse). The LED PHR is numerically smaller than the scintillator–PMT PHR because the contribution of the crystal is not present. (See Fig. 1.)

(3) ^{55}*Fe PHR for a Scintillation-Crystal-PMT Combination.* This is the PHR obtained from a scintillation crystal and an ^{55}Fe source. This PHR depends on both the crystal and the photomultiplier.

(4) *Single-Electron PHR.* The PHR of the SEPHR (single-electron spectrum) has significance only for those photomultipliers that can resolve a single-electron peak. A weak dc light is used as the source of single electrons from the photocathode. (Note that this measures only the resolution of the electron multiplier section of the device.)

(5) *Electron Resolution.* Electron resolution is a measure of the ability of the electron multiplier section of the device to resolve inputs consisting of either one or two electrons. The measurement applies only to those photomultipliers that can resolve one- and two-electron events. As an alternative to measuring the FWHM of the electron peak, the peak-to-valley ratio can be measured. (See Fig. 2.)

Measurement of ^{137}Cs PHR requires a ^{137}Cs source, an NaI(Tl) scintillation crystal of approximately the same diameter as the photocathode, a pulse-height analyzer, and the photomultiplier to be tested. The PMT is optically coupled to the scintillation crystal, for example, with the aid of silicone grease or viscous oil. The crystal housing must be at photocathode potential. The source is placed in contact with the crystal housing.

The PMT should be operated at a voltage such that linear response is obtained; that is,

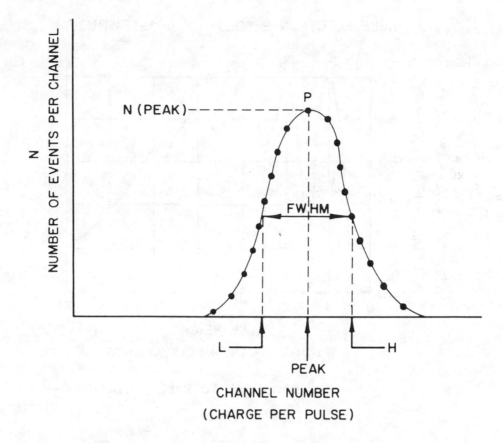

THE VALUES OF H AND L SHOULD BE INTERPOLATED AND
NOT RESTRICTED TO INTEGER CHANNEL VALUES. (H−L)≧8
R=100(H−L)/P IN PERCENT, WHERE : R=PHR
$\qquad$ H=CHANNEL HIGH
$\qquad$ L=CHANNEL LOW
$\qquad$ P=PEAK

Fig 1
^{137}Cs Equivalent LED Spectrum

output is proportional to input intensity. Improper anode bias, excessive gain (and thus excessive anode current), or improper voltage divider circuits may give rise to a compression of the output-pulse distribution, yielding an incorrect (low value) of PHR.

The scintillation-crystal–PMT combination must operate for several hours to obtain optimum PHR.

Phosphorescence of the crystal and PMT faceplate may require several hours to decay to a low enough level to permit accurate measurements to be made. Therefore, photo-

multipliers and crystals should not be exposed to ambient laboratory light for some time before measurements are made.

The test enclosure must be designed to avoid high electric fields in the region of the photocathode. If the PMT is operated at photocathode ground (positive high voltage), there is little problem with external electric fields at the photocathode. If negative high voltage is used, electric fields near the photocathode must be low. This may be accomplished by an electrostatic shield having the same potential as the photocathode. Other-

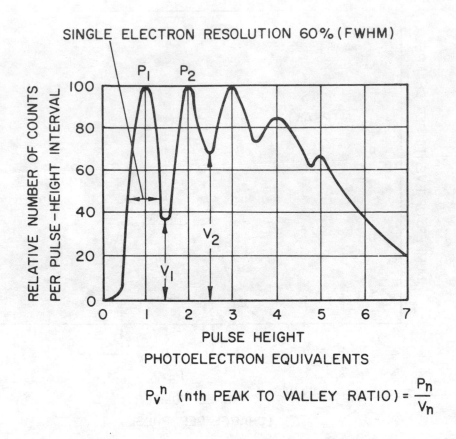

Fig 2
Typical Electron Resolution of Photomultiplier Tube Incorporating GaP First Dynode

wise, excessive noise on the output signal and electrolysis, followed by eventual loss of photosensitivity, may develop. As with other PMT measurements, a magnetic shield is required.

The ^{137}Cs distribution should be displayed on a pulse-height analyzer and so positioned that the upper half of the full-energy-peak distribution spans at least eight channels. At least 50 000 counts must be contained within the channels comprising the FWHM.

The value of PHR, in percent, is obtained from $R = 100\,(H\text{-}L)/P$, where R is the PHR, H and L are the upper and lower channels, respectively, corresponding to the half-amplitude points of the distribution, and P is the channel number (PEAK) corresponding to the peak of the distribution. A linear interpolation should be made to determine the value of H and L. Alternatively, other curve-fitting techniques may be used. The exact

method should be described.

In the case of a computer-controlled pulse-height analyzer, a different method can be employed to determine the FWHM by assuming that the full-energy peak approximates a truncated Gaussian distribution. While the observed distribution is usually slightly skewed and not truly Gaussian, PHR values determined on the basis of a Gaussian distribution will, in general, agree with PHR values obtained from the former method.

For an LED, the PHR should be stated in terms of a ^{137}Cs-equivalent flash; the LED should be adjusted in the flash intensity until the resulting pulse-height distribution exhibits a peak in the same channel as an NaI(Tl) crystal and ^{137}Cs source combination. The LED shall be positioned such that it uniformly illuminates the photocathode. PHR may be calculated from the previously discussed method.

This measurement of PHR may be independently verified with another calibrated LED, and good agreement should be obtained. This type of independent measurement is not as straightforward when NaI(Tl) crystals are used, due to variations in resolution between different crystals. The PHR obtained with a ^{137}Cs-equivalent LED may be used as a guide for estimating the PHR that could be obtained with an NaI(Tl) crystal. However, the LED flash does not simulate the spatial distribution of light that is obtained with an NaI(Tl) crystal and neither does it take account of the inherent crystal resolution, so this method can only provide an approximate limit to the ^{137}Cs resolution that can be expected.

PHR for an [55]Fe (5.9-keV X-ray) source, using an NaI(Tl) crystal coupled to the photomultiplier, provides a figure of merit for resolution of low-energy events. Measurements are made as outlined in the preceding discussion on ^{137}Cs PHR. The effect of crystal resolution is more pronounced in the [55]Fe PHR measurement than in the ^{137}Cs PHR measurement.

The [55]Fe distribution is skewed, and a Gaussian distribution does not adequately describe the observed distribution. For this reason, resolution measurements based on assumptions of Gaussian distributions are not satisfactory. The [55]Fe is placed in contact with the crystal housing.

SEPHR is the fractional FWHM of the single-electron distribution obtained from the photomultiplier. Note that this measurement applies only to those photomultipliers that exhibit a peaked single-electron distribution. (Most venetian blind photomultipliers are unable to resolve single-electron events even though their [55]Fe and ^{137}Cs PHR may be comparable to focused-type electron multipliers.)

The single-electron peak is displayed on the pulse-height analyzer system while dc (continuous) light illuminates the photocathode. The tube dark pulses may be predominantly single electron in origin and may be used to obtain a single-electron distribution; however, the dc light is required to verify the placement of the true single-electron peak on the analyzer display.

SEPHR, in percent, may be calculated from

$$S = 100(H-L)/P \qquad \text{(Eq 1)}$$

where S is the SEPHR, H and L are the upper and lower channels, respectively, corresponding to the half-amplitude points, and P is the peak channel number (PEAK). Linear interpolation should be used to locate H and L.

Some photomultipliers are able to resolve a single-electron peak, but the resulting distribution may not fall to one-half amplitude on the low-energy side due to noise contributions. In such cases, the quantity L in Eq 1 is undefined, and calculation of PHR by means of the preceding method is impossible. A statement of fractional width at some other amplitude in the distribution (75 percent, for instance) could be made, but it should be stressed that this figure is not PHR, which is defined in terms of the fractional FWHM.

Electron multiplier sections of photomultipliers are sometimes operated in saturating mode for counting experiments. Single-electron distributions for saturated-mode operation may be characterized by a fractional FWHM, but the associated value should not be termed PHR, and a statement should accompany the value to indicate that the measurement applies to nonlinear (saturated) operation.

Multiple-electron resolution (see Fig 2) is possible in certain photomultipliers. A multiple-electron display on the pulse-height analyzer can be obtained with a pulsed LED. The pulse-height analyzer should be gated on only during the time interval during which the photomultiplier output pulse is expected.

The intensity of the LED flashes can be varied while observing the distribution accumulating on the pulse-height analyzer display. The single-, double-, and other electron peaks may be built up to any relative height with respect to each other by varying the LED flash intensity.

An alternative to measuring the PHR of the 2nd-, 3rd-, ..., nth-electron peak is to measure the peak-to-valley ratio of adjacent peaks of equal amplitudes. Thus the peak-to-valley ratio could be stated for the one-to-two valley and the two-to-three valley. This measurement is simpler to perform than a calculation of PHR and adequately characterizes multiple-electron resolution.

3.1.3 *Pulse-Height-Stability Measurements.* In scintillation counting it is particularly important that the photomultiplier have very good pulse-height stability. There are two types of pulse-height stability tests which have been used to evaluate photomultipliers for this application: (1) a test of long-term drift in pulse-height amplitude measured at a constant counting rate; and (2) a measure of short-term pulse-height amplitude shift with change in counting rate.

In the time stability test, a pulse-height analyzer, a ^{137}Cs source, and an NaI(Tl) crystal are employed to measure the pulse height. The ^{137}Cs source is located along the major axis of the tube and crystal so that a count rate of about 1000 counts per second is obtained. The entire system is allowed to warm up under operating conditions for a period of ½ to 1 h before readings are recorded. Following this period of stabilization, the pulse height is recorded at 1-h intervals for a period of 16 h. The drift rate D_g is then calculated, in percent, as the mean gain deviation (MGD) of the series of pulse-height measurements as follows:

$$D_g = \frac{\sum\limits_{i=1}^{i=n} |p - p_i|}{n} \cdot \frac{100}{p}$$

where p is the mean pulse height averaged over n readings, p_i is the pulse height at the *ith* reading, and n is the total number of readings. Typical maximum MGD values for photomultipliers with high-stability CuBe dynodes are usually less than 1 percent when measured under the conditions specified above. Gain stability becomes particularly important when photopeaks produced by nuclear disintegrations of nearly equal energy are being differentiated.

In the count-rate stability test, the photomultiplier is first operated at about 10 000 counts per second. The photopeak counting rate is then decreased to approximately 1000 counts per second by increasing the source-to-crystal distance. The photopeak position is measured and compared with the last measurement made at a counting rate of approximately 10 000 per second. The count-rate stability is expressed as the percentage gain shift for the count-rate change. Example: Gain shift of —0.05 percent (10 000 to 1000 counts per second). Photomultipliers designed for pulse-height stability may be expected to have a value of no greater than 1-percent gain shift as measured by this count-rate stability test.

3.1.4 *Pulse-Height Linearity.* Pulse-height linearity in photomultipliers is normally excellent over several orders of magnitude. Deviations from linearity are usually due to the following two effects: (1) space-charge effects associated with high-current pulses; and (2) resistive divider networks which cannot supply enough current to maintain the photomultiplier dynodes and other elements at constant operating potentials. Assuming that item (2) is not the limitation, the following measurement procedures outline two methods of determining the peak linear current that can be obtained from a photomultiplier. The first method is called the two-pulse technique and requires a pulsed light source and a neutral density filter that provides about 90-percent attenuation. The pulsewidth should not exceed 1 μs, and the duty cycle should not exceed 1 percent. The output pulses are displayed on an oscilloscope (or pulse-height analyzer) with and without the filter, and the ratio of the pulse heights is noted. Next the light source intensity is increased, while noting the ratio, until the ratio deviates from that obtained at the low intensity by 10 percent. The peak current associated with the larger pulse is termed peak linear current at the given operating voltage. Note that this measurement allows peak linear current measurement as a function of tube voltage.

A simpler, but less desirable, measurement requires only the pulsed light source and an oscilloscope or pulse-height analyzer. Tube voltage is initially low enough to ensure linear operation. A plot is made of the logarithm of the peak current versus the logarithm of the supply voltage. The peak current at which the curve deviates from an extrapolated straight line by 10 percent is the peak linear current. (This measurement cannot be performed if zener diodes are employed in the voltage divider.) This test only applies to photomultipliers whose gain exhibits a simple power dependence on the applied voltage (see Fig 3).

Another simple and convenient test for the linearity of the entire system (photomultiplier, crystal, amplifier, and pulse-height analyzer) may be made by noting that the ^{60}Co high-energy peak is almost exactly twice the energy of the ^{137}Cs peak.

TWO PULSE METHOD

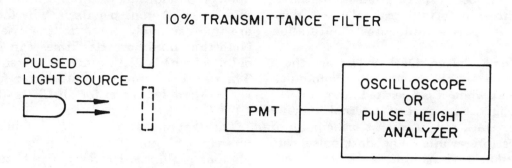

10% TRANSMITTANCE FILTER

PULSED LIGHT SOURCE

PMT

OSCILLOSCOPE OR PULSE HEIGHT ANALYZER

PEAK LINEAR CURRENT IS REACHED WHEN THE PULSE HEIGHT RATIO DEVIATES FROM 10:1 BY 10%

LOGARITHMIC PLOT METHOD

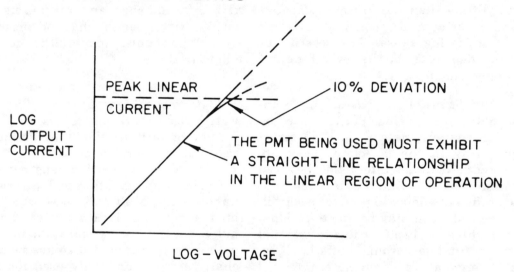

PEAK LINEAR CURRENT

10% DEVIATION

LOG OUTPUT CURRENT

THE PMT BEING USED MUST EXHIBIT A STRAIGHT-LINE RELATIONSHIP IN THE LINEAR REGION OF OPERATION

LOG – VOLTAGE

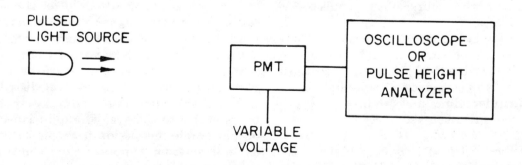

PULSED LIGHT SOURCE

PMT

OSCILLOSCOPE OR PULSE HEIGHT ANALYZER

VARIABLE VOLTAGE

Fig 3
Pulse-Height Linearity

3.2 Spurious-Pulse Characteristics. PMTs exhibit a number of spurious-pulse characteristics. Three important examples are dark pulses, afterpulses, and tube scintillation pulses.

3.2.1 *Dark Pulses*. Dark pulses are those pulses which are observed at the photomultiplier output when it is operated in total darkness and in the absence of external ionizing radiation. Dark pulses consist principally of single-electron events. The dark-pulse rate and distribution in amplitude is dependent upon the device type and operating history, that is, photocathode type and length of time since the last exposure to radiation of infrared or shorter wavelengths.

3.2.2 *Afterpulses*. Afterpulses are spurious output pulses that are time correlated with the signal pulse. Afterpulses are generally of lesser amplitude than the initiating (signal) pulse and occur at nearly constant intervals after the initiating pulse. The severity of afterpulsing depends upon the device type and the past operating history.

3.2.3 *Tube Scintillation Pulses*. Tube scintillation pulses are spurious pulses caused by light flashes that originate in the PMT structure, primarily in the faceplate. They are usually produced by an interaction between luminescent centers within the glass and radiation of sufficient energy to produce a scintillation. The radiation may be due to residual activity in the glass itself, or due to external sources or natural background radiation. Cosmic ray events are a common example of externally induced scintillation pulses. Cerenkov light is another source of these pulses.

3.3 Pulse Timing Characteristics. This section discusses delta-function light sources in order of increasing complexity and cost. In general, no single source is suitable for all photomultiplier time measurements. Applicable sources are discussed in 3.4.

3.3.1 *Light-Emitting Diodes*. The light-emitting diode (LED) is a very inexpensive delta-function light source.[1] Rise time varies from approximately 500 ps to 5 ns, depending upon chip mounting technique, material, and number of photons per flash. The fall times are approximately equal to the rise times. Note that minimum rise times can only be obtained with a delta-function current pulse. The rise time associated with a step function is considerably greater for all LEDs. (See Fig 4.)

The light output can be varied from approximately 1 photon per pulse to over 1×10^5 photons per pulse. However, rise time begins to increase when the intensity exceeds a few hundred photons per pulse.

Pulse generators for LEDs may consist of mercury-switch pulsers or avalanche transistor pulsers using either capacitors or charge lines for energy storage. Repetition rates of 400 Hz for a mercury-switch pulser and 10 kHz for avalanche transistor pulsers are typical. A trigger signal may be derived from either type of pulser to mark the occurrence of a light flash.

3.3.2 *Spark Light Sources*. The mercury-wetted relay spark source has long been used for photomultiplier time measurements. The rise time is about 500 ps, and the fall time is somewhat greater, being lengthened by an exponential-like tail that may persist for a few nanoseconds. Multiple pulses are often generated approximately 1 μs after the primary pulse; these pulses usually do not interfere with measurements. A mercury switch with a charge line is commonly used for the pulse generator, and rates of 400 Hz are typical. A charge line is commonly used for the current source. A trigger signal may be derived to mark the occurrence of the light flash.

Light output may be varied from approximately 1 photon per pulse to over 1×10^5 photons per pulse. The spectral distribution includes the ultraviolet, and the spark appears bluish to the eye. The light is useful in obtaining photoemission from photocathodes as well as from dynodes. Photoemission by spark source excitation can be obtained from BeO and GaP(Cs) dynodes. However, at the higher intensities needed for these measurements, the spark source rise time may increase to over 1 ns, and the fall time may exceed 10 ns. High-pressure spark-gap devices of equivalent performance are also suitable.

---

[1]Suitable diodes include the red-emitting Monsanto MV10A3 (peak emission at 670 nm) diode and the yellow-emitting Ferranti XP20 series diodes. Typical rise times are of the order of 500 ps for light flashes corresponding to less than 100 photons.

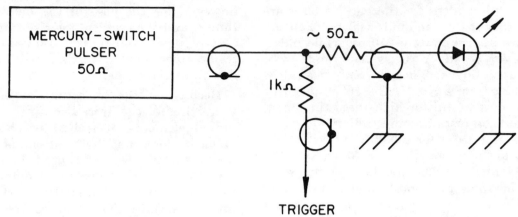

MERCURY-SWITCH
PULSER
50Ω

~ 50Ω

1kΩ

TRIGGER

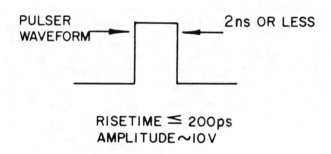

PULSER
WAVEFORM

2ns OR LESS

RISETIME ≤ 200ps
AMPLITUDE ~10V

Fig 4
LED Circuitry

3.3.3 *Cerenkov Source.* The Cerenkov light source relies on energetic particles from a small nuclear radiation source creating Cerenkov light in a high index of refraction medium. A beta source such as ^{90}Sr may be used with a thin piece of plastic (such as Lucite).

The rate is governed by the radiation source intensity, and the flashes are random in time. No electrical trigger is available to mark the occurrence of a flash.

3.3.4 *Scintillator Light Source.* The scintillator light source utilizes a fast scintillator and a nuclear radiation source to provide a light flash with a fall time longer than its rise

time. A typical rise time is 400 ps (for Naton 136 and a ^{60}Co source). The fall time (for Naton 136) is approximately 1.4 ns with a characteristic long tail of several nanoseconds.

The rate of flashing is determined by the source intensity, and the flashes are random in time. No electrical trigger is available to mark the occurrence of a light flash.

3.3.5 *Mode-Locked Laser.* Mode locking is achieved by applying a time-varying intensity or phase perturbation to the continuous-wave light signal within a laser cavity. When the perturbing frequency matches the resonant frequency $C/2L$ of the cavity, where C is the velocity of light and L is the cavity length, the modes are brought into phase, resulting in intense light pulses occurring at a rate $C/2L$.

He–Ne (633 nm), Ar (588 nm and 514.5 nm) and Nd: YAG (106 μm) lasers may be mode locked. Pulsewidths for the He–Ne laser are of the order of 1 ns, and pulsewidths for the Ar laser are approximately 250 ps, while pulsewidths for a Nd: YAG system range from about 1 to 50 ps. The shorter pulsewidths obtainable from a Nd: YAG system make this system most useful for high-speed time measurements on fast photomultipliers.

Repetition rates for a Nd: YAG system are usually in the range of 75–200 MHz. Even the 75-MHz rate may be too fast for measurements on some photomultipliers. Lower repetition rates can be obtained by using fast electrooptical modulators to gate off unwanted pulses. An electrical trigger signal is available to mark the occurrence of each pulse.

The 1.06-μm radiation can be doubled in frequency by employing a nonlinear crystal to obtain pulses at 532 nm (green). These picosecond pulses are very intense (10^8 photons per pulse) and may be used to obtain photoemission from GaP(Cs) dynodes, as well as from photocathodes. The mode-locked laser can be used to probe the surface of a dynode to map its spatial timing characteristics.

3.4 Measurement Techniques

3.4.1 *Rise-Time Measurements.*

3.4.1.1 *Device Rise Time.* Device rise time (DRT) is the mean time difference between the 10- and 90-percent amplitude points on the output waveform for full cathode illumination and delta-function excitation. DRT is measured with a repetitive delta-function light source and a sampling oscilloscope. The trigger signal for the oscilloscope may be derived from the device output pulse, so that light sources such as the scintillator light source may be employed.

The rise time as measured from an oscilloscope photograph must be corrected for the finite rise times of the individual elements comprising the system. DRT is usually calculated from the relation

$$t_d = \sqrt{(t_o)^2 - (t_s)^2 - (t_p)^2 - (t_1)^2 - (t_c)^2}$$

where t_d is the DRT, t_o is the observed rise time, t_s is the source rise time, t_p is the

splitter rise time, t_1 is the line rise time, and t_c is the oscilloscope rise time, under the assumption that the elements behave as Gaussian elements. This assumption is not valid, however, for the case of the delay-line rise time t_1 and the oscilloscope rise time, t_c. If a reactive signal splitter is employed (such as a transformer-coupled TEE), it will not behave as a Gaussian element.

Because of the uncertainties involved in correcting for the finite rise times of the individual elements, it is best to choose these elements such that their individual rise times do not exceed one third that of the DRT. The rise times of the elements comprising the measurement system shall be stated. Example: GaAsP LED light source with rise time less than 600 ps, signal-delay-line rise time of 100 ps as measured with a time-domain reflectometer, sampling oscilloscope rise time of 50 ps. A time-domain reflectometer should be employed to ensure that the system, including photomultiplier socket, exhibits minimum rise-time degradation. A time-domain reflectometer is a device that characterizes the impedance of a transmission line by sending a pulse down the line and observing the reflected pulse.

In stating the rise-time results, all relevant system parameters should be recorded, for example, number of photoelectrons per pulse, supply voltage, bleeder configuration, photomultiplier gain, and peak output current.

The spectral distribution of the light source shall be given since photoelectron energies vary with excitation wavelength, and differences in photoelectron energies can produce different measured time parameters.

The saturated anode current rise time is obtained as the time between the 10- and 90-percent points on the leading edge of the anode current waveform when the photocathode is so illuminated as to produce pulses of saturation amplitude. The value of the saturated current shall also be stated.

3.4.1.2 *Reflected-Pulse Rise Time.* Reflected-pulse rise time is that rise time measured with a time-domain reflectometer connected to the anode output connector of the device. The time-domain reflectometer should have a rise time that is short (at most one third) compared with the reflected-pulse rise time.

If the device does not employ a coaxial output connector, then the transition section between the coaxial cable and the device anode output terminals should be fabricated according to the manufacturer's instructions. The reflected-pulse rise time of the associated socket should also be stated.

3.4.1.3 *Single-Electron Rise Time.*
SERT (Single-electron rise time) is that anode-pulse rise time associated with single electrons originating at the photocathode. Measurement of SERT requires a photomultiplier having an adequate gain so that the single-electron event may be viewed on a sampling oscilloscope. (See Fig 5.)

A pulsed light source may be attenuated so that the average yield per pulse is much less than one photoelectron. Photocathode dark emission may also be used as a source of single electrons. If the dark current is too low, a dc light may be used to increase the single-electron emission rate. The use of a dc (continuous) light is required to ensure that the output pulses are single electron initiated. When dark current (or attenuated dc light) is used as a source of single electrons, the trigger signal for the sampling oscilloscope must be derived from the anode output pulse. A signal pick-off probe may be used, provided its rise time is small (one third or less) compared to the anode-pulse rise time. A resistive divider is suitable and may be fabricated in a manner to produce a rise time of less than 100 ps. Delay lines with internal trigger pick-offs (such as a Tektronix 7M11) are also available.

If the single-electron resolution of the device being measured is less than approximately 200 percent, considerable difficulty may be experienced in setting the oscilloscope trigger threshold level, and measurements may be misleading.

A description of the instrumentation and techniques used should accompany SERT data.

3.4.2 *Fall-Time Measurements.*
Fall-time measurements are made in accord with procedures outlined in 3.4.1. Device fall time is the mean time difference between the 90- and 10-percent amplitude points on the trailing edge of the output-pulse waveform for full cathode illumination and delta-function excitation. The light source used should exhibit a fall time that is less than one third the device fall time.

Single-electron fall time is measured as outlined in 3.4.1.3.

3.4.3 *Transit-Time Measurements*

3.4.3.1 *Device Transit Time.*
Device transit-time measurements require a delta-function light source having a marker pulse output synchronized in time with the light flash and a sampling oscilloscope. The device transit time is the time difference between the incidence of the light upon the photocathode (full illumination) and the occurrence of the half-amplitude point on the output-pulse leading edge.

The marker pulse and output pulse can both be positioned on the oscilloscope display by proper choice of delay cables. The device transit time is then calculated from knowledge of the delay-line delay times in the system, the time difference between marker pulse and output pulse on the oscilloscope, and the optical delay (length/C) between source and photocathode. Delay times of cables can be measured accurately with a time-domain reflectometer. Since device transit time varies with electrode potentials, these data must be specified.

3.4.3.2 *Photocathode Transit Time.*
Photocathode transit time is the time required for a packet of photoelectrons to travel from the photocathode to the first dynode. This measurement requires a delta-function light source. The device configuration must allow photoexcitation of the first dynode, as well as the photocathode, by the same source. The light source must therefore have a spectrum that provides photoemission from the first dynode. The value of photocathode transit time is obtained by first measuring device transit time and then biasing off the photocathode to allow measurement of the combined multiplier transit time and output-structure transit time. The optical delay between photocathode and first dynode must be considered in the calculation of photocathode transit time. (See Fig 6.)

The photocathode transit time is measured with full cathode illumination. The photocathode must be biased off in such a manner that the electric field pattern in the electron multiplier structure is not altered from its proper operating distribution. A suppression

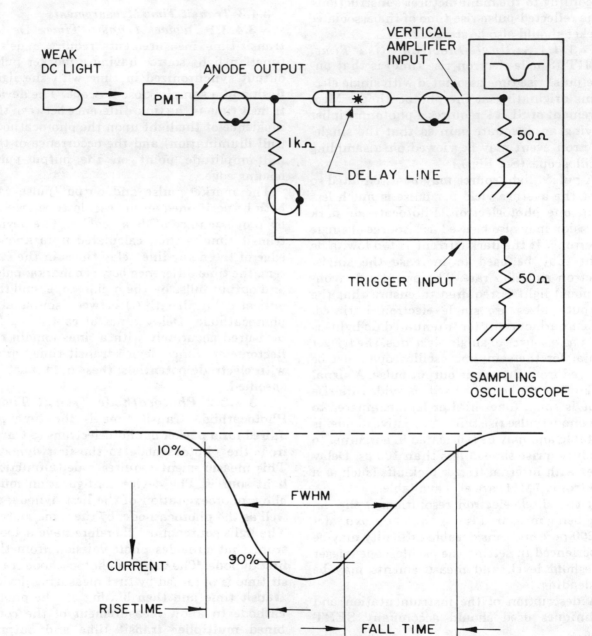

Fig 5
Single-Electron Rise Time

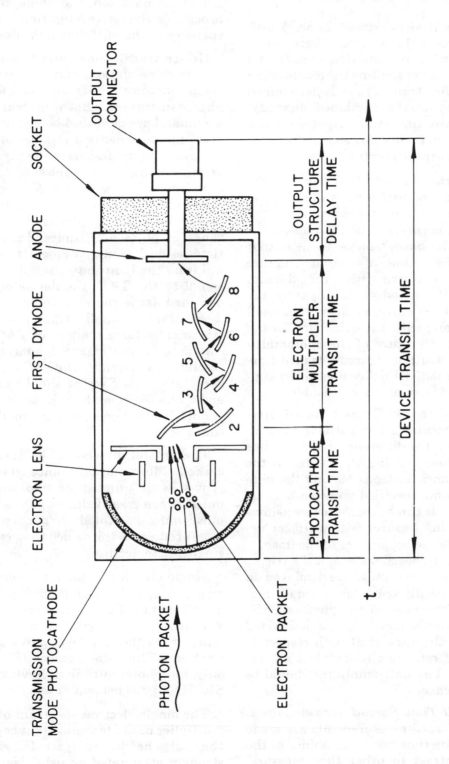

Fig 6
Photomultiplier Transit-Time Measurements

bias of 10 V is usually sufficient. Care must be exercised to insure that photoemission from all surfaces other than the first dynode is suppressed.

3.4.3.3 *Multiplier Transit Time.* Multiplier transit time is the time delay between an electron packet leaving the first dynode and the multiplied packet striking the anode of the device. Multiplier transit time is determined by first measuring the combined quantity, multiplier transit time and output-structure transit time and then subtracting the measured value of output-structure transit time.

3.4.3.4 *Output-Structure Transit Time.* Output-structure transit time is the time delay between an electron packet arriving at the anode and the occurrence of the half-amplitude point of the output pulse at the output connector. The value of output-structure transit time is measured with a time-domain reflectometer. If a wired socket is part of the external output structure, its delay time shall be specified along with the delay time of the internal output structure of the photomultiplier. Example: Output-structure transit time is 1.00 ns with 0.85-ns delay due to the supplied socket and coaxial cable assembly.

3.4.4 *Photocathode Transit-Time-Difference Measurements.* Photocathode transit-time difference is the difference in transit time between electrons leaving the center of the photocathode and electrons leaving the photocathode at some specified point on a designated diameter. It can be measured by using a delta-function light source with a trigger signal to designate the occurrence of the flash.

Delay lines are chosen such that the trigger marker and the output pulse are displayed on the sampling oscilloscope for a small spot illuminating the center of the photocathode. The spot is then displaced along a designated diameter, and the time shift with respect to the center-spot reference is stated as a function of radius. The half-amplitude should be used as a reference.

3.4.5 *Transit-Time-Spread Measurements.* Transit-time-spread measurements are made to determine the time-resolving ability of the device. In contrast to other time measurements, transit-time-spread measurements are made with the aid of a time-to-amplitude converter (TAC) instead of a sampling os-

cilloscope. The measurement involves recording the distribution of a statistically significant number of pulses, which, by virtue of the statistical nature of the photomultiplication process, arrive at varying times after the occurrence of the initiating light flash.

Device transit-time spread, also called device time resolution, can be measured for a single photomultiplier and also for a pair of photomultipliers. Both measurements are useful and are described below. In performing these measurements a number of techniques are available to designate the instant in time at which the output pulse reaches a given amplitude. The recommended technique utilizes constant-fraction timing since this leads to the best timing performance for a wide range of pulse amplitudes. In a single-tube time-resolution measurement the trigger signal from the light source supplies the "start" signal to the TAC; the device output signal obtained from the constant-fraction timing discriminator supplies the "stop" signal. A statistically large number of TAC pulses are sorted by a multichannel analyzer, and the resolution is given by the FWHM of the time spectrum. The FWHM should span at least eight channels, and at least 50 000 events should be contained within the distribution. (See Fig 7.)

The resolution should be stated for full cathode illumination. Since resolution improves with number of photoelectrons per pulse, the average value of photoelectrons per pulse must be stated. A particularly useful measurement is the device time resolution for single photoelectrons.

Single-electron time-resolution measurements using a single photomultiplier require a delta-function light source (see 3.3) with an electrical trigger signal to mark the occurrence of the flash, a TAC, and a pulse-height analyzer. The measurement (Fig 8) is valid only for photomultipliers having a value of SEPHR of 200 percent or less.

The single-electron spectrum of the photomultiplier under test must first be obtained on the pulse-height analyzer by employing a strongly attenuated dc light. Next the delta-function light source must be attenuated so that the most probable number of observed photoelectrons per flash is equal to one. The

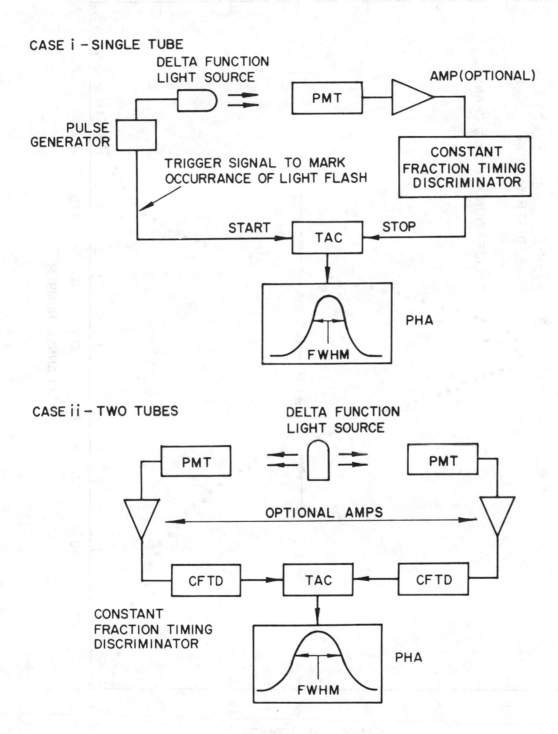

CASE i – SINGLE TUBE

DELTA FUNCTION
LIGHT SOURCE

AMP(OPTIONAL)

PMT

PULSE
GENERATOR

TRIGGER SIGNAL TO MARK
OCCURRANCE OF LIGHT FLASH

CONSTANT
FRACTION TIMING
DISCRIMINATOR

START   TAC   STOP

FWHM

PHA

CASE ii – TWO TUBES

DELTA FUNCTION
LIGHT SOURCE

PMT   PMT

OPTIONAL AMPS

CFTD   TAC   CFTD

CONSTANT
FRACTION TIMING
DISCRIMINATOR

FWHM

PHA

Fig 7
Transit-Time Spread

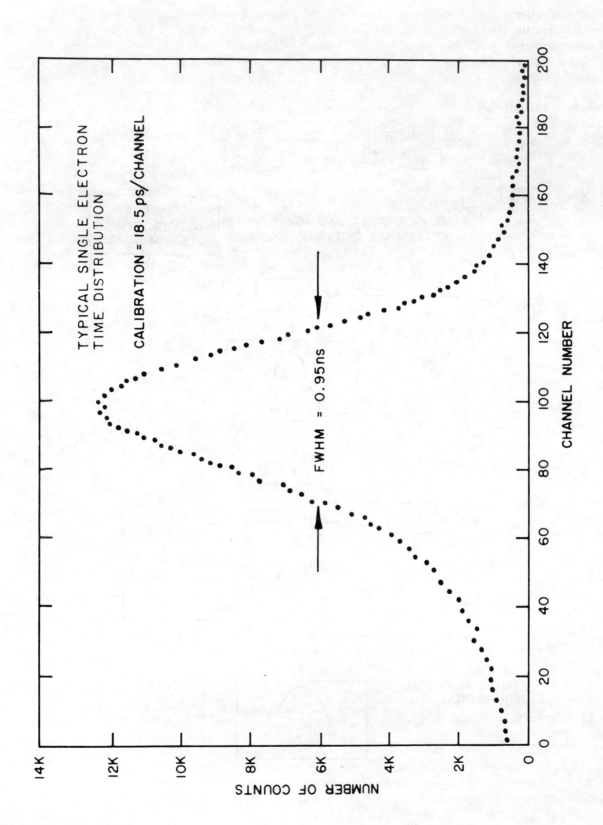

Fig 8

Single-Photoelectron Time Resolution

ratio of one-photoelectron events to two-photoelectron events should be no less than 100 to 1 as monitored with the pulse-height analyzer. This in turn means that for a significant fraction of the flashes no photoelectron will be liberated. (An upper-level discriminator could be employed to discriminate against two-electron events; however, the setting of the discriminator level is critical, and therefore this alternate method is not preferred.)

The TAC shall be calibrated according to the manufacturer's procedure. Such procedure usually includes a calibration of the time base by noting the displacement in time of a large number of events when an accurately known value of delay cable is introduced into the start or stop line.

The time-resolution distribution shall contain at least 50 000 events under the FWHM, and the FWHM shall span at least eight channels.

A second measurement of device time resolution involves the use of two photomultipliers viewing a common delta-function light source (see 3.3). This technique allows one photomultiplier to activate the TAC "start," while the other activates the TAC "stop." Note that this technique does not require a trigger signal from the source, so that either Cerenkov or scintillation sources can be employed. As with the single-tube measurement, resolution improves with the number of photoelectrons per pulse, so this figure must be stated. The FWHM of the distribution shall be stated, and the instrumentation shall be described. Calibration of the instrumentation shall be carried out according to the manufacturer's methods. (See also previous discussion.)

4. Test Conditions for Photomultipliers

Test conditions for photomultipliers are specified in terms of environmental conditions that must be met to enable accurate measurements to be made of the photomultiplier parameters discussed in the previous sections.

The test enclosure must be free of detectable light leaks. This can be verified by half-hour photon counting periods with and without bright ambient light incident on the enclosure.

The PMT should be stored in darkness for several hours prior to measurement to avoid phosphorescence effects. Cleanliness of the PMT glass and sockets is essential in preventing external noise effects. Any material near the photocathode must be at photocathode potential to prevent electroluminescence of the envelope and electrolysis of the glass.

The PMT should be degaussed before using, and a magnetic shield should be employed. Note that even the earth's magnetic field is of sufficient strength to influence measurements. Tube temperature should preferably be maintained constant within $\pm 2\,^{\circ}$C in the range 19 to 25° C. This is important in instances where the voltage divider may raise the temperature of the test enclosure.

5. Test Instrumentation

This section lists the instrumentation required to perform the measurements described in 3.4. The list is not all-inclusive, but instead serves as a guide to the type of instrumentation that is required.

All measurements require high-voltage power supplies in the range of 0 to 3 or 4 kV, with 5- to 30-mA current capability. Regulation should be 0.01 percent or better, and ripple and noise should be no more than a few millivolts.

For the tests of 3.1, the following are needed: NaI(Tl) crystals, ^{137}Cs and ^{55}Fe sources, charge-sensitive preamplifier, linear amplifier, pulse-height analyzer, precision pulser, LED, and pulser for LED.

Instrumentation for measurements in 3.2 includes the preceding, plus a TAC.

Instrumentation for 3.3 and 3.4 measurements includes fast linear amplifiers with pulse-shaping circuitry, sampling oscilloscope, delta-function light source, Cerenkov source or mode-locked laser, TAC, pulse-height analyzer, constant-fraction timing discriminator (or other fast timing discriminator).

6. Glossary for Scintillation Counting Field

accelerating electrode. An electrode to which a potential is applied to increase the velocity of the electrons.

accelerating grid. See **accelerating electrode**.

afterpulse. A spurious pulse induced in a photomultiplier by a previous pulse.

anode (electron tubes). An electrode through which a principal stream of electrons leaves the interelectrode space.

anode-reflected-pulse rise time. The rise time of a pulse reflected from the anode.

NOTE: This time can be measured with a time-domain reflectometer.

anticoincidence circuit. A circuit that produces a specified output signal when one (frequently predesignated) of two inputs receives a signal and the other receives no signal within an assigned time interval.

background counts (in radiation counters). Counts caused by ionizing radiation coming from sources other than that to be measured.

background response (in radiation detectors). Response caused by ionizing radiation coming from sources other than that to be measured.

CHANHI. Abbreviation for upper channel corresponding to the half-amplitude point of a distribution.

CHANLO. Abbreviation for lower channel corresponding to the half-amplitude point of a distribution.

coincidence circuit. A circuit that produces a specified output signal when and only when a specified number (two or more) or a specified combination of input terminals receives signals within an assigned time interval.

count (in radiation counters). A single response of the counting system.

counting efficiency (scintillation counters). The ratio of (1) the average number of photons or particles of ionizing radiation that produce counts to (2) the average number incident on the sensitive area.

count-rate meter. A device that indicates the time rate of occurrence of input pulses averaged over a time interval.

current amplification (photomultipliers). The ratio of (1) the signal output current to (2) the photoelectric signal current from the photocathode.

dark current (phototubes). That component of output current remaining when ionizing radiation and optical photons are absent.

dark pulses. Pulses observed at the output electrode when the photomultiplier is operated in total darkness. These pulses are due primarily to electrons originating at the photocathode.

decelerating electrode. An electrode to which a potential is applied to decrease the velocity of the electrons.

decelerating grid. See **decelerating electrode**.

delay circuit. A circuit that produces an output signal that is intentionally delayed with respect to the input signal.

delay coincidence circuit. A coincidence circuit that is actuated by two pulses, one of which is delayed by a specified time interval with respect to the other.

delay line. A transmission line for introducing signal delay.

delta-function light source. A light source whose rise time, fall time, and FWHM are no more than one third of the corresponding parameters of the output pulse of the photomultiplier.

device rise time. The mean time difference between the 10- and 90-percent amplitude points on the output waveform for full cathode illumination and delta-function excitation. DRT is measured with a repetitive delta-function light source and a sampling oscilloscope. The trigger signal for the oscilloscope may be derived from the device output pulse, so that light sources such as the scintillator light source may be employed.

discriminator, amplitude. See **discriminator, pulse-height**.

discriminator, constant-fraction pulse-height. A pulse-height discriminator in which the threshold changes with input amplitude in such a way that the triggering point corresponds to a constant fraction of the input pulse height.

discriminator, pulse-height. A circuit that produces a specified output signal if and only if it receives an input pulse whose amplitude exceeds an assigned value.

DRT. See **device rise time.**

dynode. An electrode which performs a useful function, such as current amplification, by means of secondary emission.

electron multiplier. That portion of the photomultiplier consisting of dynodes which produce current amplification by secondary electron emission.

electron multiplier transit time. That portion of photomultiplier transit time corresponding to the time delay between an electron packet leaving the first dynode and the multiplied packet striking the anode.

electron resolution. The ability of the electron multiplier section of the photomultiplier to resolve inputs consisting of n and $n+1$ electrons. This may be expressed as a fractional FWHM of the nth peak, or as the peak to valley ratio of the nth peak to the valley between the nth and $n \times$ 1th peaks.

focusing electrode. An electrode whose potential is adjusted to control the cross-sectional area of the electron beam.

focusing grid. See **focusing electrode.**

full width at half maximum. The full width of a distribution measured at half the maximum ordinate. For a normal distribution it is equal to $2(2 \ln 2)^{1/2}$ times the standard deviation σ.

FWHM. See **full width at half maximum.**

gain (photomultipliers). See **current amplification.**

ionizing radiation (scintillation counting). Particles or photons of sufficient energy to produce ionization in interactions with matter.

light guide. See **light pipe.**

light pipe. An optical transmission element that utilizes unfocused transmission and reflection to reduce photon losses.

MCA. See **multichannel analyzer.**

multichannel analyzer. See **pulse-height analyzer.**

multiplier phototube. See **photomultiplier.**

noise (phototubes). The random output that limits the minimum observable signal from the phototube.

opaque photocathode. See **reflection-mode photocathode.**

optical photons (scintillation counting). Photons with energies corresponding to wavelengths between approximately 120 = 1800 m.

optical pulse. A pulse of optical photons.

output-structure transit time. That portion of the photomultiplier transit time occurring within the output structure.

PEAK. Channel number corresponding to the peak of a distribution.

PHA. See **pulse-height analyzer.**

photocathode. An electrode used for obtaining photoelectric emission when irradiated.

photocathode blue response. The photoemission current produced by a specified luminous flux from a tungsten filament lamp at 2854 K color temperature when the flux is filtered by a CS 5-58 blue filter of half stock thickness (1.75-2.25 mm). This parameter is useful in characterizing response to scintillation counting sources.

photocathode luminous sensitivity. See **sensitivity, cathode luminous.**

photocathode radiant sensitivity. See **sensitivity, cathode radiant.**

photocathode spectral sensitivity characteristic. See **spectral sensitivity characteristic (photocathode).**

photocathode transit time. That portion of the photomultiplier transit time corresponding to the time for photoelectrons to travel from the photocathode to the first dynode.

photocathode transit-time difference. The difference in transit time between electrons leaving the center of the photocathode and electrons leaving the photocathode at some specified point on a designated diameter.

photocell. A solid-state photosensitive electron device in which use is made of the variation of the current–voltage characteristic as a function of incident radiation.

photomultiplier (photomultiplier tube). A phototube with one or more dynodes between its photocathode and output electrode.

photomultiplier delay time. See **photomultiplier transit time.**

photomultiplier transit time. The time difference between the incidence of a delta-function light pulse on the photocathode of the photomultiplier and the occurrence of the half-amplitude point on the output-pulse leading edge.

photomultiplier tube. See **photomultiplier.**

photon emission spectrum, scintillator material. The relative numbers of optical photons emitted per unit wavelength as a function of wavelength interval. The emission spectrum may also be given in alternative units such as wavenumber, photon energies, frequency, and so on.

phototube. An electron tube that contains a photocathode and has an output depending at every instant on the total photoelectric emission from the irradiated area of the photocathode.

PHR. See **pulse-height resolution.**

PMT. See **photomultiplier.**

pulse amplifier. An amplifier designed specifically for the ampiification of electrical pulses.

pulse counter. A device that tallies input pulses.

pulse duration. The time interval between the first and last instants at which the instantaneous amplitude reaches a stated fraction of the peak pulse amplitude.

pulse fall time. The interval between the instants at which the instantaneous amplitude last reaches specified upper and lower limits, namely, 90 percent and 10 percent of the peak pulse amplitude unless otherwise stated.

pulse-height analyzer. An instrument capable of indicating the number or rate of occurrence of pulses falling within each of one or more specified amplitude ranges.

pulse-height resolution. The fractional full width at half maximum of the pulse-height distribution curve (FWHM/$A1$) where $A1$ is the pulse height corresponding to the maximum of the distribution curve. In scintillation spectroscopy, it is customary to state PHR as a percentage.

p lse-pair resolution (photomultiplier). The time interval between two equal-amplitude delta-function optical pulses such that the valley between the two corresponding anode pulses falls to fifty percent of the peak amplitude.

pulse rise time. The interval between the instants at which the instantaneous amplitude first reaches specified lower and upper limits, namely, 10 percent and 90 percent of the peak pulse amplitude unless otherwise stated.

pulse shaper. A circuit that yields a desired pulse shape for a given input waveform.

pulse stretcher. A pulse shaper that produces an output pulse whose duration is greater than that of the input pulse and whose amplitude is proportional to that of the peak amplitude of the input pulse.

quantum efficiency (photocathode). The average number of electrons photoelectrically emitted from the photocathode per incident photon of a given wavelength.

NOTE: The quantum efficiency may be a function of the angle of incidence and of the direction of polarization of the incident radiation.

reflection mode photocathode. A photocathode wherein photoelectrons are emitted from the same surface as that on which the photons are incident.

resolving time (radiation counters). The minimum achievable pulse spacing between counts.

NOTE: This quantity is a property of the combination of the tube and recording circuit.

scintillation. The optical photons emitted as a result of the interaction of a particle or photon of ionizing radiation with a scintillator.

scintillation counter. The combination of scintillator, phototube (or photocell), and associated circuitry for detection and measurement of ionizing radiation.

scintillation-counter cesium resolution. The scintillation-counter energy resolution for the gamma ray or conversion electron from [137] Cs.

scintillation-counter energy resolution. The FWHM, in units of energy, of the scintillation counter spectrum at a stated energy.

scintillation decay time. The time required for the rate of emission of optical photons of a scintillation to decrease from 90 percent to 10 percent of its maximum value.

scintillation duration. The time interval from the emission of the first optical photon of a scintillation until 90 percent of the optical photons of the scintillation have been emitted.

scintillation rise time. The time required for the rate of emission of optical photons of a scintillation to increase from 10 percent to 90 percent of its maximum value.

scintillator. The body of scintillator material together with its container.

scintillator material. A material which emits optical photons in response to ionizing radiation.

NOTE: There are five major classes of scintillator materials, namely,
(1) Inorganic crystals [for example, NaI(T) single crystals, ZnS(Ag) screens]
(2) Organic crystals (for example, anthracene, *trans*-stilbene)
(3) Solution scintillators: (a) liquid, (b) plastic, (c) glass
(4) Gaseous scintillators
(5) Cerenkov scintillators

scintillator photon distribution (in number). The statistical distribution of the number of optical photons produced in the scintillator by total absorption of monoenergetic particles. (See Fig 9.)

scintillator material total conversion efficiency. The ratio of the total energy of the optical photons emitted by a scintillator to the incident energy of a particle or photon of ionizing radiation. (The efficiency is generally a function of the type and energy of the ionizing radiation.)

semitransparent photocathode. See **transmission mode.**

sensitivity, cathode luminous (photocathodes, scintillation counting). The quotient of photoelectric emission current from the photocathode by the incident luminous flux under specified conditions of illumination.

NOTE (1): Since cathode luminous sensitivity is not an absolute characteristic but depends on the spectral distribution of the incident flux, the term is commonly used to designate the sensitivity to radiation from a tungsten filament lamp operating at a color temperature of 2854 K.
NOTE (2): Cathode luminous sensitivity is usually measured with a beam at near normal incidence.

sensitivity, cathode radiant (photocathodes, scintillation counting). The quotient of the photoelectric emission current from the photocathode by the incident radiant flux at a given wavelength under specified conditions of irradiation.

NOTE: Cathode radiant sensitivity is usually measured with a beam at near normal incidence.

SEPHR. See **single-electron pulse-height resolution.**

SERT. See **single-electron rise time.**

single electron distribution. The pulse-height distribution associated with single electrons originating at the photocathode.

single-electron PHR. The fractional FWHM of the single-electron distribution of a photomultiplier.

single-electron rise time. The anode-pulse rise time associated with single electrons originating at the photocathode.

Fig 9
Scintillator Photon Distribution

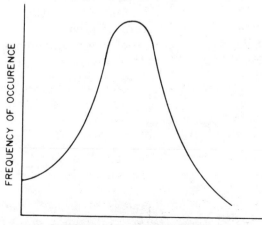

FREQUENCY OF OCCURENCE

PHOTONS PER SCINTILLATION

single-electron time jitter. See single-electron transit-time spread.

single-electron time resolution. See single-electron transit-time spread.

single-electron transit-time spread. Transit-time spread measured with single-electron events.

spectral quantum yield (photocathode). See quantum efficiency (photocathode).

spectral sensitivity characteristic (photocathode). The relation between the radiant sensitivity and the wavelength of the incident radiation under specified conditions of irradiation.

NOTE: Spectral sensitivity characteristic is usually measured with a beam at near normal incidence.

spurious count. A count from a scintillation counter other than (1) one purposely generated, or (2) one due directly to ionizing radiation.

spurious pulse. A pulse in a scintillation counter other than (1) one purposely generated, or (2) one due directly to ionizing radiation.

TAC. See time-to-amplitude converter.

time jitter. See transit-time spread.

time resolution. See transit-time spread.

time-to-amplitude converter. An instrument producing an output pulse whose amplitude is proportional to the time difference between start and stop pulses.

transit-time spread. The FWHM of the time distribution of a set of pulses each of which corresponds to the photomultiplier transit time for that individual event.

transmission-mode photocathode. A photocathode in which radiant flux incident on one side produces photoelectric emission from the opposite side.

tube scintillation pulses (photomultipliers). Dark pulses caused by scintillations within the photomultiplier structure. *Example:* Cosmic-ray-induced events.

wavelength shifter (scintillator). A photofluorescent compound used with a scintillator material to absorb photons and emit related photons of a longer wavelength.

Common Acronyms

CHANHI	Upper channel corresponding to the half-amplitude point of a distribution
CHANLO	Lower channel corresponding to the half-amplitude point of a distribution
DRT	Device rise time
FWHM	Full width at half maximum
LED	Light-emitting diode
MGD	Mean gain deviation
PEAK	Pulse-height analyzer channel corresponding to the peak of a distribution
PHR	Pulse-height resolution
PMT	Photomultiplier
SEPHR	Single-electron pulse-height resolution
SERT	Single-electron rise time
TAC	Time-to-amplitude converter

415

IEEE Guide for Planning of Pre-Operational Testing Programs for Class 1E Power Systems for Nuclear-Power Generating Stations

Sponsor

**Nuclear Power Subcommittee
of the
IEEE Power Generation Committee**

Foreword

(This foreword is not a part of IEEE Std 415-1976, Guide for Planning of Pre-Operational Testing Programs for Class 1E Power Systems for Nuclear-Power Generating Stations.)

This document was prepared by the Working Group of Pre-Operational Testing Criteria of the Nuclear Power Subcommittee of the IEEE Power Generation Committee. At the time of final Working Group approval, the members were:

R.A. Clark, *Chairman*

S. Cavallaro	E. P. Peabody
J. W. Colwell	J. R. Reesy
K. K. Khanna	F. D. Robbins
J. R. Kesler	J. J. Seibert
J. T. Lence	E. J. Warchol, Chairman WG

D. C. McClintock

Comments, suggestions, and requests for interpretations should be addressed to:

Mr. E. J. Warchol
Bonneville Power Administration
P. O. Box 3621 — EIC
Portland, Ore. 97208

with copies to:

Secretary, IEEE Standards Committee
IEEE Headquarters
345 East 47th Street
New York, N. Y. 10017

Contents

IEEE Guide for Planning of Pre-Operational Testing Programs for Class 1E Power Systems for Nuclear-Power Generating Stations

1. Scope

This document provides guidance for pre-operational testing of Class 1E Power systems for nuclear-power generating stations. The extent of the system shall be that covered by IEEE Std 308-1974, "Standard Criteria for Class 1E Power Systems for Nuclear Power Generating Stations." A system is illustrated by Fig 1. The power systems include both ac and dc supplies but do not include the equipment which utilizes the ac and dc power.

2. Purpose

The purpose of this document is to provide direction for establishing an acceptable pre-operational testing program for Class 1E power systems in nuclear-power generating stations. The pre-operational tests are performed after the appropriate construction tests have been completed. The purpose of the pre-operational tests is to verify that the functioning of Class 1E power systems performs their required function as described in the Final Safety Analysis Report (FSAR) of the station.

3. Referenced Documents

The following Standards and Criteria shall be applicable.

3.1 IEEE Std 308-1974, Standard Criteria for Class 1E Power Systems for Nuclear Power Generating Stations. (ANSI N41.12-1975)

3.2 IEEE Std 336-1971 (ANSI N45.2.4-1972), IEEE Standard Installation, Inspection, and Testing Requirements for Instrumentation and

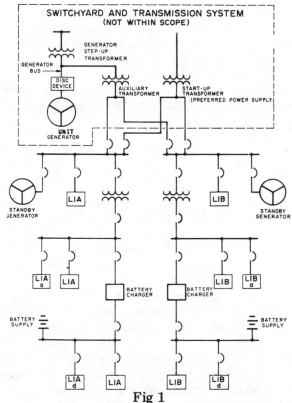

Fig 1
Example Class 1E Power System

Electric Equipment During the Construction of Nuclear Power Generating Stations.

3.3 IEEE Std 380-1972, Definitions of Terms Used in IEEE Nuclear Power Generating Station Standards.

3.4 IEEE Std 450-1972, Recommend Practice for Maintenance, Testing, and Replacement of Large Stationary Type Power Plant and Substation Lead Storage Batteries.

3.5 IEEE Std 387-1972, Criteria for Diesel-Generator Units Applied as Standby Power Supplies for Nuclear Power Generating Stations.

4. Definitions

For convenience, the following specific definitions are listed herein.

4.1 construction tests. A construction test is defined as a test to verify proper installation and operation of individual components in a system prior to operation of the system as an entity.

4.2 pre-operational system tests. A pre-operational test is defined as a test to confirm that all individual component parts of a system function as a system and the system functions as designed.

The above tests shall be performed following both the initial construction of the facility and subsequent modifications or additions made to the facility at later dates.

For other definitions, refer to the IEEE Standards listed in Section 3 for applicable definition of terms.

5. General Requirements

5.1 Testing Program, Planning, and Administrative Procedures. Planning, administrative procedures, and work instructions for the total plant testing program shall have been established. Detailed planning, work instructions, and administrative procedures for pre-operational testing of Class 1E power systems shall be prepared. Appropriate documentation for pre-operational testing of Class 1E power systems shall be prepared and coordinated with other phases of the overall plant testing program.

5.2 Prerequisites. The Class 1E power systems components shall have met all requirements of design, fabrication, installation, and all tests in compliance with applicable codes, standards, and Quality Assurance Programs. The records of construction tests shall be reviewed and compared with requirements as part of the Quality Assurance Program and checked for appropriate certification.

5.3 Performance. After meeting the requirements of Sections 5.1 and 5.2 above, the tests outlined in Section 6 and other referenced documents shall be performed.

6. Testing Guidelines

6.1 Per-Operational Tests. The tests shall be of sufficient number and diversity of modes to demonstrate that the Class 1E power systems meet design specifications under the various design conditions. These tests should be conducted with bus loads and sequencing times, as close as possible to actual system operating conditions. If actual full loads are not practical simulated full loads should be obtained by load boxes, by feeding power back into the power line, or by other suitable means. The loads themselves are not part of the system being tested.

6.2 Pre-Operational Test Categories. This section describes some pre-operational test categories for Class 1E power system. Table 1 shows a matrix of minimum applicability of typical categories to Class 1E power systems.

6.2.1 Starting Tests. Starting tests shall demonstrate the capability to attain and stabilize frequency and voltage within the design limits and time.

6.2.2 Load Acceptance Test. Load acceptance tests shall demonstrate the capability to accept the individual loads that make up the design load, in the desired sequence and time duration, and to maintain the voltage and frequency within design limits.

6.2.3 Design Load Test. Design load tests shall demonstrate the capability to carry the maximum design load (magnitude and sequence) for a time required to reach a temperature equilibrium plus one hour, without exceeding the specified design limits.

6.2.4 Load Rejection Test. Load rejection tests shall demonstrate the capability of rejecting the maximum required load without reaching speeds or voltages which will cause tripping, mechanical damage, or harmful overstresses.

6.2.5 Electric Test. Electric tests shall demonstrate that all electric requirements of the diesel generator, excitation system, voltage regulation system, engine governor system, transformers, switchgear, bus transfer schemes, control interlocks, and surveillance systems are acceptable for the intended application.

6.2.6 Functional Test. Functional tests shall demonstrate the capability of the control, surveillance, and protective switching devices to function in accordance with the requirements of the intended application.

Table 1
Matrix of Minimum Applicability of Typical Categories

	Starting Test	Load Acceptance Test	Design Load Test	Load Rejection Test	Electric Test	Functional Test	Independence	Battery Capacity Test	Battery Charging System
Standby diesel generator[1]	X	X	X	X	X	X	X		
Preferred ac power supply		X	X		X	X	X		
AC power distribution systems		X	X		X	X	X		
Battery supply[2]								X	
Battery charger									X
DC power distribution system		X	X		X	X	X		
Vital instrumentation and control power supplies	X	X	X	X	X	X	X		
Vital instrumentation and control distribution system		X	X		X	X	X		

[1] For detailed testing procedure refer to Ref 3.5.
[2] For detailed testing procedure refer to Ref 3.4.

6.2.7 Test for Independence. These tests assure that each Class 1E power system including its load group and each available power source can function independent of any other system.

6.2.8 Battery Capacity Test. Battery capacity test shall be performed in accordance with paragraph 5.6 of IEEE Std 450-1972 (Ref 3.4). This is a capacity test which determines if the battery will meet the design requirements.

6.2.9 Battery Charging System Capability Test.

6.2.6.1 These tests shall demonstrate the capability of the charging system to maintain battery voltage within design limits while carrying the required steady-state load.

6.2.6.2 These tests shall demonstrate the capability of the charging system to maintain battery voltage within design limits while carrying the required steady-state load.

7. Documentation

7.1 Contents. The written pre-operational test procedures should include:

(1) purpose
(2) references and drawings
(3) prerequisites
(4) special precautions
(5) detailed test procedures
(6) test equipment required with serial numbers
(7) forms and data sheets
(8) return to normal
(9) acceptance criteria
(10) verification of successful completion of each test
(11) deviation report and resolution procedure.

7.2 Test Report. The compilation of test procedures and the completed data sheets shall constitute the test report. Each phase of the tests and the test data shall be initialed by the tester to certify completion, and shall have approval signature and date. The documents shall include dates for the beginning and end of the tests, conditions as found, and corrections made. Detailed documentation of simulated load or operating condition in lieu of actual load or operating condition shall be maintained.

7.3 Retention. The procedure should state that the completed procedure, along with the collected data, are part of the plant historical record and shall be retained. Administrative procedures should provide for the length of retention of any required summarization of test results, and permanent retention of documented summaries and evaluations.

ANSI/IEEE Std 420-1973
(ANSI N41.17)

420

IEEE Trial-Use Guide for Class 1E Control Switchboards for Nuclear Power Generating Stations

Sponsor

Power Generation Committee of the IEEE Power Engineering Society

Foreword

(This Foreword is not part of IEEE Std 420-1973, Trial-Use Guide for Class 1E Control Switchboards for Nuclear Power Generating Stations.)

This document offers guidance for the design, manufacture, and installation of Class 1E control boards and is a matter of concern for manufacturers, users, and for those who are responsible for licensing and regulating nuclear power generating stations. This guide is an industry consensus of an acceptable approach to assessing the adequacy of Class 1E switchboards at this time. Work to expand the scope of this document is under way and will be reflected in modifications following the period of *trial use*.

This guide is not intended to define the selection, design, or qualification of piping or modules mounted on the Class 1E control switchboards. The guide is concerned with the effect of the mounted piping and modules on the integrity of the Class 1E control switchboard in the areas listed in the guide. The selection, design, and qualification of piping and modules are covered by other documents.

Work remains to be done in the separation area to progress from the trial-use basis on the subject of physical separation of redundant Class 1E circuits and devices or other components. Fire tests are urgently needed to determine the minimum separation within the control board for maintained air space or for a fire-retardant barrier. Tests involving different types of barriers such as metallic conduit, two sheets of retardant material with air space or thermal insulating material, need to be carried out. The sufficiency of a single barrier with air space or thermal insulation needs evaluation from such tests.

Terms such as *flame retardance, self-extinguishing,* and *nonpropagating* have been used in the guide without definitions being given. Nor do the definitions exist. The committee feels that much more work is required to arrive at satisfactory terminology and the consensus is that it should not be in the scope of this document. Coordination with work on *separation criteria* documents, presently in preparation by an *ad hoc* committee of the Nuclear Power Engineering Committee of the IEEE Power Engineering Society needs to be done and the differences resolved.

In the area of quality assurance, the differences between the quality assurance classifications for electric equipment proposed by ANSI and those currently advocated by IEEE must be resolved. In addition, work is currently in progress on the formulation of ANSI standards which supplement the basic ANSI quality assurance document, Requirements for Nuclear Power Plants, Quality Assurance Program, ANSI N45.2-1971. When available, the results of this effort must be reviewed by the working group to determine the extent of its applicability to the Class 1E control switchboard and incorporate the appropriate portions of the requirements in the quality assurance section of this guide.

The Institute of Electrical and Electronics Engineers (IEEE) undertook the development of this guide under the auspices of the Power Generation Committee. Members of the Nuclear Power Plant Control and Protection Working Group at the time of generation of the Guide were:

A.J. Spurgin, *Chairman*

C. Burns	J.I. Martone	W.G. Schwartz
C.L. Cobler	R. Mattson	R.A. Schmitter
R.S. Darke	M.I. Olken	C.H.L. Sprigg
B.P. Grim	R.J. Reiman	J.V. Stephens
J.R. Hall	H.F. Reischel	D. Tondi
D.A. Hansen	W.P. Robinson, Jr.	J.R. Ward
W. Kerchner	R. Saya	L.C. Wilmot
J.R. Kesler		W.E. Wilson

The Chairman for much of the time of preparation of the document was M. I. Olken.

Contents

IEEE Trial-Use Guide for Class 1E Control Switchboards for Nuclear Power Generating Stations

1. Scope

This document applies to all Class 1E control switchboards (for example, main control room assemblies, auxiliary panels, local panels or racks, etc) as defined in Section 3, regardless of their application or location. It does not apply to the non-Class 1E control switchboards, except as they affect the Class 1E control switchboards, but recognizes that Class 1E control switchboards may be adjacent to, or joined with, other non-Class 1E control switchboards.

This document includes, but is not necessarily limited to, design, construction, wiring practices, shipping, handling, storage, installation, testing, and quality assurance.

General design practices which are not unique to Class 1E control switchboards are not addressed in this document. This document applies to piping systems and Class 1E equipment only as they may affect the integrity of the Class 1E control switchboard.

2. Purpose

The purpose of this document is to provide guidance for the design, manufacture and installation of Class 1E control switchboards in nuclear power generating stations to conform to IEEE Std 279-1971, Criteria for Protection Systems for Nuclear Power Generating Stations, ANSI N42.7-1972 and other documents as referenced.

3. Definitions

control switchboard. A type of switchboard including control, instrumentation, metering, protective (relays), or regulating equipment for remotely controlling other equipment. (See IEEE Std 27-1969, Switchgear Assemblies, including Metal-Enclosed Bus, ANSI C37.20-1969.)

Class 1E. The safety classification of the electric equipment and systems that are essential to emergency reactor shutdown, containment isolation, reactor core cooling, and containment and reactor heat removal, or are otherwise essential in preventing significant release of radioactive material to the environment.

Class 1E control switchboard. A rack panel, switchboard, or similar type structure fitted with any Class 1E equipment.

4. Criteria

4.1 Seismic

4.1.1 *Design Considerations.* A Class 1E control switchboard should provide sufficient support and physical protection to its Class 1E equipment to enable the equipment to meet its essential performance requirements during and after a safe shutdown earthquake. The safe shutdown earthquake is defined in IEEE Std 344-1971, Trial-Use Guide for the Seismic Qualification of Class 1 Electric Equipment for Nuclear Power Generating Stations.

The Class 1E control switchboard shall be so designed, that at the frequencies and accelerations of the floor or equipment mounting resulting from the safe shutdown earthquake, they do not amplify the forces beyond the level at which the equipment is qualified to function properly. The equipment used in Class 1E systems should be seismically qualified per the requirements of IEEE Std 344-1971.

Class 1E control switchboards may be designed in a variety of ways to meet the seismic requirements of the nuclear station. One approach is to design the control switchboard with sufficient rigidity so that no natural frequencies or resonances can exist at a frequency less than 33 Hz. In selecting equipment, its ability to withstand such accelerations must be a prime consideration. The methods most commonly used to achieve the necessary degree of rigidity are:

(1) Welded stiffeners every several feet
(2) Diagonal braces to the base
(3) Thick plate for the *skin*

The above methods may be used singly, or in any combination that will satisfy the rigidity requirement.

An alternative approach is to design flexibility into the control switchboards. This approach necessitates that the equipment chosen for installation in the board be capable of operating properly at the amplified acceleration levels. Flexible, as defined herein, means a natural frequency less than 33 Hz. Flexible boards are generally made of lighter materials and include bracing only as needed to insure structural and mechanical integrity.

4.1.2 *Other Considerations.* The control switchboard design should include provisions for securely mounting the board to its support. A dynamically equivalent support should be used in the seismic testing of the switchboard. If a calculation is used to show the seismic capability of the board, then the actual mounting should be modeled.

Plug-in or slide-mounted equipment should be provided with mechanical restraints, if needed, to maintain positional integrity.

All equipment within, attached to, or adjacent to, a Class 1E control switchboard shall be mounted such that the structural failure of this equipment cannot damage Class 1E equipment or wiring.

Where relative motion exists between equipment and the control board, the maximum range of such motion shall be considered in the physical arrangement and design parameters for wiring and adjacent equipment. Wiring to equipment and terminal boards shall have sufficient slack or be installed such that relative motion within the control switchboard will not cause failure.

4.2 Separation

4.2.1 *Design Considerations.* Physical separation of redundant Class 1E circuits and devices or other components should be provided within each Class 1E control switchboard, so that no single credible event as identified below can prevent the proper functioning of any Class 1E system.

Separation may be accomplished by mounting the redundant Class 1E equipment on physically separated control switchboards, if feasible from a plant operational point of view (see Section 4.4). When operational design dictates that redundant Class 1E equipment be in close proximity, separation shall be achieved by a fire-retardant barrier or an air space. Wiring shall be supported in a manner, such that the designed air space would be maintained throughout the entire life of the switchboard. Such Class 1E control switchboards shall be located and protected in a manner such that a single credible event is limited to an internally generated fire, a low energy missile, or an event caused by an improper maintenance procedure.

Examples of acceptable separation barriers are:

(1) Metallic conduit
(2) Two sheets of fire-retardant material separated by an air space or thermal insulating material
(3) A single barrier with a one inch maintained air space or thermal insulating material between the components or devices and the barrier

NOTE: At present, there is no industry consensus of what is an acceptable air gap, however, a minimum separation of 6 in of unobstructed air space appears prudent.

4.2.2 *Circuitry Considerations.* Each circuit of a group of redundant Class 1E circuits which serve the same protective function shall enter the control switchboard through separated apertures and terminate on separate blocks or connectors, which must also be separated as previously described under Section 4.2.1.

Where redundant Class 1E circuits ultimately terminate on the same device, barriers shall be provided between the device terminations to insure circuit separation.

Circuits for non-Class 1E functions may be run in the same wire bundles or wireways used by one of a group of redundant Class 1E circuits. When this is done, the circuits for non-Class 1E functions must not be subsequently routed with any other Class 1E circuits.

4.3 Flame Retardance

4.3.1 *Design Considerations.* Inherent flame-retardant characteristics and properties should be an important consideration in the selection, design, and fabrication of components and materials for the control switchboards. Normally metallic, the framework and surfaces of control switchboards inherently offer a high degree of flame resistance.

Where Class 1E and non-Class 1E equipment are located on the same control board, the non-Class 1E equipment should have the same fire-retardant qualities as the Class 1E circuits.

Any nonmetallic components and devices such as terminal blocks, raceways, wire troughs, wire cleats, cable ties, receptacles, indicating lamp lenses, etc, should be manufactured from self-extinguishing materials as classified by ASTM Standard D635-72, Standard Method of Test for Flammability of Self-Supportive Plastics.

NOTE: *Self-extinguishing* only refers to definitions in this specific test, and correlation with flammability under actual use condition is not implied.

Wire and cable should be flame retardant with self-extinguishing nonpropagating characteristics. Consideration should be given to the release of toxic or corrosive gases and dense smoke and their effects upon personnel and equipment.

Paints or other applied surface preparations should contribute minimally, relative to the total combustible potential of materials or components in or on the control switchboard. Such preparations shall not propagate flames when heated or ignited. Consideration should be given to the release of toxic or corrosive gases and dense smoke and their effects upon personnel and equipment.

4.4 Environment and Location. Class 1E control switchboards and Class 1E components mounted thereon should be designed and fabricated to withstand the environmental conditions as stated in IEEE Std 279-1971 and IEEE Std 323-1971, IEEE Trial-Use Standard, General Guide for Qualifying Class 1 Electric Equipment for Nuclear Power Generating Stations. The control boards should be located so that hazards such as fires, missiles, vibrations, pipe whips, water sprays, etc, shall not cause common failure of redundant Class 1E systems (see IEEE Std 279-1971).

4.5 Wire Selection

4.5.1 *Design Considerations.* All control switchboard wiring should have sufficient current capacity[1], mechanical strength, thermal rating, and insulation characteristics to meet the circuit and installation requirements established by the plant design. In selecting conductor size, consideration should be given to derating factors for grouped conductors without maintained spacing.

4.5.2 *Flame Retardance.* Flame retardance (see Section 4.3) and the possible release of toxic or corrosive gases from cable insulation or from jackets exposed to heat and flame must be considered.

4.5.3 *Radiation.* Where required by location, the wire insulation should have a gamma and neutron flux environment classification suitable for the calculated total integrated dose accumulated throughout the design life of the control switchboard plus the total integrated dose resulting from the postulated design basis event.

4.6 Wiring Practices

4.6.1 *Design Considerations.* All wiring should be done in a neat organized manner and in accordance with requirements and specifications as set forth by both user and manufacturer. However, on Class 1E wiring, specific requirements as detailed below must be complied with.

4.6.1.1 Wire bundles shall be supported (generally at 15 to 24 in intervals) by such devices as clamps, straps, ties, wire ducts, etc.

NOTE: Metallic wire ties are not acceptable.

4.6.1.2 Wire splices are not allowed.

4.6.1.3 Wires forming a hinge cable shall be multistrand flexible wire, secured on each side of the hinge and formed in a loop to minimize strand fatigue.

4.6.2 *Testing.* Tests should be conducted to confirm continuity and correctness of all circuits.

4.6.3 *Identification.* All equipment should be permanently marked and identified on the interior of Class 1E control switchboards.

---

[1]New current capacity tables and derating factors for these conditions are under development by the Insulated Power Cable Engineers Association. The current capacity of wiring in control switchboards may be considered identical to cables installed in ventilated cable trays.

Such identification should be conspicuously placed to minimize the possibility of maintenance or repair personnel incorrectly identifying the device. Identification should not be placed on removable covers or parts which might easily become interchanged during maintenance or inspection procedures, or both. In addition, Class 1E equipment and its wiring should be identified as such, so that personnel may confirm its independence from its redundant Class 1E equipment and wiring. Tags, color codes, colored tags, marking tapes, or other suitable means are recommended for identification.

4.6.4 *Grounding.* Caution should be exercised to insure that the method of grounding does not compromise the independence of the redundant Class 1E equipment. The control switchboard and its appurtenant equipment should be grounded for personnel safety, and a method of terminating the user's station ground should be provided.

4.6.5 *Tools.* All tools used in wiring operations should be a high quality type designed for the intended purpose, and checked at predetermined intervals to verify their proper operation.

4.7 Other Design Considerations

4.7.1 *Periodic Testing.* The requirement of IEEE Std 279-1971 and IEEE Std 338-1971, Trial-Use Criteria for the Periodic Testing of Nuclear Power Generating Station Protection Systems relating to periodic testing should be reflected in the design of Class 1E control switchboards. Periodic testing should be accomplished without the use of jumpers or the disconnection of permanent wiring.

4.7.2 *Accessibility.* When a control board has both electrical components and piping equipment, the physical relationship of these components should permit accessibility for required testing and maintenance. Testing and maintenance of piping systems should not degrade the ability of the Class 1E equipment to properly perform its function.

4.8 Shipping, Handling, and Storage

4.8.1 *Introduction.* Class 1E equipment demands extreme care in preparation for shipment, in method of securing and protecting equipment while in transit, in unloading, and during storage at the job site.

The method of shipment (air, rail, truck, ship, etc) is an important consideration. Shipping by exclusive-use vans (specially equipped vans used to transport highly sensitive electronic equipment and devices) is desirable. However, if this method is unfeasible, then packaging should be designed for environmental protection such as to avoid shock and vibrations, or to control temperature and humidity within the specified limits, or both. The package design and instruction data on procedures for proper storage, handling and protection of equipment, from time of shipment until it is in operation, should be agreed upon by manufacturer and user, in conjunction with provisions stated in IEEE Std 336-1971.

4.8.2 *Considerations Prior to Shipment*

4.8.2.1 The interior of all equipment should be free of foreign matter prior to shipment, except as required for packing and shipping.

4.8.2.2 Items not fixed in place on the control switchboard at the time of shipment should be separately barrier bagged, boxed, crated, or otherwise protected against contamination, damage, or loss.

4.8.2.3 Heavy or cumbersome control switchboards should be prepared for shipment so that mechanical handling equipment can be used, but will not damage the control switchboards.

4.8.2.4 Assemblies to be shipped in sections should be clearly identified and include complete instructions for field assembly.

4.8.3 *Considerations at Job Site*

4.8.3.1 Upon receipt of equipment at job site, visual inspection should be made for apparent damage incurred during shipment.

4.8.3.2 All equipment should be stored and installed as outlined under Section 4.8.1 and protected from damage, dust, debris, and harmful environments.

4.8.3.3 Except as necessary for receiving inspection, protective coverings, shipping stops, etc. should not be removed until the equipment is located at its permanent station and the final assembly or connecting operation is initiated.

4.8.3.4 If necessary to energize equipment during storage, means should be provided to apply power and remove heat during the power application. If equipment is crated, power connections should be accessible without opening the crate. If flammable, all protective coverings, temporary braces, vibration and shock snubbers, etc, should be arranged such that they will not initiate or support fire due to their proximity to energized equipment.

5. Quality Assurance

5.1 Specifications. Written specifications for Class 1E control switchboards should include, but not be limited to:

  (1) Design requirements
  (2) Fabrication practices
  (3) Equipment mounting procedures
  (4) Wiring procedures
  (5) Approved terminal equipment
  (6) Grounding requirements
  (7) Installation (including anchors and corresponding board attachment points)
  (8) Testing requirements
  (9) Seismic requirements
  (10) Separation requirements
  (11) Packaging and shipping requirements
  (12) Receiving, handling, and storage requirements
  (13) Documentation requirements
  (14) Wiring and component identification requirements
  (15) Environment and location requirements
  (16) Flame-retardance requirements

5.2 Documentation. Documentation indicating compliance with specifications shall be provided.

6. References

[1] IEEE Std 27-1969, Switchgear Assemblies, including Metal-enclosed Bus, ANSI C37.20-1969

[2] IEEE Std 279-1971, Criteria for Protection Systems for Nuclear Power Generating Stations, ANSI N42.7-1972

[3] IEEE Std 308-1971, Criteria for Class 1E Electric Systems for Nuclear Power Generating Stations

[4] IEEE Std 323-1971, IEEE Trial-Use Standard, General Guide for Qualifying Class 1 Electric Equipment for Nuclear Power Generating Stations ANSI N41.5

[5] IEEE Std 336-1971, IEEE Standard for Installation, Inspection, and Testing Requirements for Instrumentation and Electric Equipment during the Construction of Nuclear Power Generating Stations, ANSI N45.2.4-1972

[6] IEEE Std 338-1971, Trial-Use Criteria for the Periodic Testing of Nuclear Power Generating Station Protection Systems ANSI N41.3

[7] IEEE Std 344-1971, Trial-Use Guide for the Seismic Qualification of Class 1 Electric Equipment for Nuclear Power Generating Stations ANSI N41.7

[8] ASTM Standard D635-1972, Standard Method of Test for Flammability of Self-Supportive Plastics

IEEE Recommended Practice for Maintenance, Testing, and Replacement of Large Lead Storage Batteries for Generating Stations and Substations

450

Sponsor

**Power Generation Committee
of the
IEEE Power Engineering Society**

Foreword

(This foreword is not a part of IEEE Std 450-1975, Recommended Practice for Maintenance, Testing, and Replacement of Large Lead Storage Batteries for Generating Stations and Substations.)

Large stationary lead storage batteries play an ever increasing role in substation and generating station control systems and in providing the backup energy for emergencies. A definite need exists within the industry for an application guide and testing procedure to provide a common or standard method for selecting, applying, and installing batteries to meet station requirements. However, to provide a guide that can be used for determining the available capacity of the battery (especially for nuclear stations), this document will limit its scope to recommended practice for lead storage battery maintenance, testing, and replacement.

The IEEE will maintain this document current with the state of the technology. Comments on this document and suggestions for additional material that should be included are invited. These should be addressed to:

> Secretary
> IEEE Standards Board
> The Institute of Electrical and Electronics Engineers, Inc
> 345 East 47 Street
> New York, NY 10017

This document was prepared by a Working Group on Batteries, Station Design Subcommittee of the Power Generation Committee of the IEEE Power Engineering Society. The members of the Working Group were:

J. H. Bellack, *Chairman*

J. W. Anderson	C. W. Jordan
J. L. Giambalvo	K. C. Lockwood
W. E. Golde	H. L. McCloud
A. P. Grande	M. W. Migliaro
F. R. Greenwood	H. K. Reid
E. C. Haupt	M. Srinivasan
R. W. Hopewell	B. G. Treece
W. F. Hurley	L. D. Zachau, Jr

Liaison Representative

K. C. Andrus, *Substation Committee*

The Station Design Subcommittee, at the time that it reviewed and approved this document, had the following membership:

J. B. Sullivan, *Chairman*

J. C. Appiarius	A. Foss	W. B. Raley
M. S. Baldwin	J. J. Garland	J. R. Reesy
R. T. Barnum	W. E. Golde	M. P. Roller
J. H. Bellack	W. F. Gundaker	W. J. Rom
I. B. Berezowski	J. J. Heagerty	J. D. Rosenblatt
F. W. Brandt	B. R. Jessop	J. M. Sappington
E. W. Brunton	C. W. Jordan	M. N. Sprouse
F. D. Burton	E. F. Kratz	B. J. Stables
Richard S. Coleman	A. Lehrkind	C. E. Stine
John W. Colwell	O. S. Mazzoni	R. E. Strasser
C. C. Coppin	M. W. Migliaro	S. Tjepkema
R. E. Cotta	J. L. Mills	C. J. Wylie
S. M. Denton	P. M. Niskode	H. E. Yocom, Jr
J. D. Farber	R. E. Penn	R. Zweigler
	J. D. Plaxco	

Contents

IEEE Recommended Practice for Maintenance, Testing, and Replacement of Large Lead Storage Batteries for Generating Stations and Substations

1. Scope

This document is limited to providing recommended practices of maintenance, test schedules, and testing procedures that can be used to optimize life and performance from large lead storage batteries. It also provides guidance to determine when batteries should be replaced. There are other test procedures and replacement techniques used within the industry (especially for smaller substation batteries) that are equally as effective but are beyond the scope of this document. Sizing, installation, other battery types, and application are also beyond the scope.

This document does not include surveillance and testing of the dc system even though the battery is part of that system.

2. Definitions

The following definitions apply specifically to the subject matter of this standard. For other definitions, see IEEE Std 380-1972, Definitions of Terms Used in IEEE Nuclear Power Generating Station Standards, and IEEE Std 100-1972, Dictionary of Electrical and Electronics Terms (ANSI C42.100-1972).

acceptance test (lead storage batteries). A capacity test made on a new battery to determine that it meets specifications or manufacturer's ratings.

battery rack. A rigid structure used to accommodate a group of cells.

performance test (lead storage batteries). A capacity test made on a battery, as found, after being in service to detect any change in the capacity determined by the acceptance test.

service test (lead storage batteries). A special capacity test made to demonstrate the capability of the battery to meet the design requirements of the system to which it is connected.

terminal connection detail (lead storage batteries). Connections made between rows of cells or at the positive and negative terminals of the battery, which may include lead-plated terminal plates, cables with lead-plated lugs, and lead-plated rigid copper connectors.

3. Maintenance

3.1 General. Proper maintenance will prolong the life of a battery and will aid in assuring that it is capable of supplying its design power requirements. A good battery maintenance program will serve as a valuable aid in determining the need for battery replacement. Station battery maintenance should be performed by personnel knowledgeable of batteries and the safety precautions involved.

3.2 Safety. The safety precautions listed herein should be followed in station battery maintenance. Work performed on batteries should be done only if the proper and safe tools are available with the protective equipment listed.

3.2.1 *Methods.* Work performed on a battery in service should use methods to preclude circuit interruption or arcing in the vicinity of the battery.

3.2.2 *Protective Equipment.* The following protective equipment should be available to personnel who perform battery maintenance work:

(1) Goggles
(2) Acid-resistant gloves
(3) Protective aprons
(4) Portable or stationary water facilities for rinsing eyes and skin in case of acid spillage

(5) Bicarbonate of soda or other suitable neutralizing agent for acid spillage

3.2.3 *Precautions.* The following protective procedures should be followed:

(1) Insulate the handles of tools used for tightening connector bolts

(2) Prohibit arcing, smoking, and an open flame in the immediate vicinity of the battery

(3) Ensure that the test leads shall be connected with sufficient length of cable to prevent accidental arcing in the vicinity of the battery

(4) Use fuses on all connections for test equipment

(5) Ensure that battery area ventilation is operable

(6) Ensure unobstructed egress from the battery area

3.3 Inspections

3.3.1 *General.* Inspection of the battery on a regular scheduled basis should include a check and record of the following:

(1) General cleanliness of the battery and battery area

(2) Float voltage

(3) Cells for cracks or electrolyte leakage

(4) Plates of cells (plates buckling, discoloring, grid cracks, or plate growth)

(5) Ambient temperature and ventilation equipment

(6) Pilot cell (if used) voltage, specific gravity, and electrolyte temperature and level; specific gravity readings should be taken in accordance with manufacturer's instructions

(7) Terminals and connectors for evidence of corrosion

3.3.2 *Quarterly Inspections.* Quarterly inspections should include a check and record of the following:

(1) Specific gravity readings of each cell

(2) Voltage reading of each cell and total battery terminal voltage (cell voltages shall be post to post to include intercell connector)

(3) Electrolyte level of each cell

(4) Float voltage

(5) Temperature of electrolyte of representative cells (suggested every sixth cell)

(6) Battery load with battery on float charge (charger current)

3.3.3 *Yearly Inspections.* Yearly inspections should include a check and record of the following:

(1) Cell condition (detailed visual inspection)

(2) Cell-to-cell and terminal connection detail resistance

(3) Integrity of battery rack

(4) Tightness of bolted battery connections

3.3.4 *Special Inspection.* Under any abnormal conditions or severe circumstances, the inspection procedures listed in Sections 3.3.2 and 3.3.3 should be repeated.

NOTE: A regular schedule for a nuclear plant Class 1E battery should be consistent with the inspection intervals described in IEEE Std 308-1974, Standard Criteria for Class 1E Power Systems for Nuclear Power Generating Stations.

3.4 Corrective Actions

(1) An equalizing charge should be given if the specific gravity of an individual cell drops more than 0.010 from the average of all cells at time of inspection.

(2) An equalizing charge should be given if the average specific gravity of all cells drops more than 0.010 from the acceptance test (Section 4.1) value when corrected for temperature and electrolyte levels.

(3) An equalizing charge should be given if any cell voltage deviates more than 0.04 V from the average at time of inspection.

(4) If not required by Section 3.4 (1), (2), or (3), an equalizing charge should be given at least once a year. This equalizing charge can be waived for certain batteries based on analysis of records of operation and maintenance inspections (Section 7).

(5) Add water when any cell electrolyte reaches the low-level line. Water should be added to bring all cells to the high-level line. Water quality should be in accordance with manufacturer's instructions.

(6) Any intercell connection and terminal connection detail should be taken apart, cleaned, and refastened if its measured resistance value is more than 20 percent above the average at the time of installation.

(7) If cell temperatures deviate more than 3°C (5°F) from each other during a single inspection, determine the cause and correct.

(8) When excessive dirt is noted on cells or connectors, wipe with water-moistened clean wiper. Remove electrolyte spillage on cell covers and containers with a bicarbonate of soda and water-moistened wiper. All wipers used should be free of oil distillates.

(9) Correct any other abnormal conditions noted.

4. Capacity Test Schedule

The following schedule of capacity tests is used to (1) test the battery to determine whether the battery meets its specification or manufacturer's rating, or both, (2) periodically test to determine whether the rating of the battery, as found, is holding up, and (3) if required, test to determine whether the battery meets the design requirements of the system to which it is connected.

4.1 Acceptance. An acceptance test of battery capacity (Section 5) should be made either at the factory or upon initial installation as determined by the user. The test should meet a specific discharge rate and duration relating to the manufacturer's rating.

4.2 Performance

(1) A performance test of battery capacity (Section 5) should be made within the first 2 years of service. Initial conditions shall be as described under Section 5.1, omitting requirements (1) and (2). Results of this test reflect all factors including maintenance that determine battery capability. It is desirable for comparison purposes that the performance tests be similar in duration to the battery acceptance test (Section 4.1). If on a performance test the battery does not deliver its expected capacity, the test should be repeated after completing the requirements of Sections 5.1 (1) and (2).

(2) Additional performance tests should be given to each battery at 5 year intervals until it shows signs of degradation as outlined in Section (3).

(3) Annual performance tests of battery capacity should be given to any battery that shows signs of degradation or has reached 85 percent of the service life expected for the application. Degradation is indicated when the battery capacity drops more than 10 percent of rated capacity from its average on previous performance tests, or is below 90 percent of the manufacturer's rating.

4.3 Service. A service test of battery capacity (Section 5.6) may be required by the user to meet a specific application requirement upon completion of the installation. This test is a test of the battery to deliver the design requirements of the dc system. This test is performed for Class 1E nuclear power generating station batteries as part of the preoperational and periodic dc system tests described in IEEE Std 308-1974. If the system design changes, the user may be required to repeat this test to meet requalification requirements.

5. Procedure for Battery Capacity Tests

This procedure describes the recommended practice of capacity testing by discharging the battery. For nuclear power stations use of Class 1E batteries; also refer to IEEE Std 323-1974, Standard for Qualifying Class 1E Equipment for Nuclear Power Generating Stations. All testing should follow the safety precautions listed in Section 3.2.

5.1 Initial Conditions. The following list gives the initial requirements for all battery capacity tests. Requirements (1) and (2) should not be completed for "performance" and "service" tests as discussed in Sections 4.2, 4.3, and 5.6.

(1) Verify that the battery has had an equalizing charge completed 3 days to a week prior to the start of the test.

(2) Check all battery connections and make sure all connectors are clean, tight, and free of corrosion.

(3) Read and record the specific gravity and voltage of each cell just prior to the test.

(4) Read and record the temperature of battery electrolyte for an average temperature (suggested every sixth cell).

(5) Read and record the battery terminal voltage.

(6) Disconnect the charger from the battery.

(7) Take adequate precautions (such as isolating battery to be tested from other batteries and critical loads) to ensure that a failure will not jeopardize other systems or equipment.

5.2 Test Length. The recommended procedure is to make a capacity test for approximately the same length of time as the critical period for which the battery is sized. See Section 5.6 for test length of the service test.

5.3 Test Discharge Rate. The discharge rate depends upon the type of test selected. For the acceptance test or performance test, the discharge rate should be a constant current load equal to the manufacturer's rating of the battery for the selected test length. See Section 5.6 for the test discharge rate of the service test.

5.4 Acceptance and Performance Test Description

5.4.1 *Discharge: 1 Hour or Longer.* Set up a load with an ammeter and voltmeter with the provisions that the load be varied to maintain a constant current drain equal to the rating of the battery at the selected rate (see Section 5.3).

(1) Connect the load to the battery, start the timing, and continue to maintain the correct discharge rate.

(2) Maintain the discharge rate until the battery terminal voltage falls to a value equal to the minimum specified average voltage per cell (usually 1.75) times the number of cells.

(3) Read and record individual cell voltages (Section 3.3.2 (2)) and battery terminal voltage. The readings should be taken while the load is applied at the beginning and completion of the test and at specified intervals. There should be a minimum of three readings.

(4) If an individual cell is approaching reversal of its polarity (plus 1 V or less) but the terminal voltage has not yet reached its lower test limit, the test should be continued with a jumper across the weak cell. Complete the jumper connection away from the cell to avoid arcing near the cell. The new minimum terminal voltage should be determined based on the remaining cells (see item (2)).

(5) Observe the battery for intercell connector heating.

(6) At the conclusion of the test, determine the battery capacity according to the procedures outlined in Section 5.5.

5.4.2 *Discharge: Less Than 1 Hour.* This test is similar to the test described in Section 5.4.1, except in the use of a higher discharge rate and the temperature correction factor. When the battery under test is not at the standard temperature of 25°C (77°F) the discharge rate is corrected for the temperature difference. Consult the manufacturer for temperature correction factors K_2 and refer to Section 5.5.2.

5.5 Determining Battery Capacity

5.5.1 *Test Length: 1 Hour or Longer.* For an acceptance or performance test that runs for 1 hour or longer, use the following equation to determine battery capacity:

percent capacity at 25°C (77°F) =

$$\frac{T_a}{T_s K_1} \times 100$$

where the test discharge current is at rating and where

T_a = Actual duration of test to minimum specified terminal voltage (see Section 5.4.1 (2))

T_s = Rated time to final voltage

K_1 = Capacity correction factor (see Table 1) relating to cell temperature at start of test; for other values consult the manufacturer

The method described in Section 5.4.2 is also applicable.

5.5.2 *Test Length: Less Than 1 Hour.* For an acceptance or performance test that runs for less than 1 hour, use the following equation to determine battery capacity:

percent capacity at 25°C (77°F) =

$$\frac{T_a}{T_s} \times 100$$

where

T_a = Actual duration of test to minimum specified terminal voltage (see Section 5.4.1 (2))

T_s = Rated time to final voltage

Note that the test discharge current is equal to the rated discharge current times K_2, where K_2 is the current discharge rate correction factor related to temperature. See Section 5.4.2.

5.6 Service Test Description. A service test is a special battery capacity test which may be required to determine if the battery will meet the design requirements of the connected load. The system designer must establish the test procedure and acceptance criteria prior to the test. Procedure for the test is:

Table 1
Capacity Correction Factors K_1 for Temperatures at Variances to Standard 25°C (77° F) at 1 Hour Through 8 Hour Rate Discharges (1.210 Specific Gravity)

Temperature (°C)	(°F)	Factor K_1
16.7	62	0.936
17.2	63	0.941
17.8	64	0.945
18.3	65	0.950
18.9	66	0.954
19.4	67	0.959
20.0	68	0.963
20.6	69	0.968
21.1	70	0.972
21.7	71	0.976
22.2	72	0.980
22.8	73	0.984
23.4	74	0.988
23.9	75	0.992
24.5	76	0.996
25	77	1.000
25.6	78	1.003
26.1	79	1.006
26.7	80	1.008
27.2	81	1.011
27.8	82	1.014
28.3	83	1.017
28.9	84	1.019
29.4	85	1.022
30.0	86	1.025
30.6	87	1.027
31.1	88	1.030
31.6	89	1.032
32.2	90	1.035
32.8	91	1.037
33.4	92	1.039

(1) The initial conditions shall be as identified in Section 5.1.

(2) The discharge rate and test length should correspond as closely as is practical to the load the battery will be subjected to during the critical period.

(3) If the battery does not meet its service load in accordance with its system design criteria, review its rating to see if it is properly sized, equalize the battery, and if necessary, inspect the battery as discussed in Section 3.3.4 and take necessary corrective actions. A battery performance test (Section 4.2) may also be required to determine whether the problem is the battery or the application.

5.7 Restoration. Disconnect all test apparatus. Equalize the battery if necessary. Return to normal service.

6. Battery Replacement Criteria

The recommended practice is to replace the battery if its capacity as determined by the tests described in Section 5.5.1 or 5.5.2 is below 80 percent of the manufacturer's rating. The timing of the replacement is a function of the sizing criteria utilized and the capacity margin available, compared to the load requirements. Whenever replacement is required, the recommended maximum time for replacement is 1 year. A capacity of 80 percent shows that the battery rate of deterioration is increasing even if there is ample capacity to meet the load requirements. Other factors, such as unsatisfactory battery service test results (Section 5.6) require battery replacement unless a satisfactory service test can be obtained following corrective actions.

Physical characteristics, such as plate condition together with age, are often determinants for complete battery or individual cell replacements. Reversal of a cell as described in Section 5.4.1 (4) is also a good indicator for further investigation into the need for individual cell replacement. Replacement cells, if used, should be compatible with existing cells and should be tested prior to installation. Replacement cells are not usually recommended as the battery nears end of life.

Failure to hold a charge, as shown by cell voltage and specific gravity readings, is a good indicator for further investigation into the need for replacement.

7. Records

Data obtained from inspections and corrective actions are important to the operation and life of the batteries. Data such as indicated in Section 3.3 should be recorded at the time of installation and as specified during each inspection. Data records should also contain reports on corrective actions (Section 3.4) and on capacity and other tests indicating the discharge rates, their duration, and results.

At nuclear stations, records of Class 1E batteries shall include a written test procedure and documentation adequate to meet the requirements of IEEE Std 308-1974.

It is recommended that forms be prepared to record all data in an orderly fashion and in such a way that comparison with past data is convenient. For a suggested format see IEEE Std 323-1974, Section 8. A meaningful comparison will require that all data be converted to a standard base in accordance with the manufacturer's recommendations.

8. Standards References

IEEE Std 380-1972, Definitions of Terms Used in IEEE Nuclear Power Generating Station Standards

IEEE Std 100-1972, Dictionary of Electrical and Electronics Terms (ANSI C42.100-1972)

IEEE Std 308-1974, Standard Criteria for Class 1E Power Systems for Nuclear Power Generating Stations

IEEE Std 323-1974, Standard for Qualifying Class 1E Equipment for Nuclear Power Generating Stations

IEEE Std 484-1975, Recommended Practice for Installation Design and Installation of Large Lead Storage Batteries for Generating Stations and Substations

ANSI/IEEE Std 484-1975
(ANSI N41.24-1976)

An American National Standard

484

IEEE Recommended Practice for Installation Design and Installation of Large Lead Storage Batteries for Generating Stations and Substations

Sponsor

**Power Generation Committee
of the
IEEE Power Engineering Society**

Approved January 15, 1976
American National Standards Institute

Foreword

(This foreword is not a part of IEEE Std 484-1975, Recommended Practice for Installation Design and Installation of Large Lead Storage Batteries for Generating Stations and Substations.)

Proper installation design and installation procedures are prerequisite to long and reliable service of large stationary lead storage batteries. A definite need exists within the industry for recommendations that provide a uniform method for the design of the installation and the installation procedures of all batteries to meet station requirements. However, to provide practices that can be used immediately (especially for nuclear power generating stations), this document will limit its scope to the installation design and installation procedures of large stationary lead storage batteries. Included will be the seismic considerations for new battery installations. Work is being initiated on determination of qualification for batteries. See IEEE Std 450-1972, Recommended Practice for Maintenance, Testing, and Replacement of Large Stationary Type Power Plant and Substation Lead Storage Batteries. Fuure documents that will describe sizing and determination of qualification for batteries are being prepared.

Prepared by the Working Group on Batteries, Station Design Subcommittee of the Power Generation Committee, this document correlates and summarizes industry practices; it is not intended to be an exhaustive compilation or a rigid procedure manual.

The members of the Working Group, at the time this document was approved, were:

J. H. Bellack, *Chairman*

J. W. Anderson	W. F. Hurley
K. C. Andrus*	C. W. Jordan
J. A. Calvo	K. C. Lockwood
J. L. Giambalvo	H. L. McCloud
W. E. Golde	M. W. Migliaro
A. P. Grande	H. K. Reid
F. R. Greenwood	B. G. Treece
E. C. Haupt	L. D. Zachau, Jr

R. W. Hopewell

*Liaison representative from the Substation Committee of the Power Engineering Society.

At the time this document was approved the members of the Station Design Subcommittee were:

J. B. Sullivan, *Chairman*

J. C. Appiarius	A. Foss	W. B. Raley
M. S. Baldwin	J. J. Garland	J. R. Reesy
R. T. Barnum	W. E. Golde	M. P. Roller
J. H. Bellack	W. F. Gundaker	W. J. Rom
I. B. Berezowski	J. J. Heagerty	J. D. Rosenblatt
F. W. Brandt	B. R. Jessop	J. M. Sappington
E. W. Brunton	C. W. Jordan	M. N. Sprouse
F. D. Burton	E. F. Kratz	B. J. Stables
Richard S. Coleman	A. Lehrkind	C. E. Stine
John W. Colwell	O. S. Mazzoni	R. E. Strasser
C. C. Coppin	M. W. Migliaro	S. Tjepkema
R. E. Cotta	J. L. Mills	C. J. Wylie
S. M. Denton	P. M. Niskode	H. E. Yocom, Jr
J. D. Farber	R. E. Penn	R. Zweigler
	J. D. Plaxco	

At the time this document was approved the members of the Power Generation Committee were:

Comments, suggestions, and requests for interpretations should be addressed to:

Secretary
IEEE Standards Board
The Institute of Electrical and Electronics
345 East 47th Street
New York, NY 10017.

Contents

An American National Standard

IEEE Recommended Practice for Installation Design and Installation of Large Lead Storage Batteries for Generating Stations and Substations

1. Scope

This document provides recommended design practices and procedures for storage, location, mounting, ventilation, instrumentation, preassembly, and assembly of large lead sealed cell storage batteries. Required safety practices are also included. This document applies particularly to installations at generating stations (including nuclear) and large substations.

The portions of this document that will specifically relate to personnel and nuclear plant safety are mandatory and are designated by the word "shall"; all other portions are recommended practices and are designated by the word "should."

Battery sizing, maintenance, capacity testing, charging equipment, and consideration of other types of batteries are beyond the scope of this document.

2. Definitions

The following definitions apply specifically to the subject matter of this document. For other definitions, see IEEE Std 380-1972, Definitions of Terms Used in IEEE Nuclear Power Generating Stations Standards, and IEEE Std 100-1972 (ANSI C42.100-1972), Dictionary of Electrical and Electronics Terms.

freshening charge. The charge given to a storage battery following nonuse or storage.

sealed cell. A cell in which the only passage for the escape of gases from the interior of the cell is provided by a vent of effective spraytrap design adapted to trap and return to the cell particles of liquid entrained in the escaping gases.

shall. Intended to indicate requirements where those recommendations are actually standards. (See Section 1.)

should. Intended to indicate that which, at the present time, is considered a recommendation, that is, advised but not required. (See Section 1.)

3. Safety

The safety precautions listed herein shall be followed during station battery installation. Work performed on batteries shall be done only if the proper and safe tools are available in addition to the protective equipment listed. Station battery installation shall be performed or supervised by personnel knowledgeable of batteries and the safety precautions required.

3.1 Protective Equipment. The following equipment for safe handling of the battery and protection of personnel shall be available:
  (1) Goggles
  (2) Acid-resistant gloves
  (3) Protective aprons and overshoes
  (4) Portable or stationary water facilities for rinsing eyes and skin in case of acid spillage
  (5) Bicarbonate of soda or other suitable neutralizing agent for acid spillage
  (6) Lifting devices of adequate capacity, when required
  (7) Connector bolt tools with insulated handles

3.2 Procedures. The following safety procedures shall be followed prior to, and during installation:
  (1) Connect metal racks to station ground system

7

(2) Inspect all lifting equipment for functional adequacy

(3) Restrict all unauthorized personnel from the battery area

(4) Prohibit smoking, arcing, and open flame in the immediate vicinity of the battery

(5) Keep the top of the battery clear of all tools and other foreign objects at all times

(6) Ensure that illumination requirements are met

(7) Ensure unobstructed egress from the battery area

4. Installation Design Criteria

Considerations that should be included in the design of the battery installation depend upon the requirements or function of the system of which the battery is a part.

4.1 General (All Installations). The following are the general installation design criteria for all large generating station and substation batteries, including Class 1E batteries for nuclear power generating stations.

4.1.1 Location

(1) Space and floor supports allocated for the battery and associated equipment should allow for present and future needs

(2) The area selected should be clean, dry, well ventilated (see Section 4.1.4), and provide aisle space for inspection, maintenance, testing, and cell replacement. Space should also be provided above the cells to allow for operation of lifting equipment, addition of water, and taking measurements (for example, temperature, specific gravity, etc)

(3) The optimum cell electrolyte temperature is 77°F (25°C), which is the basis for rated performance. A location where this temperature can be maintained will contribute to optimum battery life, performance, and cost of operation; extreme ambient temperatures should be avoided because low temperatures decrease battery capacity while prolonged high temperatures shorten battery life

NOTE: Installation in a location with an ambient below the optimum temperature will affect sizing.

(4) The location or arrangement should result in no greater than a 5°F (3°C) temperature differential between cells at a given

time; localized heat sources such as direct sunlight, radiators, steam pipes, and space heaters should be avoided

(5) Portable or stationary water facilities should be provided for rinsing spilled electrolyte. Provisions for neutralizing, containing, or safely dispersing spillage should be included

(6) The charger and main power distribution center should be as close as practical to the battery

(7) Illumination in the battery area should equal or exceed the interior lighting recommendations in the *Illuminating Engineering Society Handbook*, Fig 9-80 for central station battery rooms

(8) Nearby equipment with arcing contacts shall be located in such a manner as to avoid those areas where hydrogen pockets could form.

4.1.2 Mounting

(1) The most common practice is to mount cells on a steel rack with acid-resistant insulation between the cells and the steel of the rack. Metal racks, if used, shall be connected solidly to the station grounding system. The cells may also be mounted on insulated supports secured to a floor or base; the insulation used should be rated for full battery voltage

(2) Not more than two tiers or two steps should be considered for these batteries; this choice of rack results in a minimum temperature differential between cells and will facilitate maintenance

(3) Cells in clear jars should be mounted so that the edge of all plates on one side are plainly visible for inspection

4.1.3 Seismic. Although seismic consideration is a regulatory design requirement for nuclear power generating stations, fossil fuel generating stations and substations may appropriately follow the provisions of Section 4.2 when the installation is to be in a location subject to a high probability of seismic disturbance.

4.1.4 Ventilation. The battery area shall be ventilated, either by a natural or induced ventilation system, to prevent accumulation of hydrogen and to maintain design temperature. The ventilation system shall limit hydrogen accumulation to less than two percent of the total volume of the battery area. Maximum hydrogen evolution rate is 0.000269 cubic feet per minute per charging ampere per cell at 77°F (25°C), one atmosphere. The worst expected condition is forcing maximum current into a fully charged battery.

4.1.5 *Instrumentation and Alarms.* The following general recommendations for instrumentation and alarms apply to the battery installation only. Requirements for the charger, dc system design, etc, are beyond the scope of this document.

Each battery installation should include the following instrumentation and alarms:

(1) Voltmeter

(2) High and low battery voltage alarm

(3) Ground detector (for ungrounded systems)

NOTE: The preceding recommendations for instrumentation and alarms could be satisfied by equipment in the dc system.

4.2 Nuclear Power Generating Station Class 1E Batteries. In addition to the general requirements presented in Section 4.1, a Class 1E battery, located within a nuclear power generating station, is subject to a number of additional design criteria. See IEEE Std 308-1974, Criteria for Class 1E Power Systems for Nuclear Power Generating Stations, and IEEE Std 323-1974, Standard for Qualifying Class 1E Equipment for Nuclear Power Generating Stations. When the battery performs a Class 1E function, a quality assurance program shall be adopted to control and document all activities related to such functions.

4.2.1 *Location*

(1) The battery shall be protected against natural phenomena such as earthquakes, winds, and flooding, as well as induced phenomena such as fire, explosion, missiles, pipe whips, discharging fluids, CO_2 discharge, and other environmental hazards

(2) To minimize the effect of seismic forces, the battery should be located at as low an elevation as practical

(3) Where batteries are required in redundant systems, the batteries shall be separated as specified by IEEE Std 384-1974 (ANSI N14.14), Trial-Use Standard Criteria for Separation of Class 1E Equipment and Circuits

4.2.2 *Mounting*

(1) All cells shall be restrained; side and end rails with spacers between cells is one method that can be used to prevent loss of function due to a seismic event

(2) Cells shall be mounted in accordance with manufacturer's recommended separation distance; any spacers used shall be moisture and acid resistant to avoid cell damage caused by deformation

(3) Where more than one rack section is used, they shall be rigidly joined, or adjacent end cells joined with flexible connectors as provided by or recommended by the manufacturer

(4) Racks shall be firmly connected to the building structure as specified by using approved fastening techniques such as embedded anchor bolts or racks welded to structural steel faceplates (sized to accommodate a range of battery rack sizes)

CAUTION: Anchoring a rack to both the floor and wall may cause stress due to conflicting modes of vibration.

(5) The racks, anchors, and installation thereof shall be able to withstand the force calculated for a safe shutdown earthquake to allow continuous battery service during and following that event, in accordance with IEEE Std 344-1975, Seismic Qualification of Class 1 Electric Equipment for Nuclear Power Generating Stations

5. Installation Procedures

See Section 3 for safety precautions to be followed. In the case of Class 1E installation, refer to IEEE Std 336-1971 (ANSI N45.2.4-1972), Installation, Inspection, and Testing Requirements for Instrumentation and Electrical Equipment During the Construction of Nuclear Power Generating Stations.

5.1 Receiving and Storage

5.1.1 *Receiving Inspection.* Upon receipt, and at the time of actual unloading, each package should be visually inspected for apparent damage and electrolyte leakage. If either is evident, a more detailed inspection of the entire shipment should be conducted and cell repair or replacement instituted as required. Record receipt date and inspection data.

5.1.2 *Unpacking*

(1) When lifting cells, a strap and strap spreader should be used, if applicable

(2) Always lift cells by the bottom, never by the cell posts

(3) Check electrolyte levels for evidence of leakage and that the plates are covered; any cell should be replaced if the electrolyte level is 1/2 inch or more below the top of the plates

(4) All cells with visible defects such as cracked jars, loose terminal posts, or improperly aligned plates shall be repaired or replaced

5.1.3 *Storage*

(1) Cells should be stored indoors in a clean, level, dry, and cool location; extremely low ambient temperatures or localized sources of heat should be avoided

(2) Cells should not be stored for more than the time period recommended by the manufacturer, without applying a charge to the battery; in all cases, a period of three months storage is allowable between charges if the recommendations of (1) are followed

(3) For charging during storage or special conditions, the battery manufacturer should be consulted. Record dates and conditions for all charges during storage

5.2 Assembly

5.2.1 *Rack Assembly.* The assembly of the rack should be in accordance with the manufacturer's recommended procedure.

5.2.2 *Cell Mounting and Connections*

(1) Lift the individual cells onto the rack following the procedures outlined in Section 5.1.2 (1) and (2); mount the cells in accordance with the manufacturer's recommendations

(2) Check cell polarity for positive to negative connections throughout the battery

(3) Battery manufacturers ship large lead cells with corrosion-inhibiting grease applied to all terminal posts and connecting hardware; clean any area showing evidence of corrosion, dirt, or acid, then recoat that area with a thin film of the corrosion-inhibiting grease

(4) The intercell connector contact surfaces should be cleaned by rubbing gently with a brass suede brush or fine emery paper; care should be exercised in cleaning to prevent removal of the lead plating. Apply a thin film of corrosion-inhibiting grease to all contact surfaces

(5) Make intercell connections using manufacturer approved connectors (normally furnished with the battery)

(6) When more than one intercell connector per cell post is required, mount on opposite sides of post for maximum surface contact

(7) Tighten both ends of connection bolts to battery manufacturer's recommended torque value

(8) Clean all cell covers and containers; for dust and dirt use a water moistened clean wiper; for electrolyte spillage, use a bicarbonate of soda and water moistened wiper. All wipers used should be free of oil distillates

(9) Where specified, install explosion-resistant vent plugs

(10) For future identification, apply individual cell numbers in sequence beginning with number 1 at the positive end of the battery; also add any required operating identification

(11) Read the voltage of the battery to ensure that individual cells are connected correctly, that is, the total voltage should be approximately equal to the number of cells times the measured voltage of one cell. If the measurement is less, recheck the individual cell polarities

(12) Read and record intercell connection resistance to determine adequacy of initial installation and as a reference for future maintenance requirements (see Section 6, Records). Review records of each intercell connection resistance measurement; remake and remeasure any connection that has a resistance measurement more than 10 percent over the average

(13) When (1) through (12) have been satisfactorily completed, make final connections from the battery to the charger and dc system

5.2.3 *Preoperational Care.* After battery assembly is complete to the stage of intercell connections (Section 5.2.2 (5)), the battery should be maintained in accordance with IEEE Std 450-1972, Recommended Practice for Maintenance, Testing, and Replacement of Large Stationary Type Power Plant and Substation Lead Storage Batteries. Protect the battery from construction dirt and debris.

5.3 Freshening Charge and Testing. Since a battery loses some of its charge during shipment and storage, a freshening charge should be applied after installation.

5.3.1 *Freshening Charge*

(1) Read and record the temperature, voltage and specific gravity of each cell prior to

applying the charge; select the cell with the lowest specific gravity as a pilot cell

(2) Inspect all cells and add approved water as necessary to bring electrolyte to the "low level" line

(3) Follow manufacturer's recommendation for applying a freshening charge

(4) Read and record the specific gravity and temperature of electrolyte of the pilot cell at least once daily

(5) After pilot cell specific gravity readings become constant (with correction for temperature), add approved water to all cells to bring electrolyte up to the "high level line"; continue charge for 3 hours

(6) Return charger to float voltage

(7) At the end of 72 hours, read and record all individual cell voltages, temperatures, and specific gravities (see Section 6, Records)

(8) Any cell that shows a specific gravity (corrected to 77 °F (25 °C)) less than 1.200 or more than 1.220, or a voltage lower than 0.04 V below the average, requires corrective action in accordance with the manufacturer's instruction

5.3.2 *Acceptance Test.* When required, a capacity discharge test shall be conducted in accordance with IEEE Std 450-1972.

6. Records

Data obtained from receiving, storage, and assembly are pertinent to the maintenance and operational life of the battery. The data, as applicable, that should be recorded and maintained in a suitable permanent file for record purposes and future reference are:

(1) Date of receipt of battery and receiving inspection data; dates and conditions of charge (see Section 5.1.3 (3))

(2) Initial resistance values of the intercell connections (see Section 5.2.2 (12))

(3) Individual cell specific gravity (corrected for temperature) and voltage measure-

ments (see Section 5.3.1 (1), (4), (5), and (7))

(4) Acceptance test data (see Section 5.3.2)

(5) Quality assurance records for class 1E batteries.

NOTE: The preceding records should be in accordance with cell identification (see Section 5.2.2 (10)).

7. References

IEEE Std 100-1972 (ANSI C42.100-1972), Dictionary of Electrical and Electronics Terms

IEEE Std 308-1974, Criteria for Class 1E Power Systems for Nuclear Power Generating Stations

IEEE Std 323-1974, Standard for Qualifying Class 1E Equipment for Nuclear Power Generating Stations

IEEE Std 336-1971 (ANSI N45.2.4-1972), Installation, Inspection, and Testing Requirements for Instrumentation and Electric Equipment During the Construction of Nuclear Power Generating Stations

IEEE Std 344-1975, Recommended Practices for Seismic Qualification of Class 1 Electric Equipment for Nuclear Power Generating Stations

IEEE Std 380-1972, Definitions of Terms Used in IEEE Nuclear Power Generating Station Standards

IEEE Std 384-1974 (ANSI N14.14), Trial-Use Standard Criteria for Separation of Class 1E Equipment and Circuits

IEEE Std 450-1972, Recommended Practice for Maintenance, Testing, and Replacement of Large Stationary Type Power Plant and Substation Lead Storage Batteries

Illuminating Engineering Society Lighting Handbook, 5th ed, Illuminating Engineering Society, New York, NY, 1972

IEEE Recommended Practice for Sizing Large Lead Storage Batteries for Generating Stations and Substations

485

Sponsor

**Power Generation Committee of the
IEEE Power Engineering Society**

© Copyright 1978 by

The Institute of Electrical and Electronics Engineers, Inc.

Foreword

(This Foreword is not a part of IEEE Std 485-1978, Recommended Practice for Sizing Large Lead Storage Batteries for Generating Stations and Substations.)

Although the storage battery is of primary importance in assuring the satisfactory operation of generating stations and substations, no single up-to-date guide exists to aid engineers in sizing the battery for a particular installation. This Recommended Practice is based on commonly accepted methods used to define the load and to ensure adequate battery capacity. The method described is applicable to all installations and battery sizes.

The installations considered herein are designed for "full float" operation with a battery charger serving to maintain the battery in a charged condition as well as to supply the normal dc load.

This Recommended Practice was prepared by the Working Group on Batteries of the Station Design Subcommittee of the IEEE Power Generation Committee. It may be used separately, but, when combined with ANSI/IEEE Std 450-1975, IEEE Recommended Practice for Maintenance, Testing, and Replacement of Large Lead Storage Batteries for Generating Stations and Substations, and ANSI/IEEE Std 484-1975, IEEE Recommended Practice for Installation Design and Installation of Large Lead Storage Batteries for Generating Stations and Substations, it will provide the user with a general guide to designing, placing in service, and maintaining a large storage battery installation. At the time of final working group approval, the members were:

J. H. Bellack, *Chairman*

J. W. Anderson	E. C. Haupt
D. B. Brandt	R. W. Hopewell
V. E. Dalke	W. F. Hurley
R. DeBlasio	C. W. Jordan
G. Endicott	H. L. McCloud
G. Fligg	M. W. Migliaro
J. L. Giambalvo	L. A. Nienaber
W. E. Golde	H. K. Reid
A. P. Grande	B. G. Treece

L. D. Zachau, Jr

Comments, suggestions, and requests for interpretations should be addressed to:
Secretary
IEEE Standards Board
Institute of Electrical and Electronics Engineers, Inc
345 East 47th Street
New York, NY 10017

Contents

IEEE Recommended Practice for Sizing Large Lead Storage Batteries for Generating Stations and Substations

1. Scope

This Recommended Practice describes methods for defining the dc load in a generating station or a substation and for sizing a lead storage battery to supply that load. Some factors relating to cell selection are provided for consideration. Installation, maintenance, qualification, testing procedures, and consideration of battery types other than lead storage are beyond the scope of this Recommended Practice. Design of the dc system and sizing of the battery charger(s) are also beyond the scope of this Recommended Practice.

2. Definitions

The following definitions apply specifically to this Recommended Practice. For other definitions, see IEEE Std 380-1975, Definitions of Terms Used in IEEE Nuclear Power Generating Station Standards, and ANSI/IEEE Std 100-1977, Dictionary of Electrical and Electronics Terms.

battery duty cycle. The load currents a battery is expected to supply for specified time periods.

cell size. The rated capacity of a lead storage cell or the number of plates in the cell.

full float operation. Operation of a dc system with the battery, battery charger, and load all connected in parallel and with the battery charger supplying the normal dc load plus any self-discharge or charging current, or both, required by the battery. (The battery will deliver current only when the load exceeds the charger output.)

period. An interval of time in the battery duty cycle during which the current is assumed to be constant for purposes of cell sizing calculations.

rated capacity. The ampere-hour capacity assigned to a lead storage cell by its manufacturer for a given discharge time, at a specified electrolyte temperature and specific gravity, to a given end-of-discharge voltage.

3. References

[1] IEEE Std 380-1975, Definitions of Terms Used in IEEE Standards on Nuclear Power Generating Stations.
[2] ANSI/IEEE Std 100-1977, Standard Dictionary of Electrical and Electronics Terms.
[3] ANSI/IEEE Std 484-1975, Recommended Practice for Installation Design and Installation of Large Lead Storage Batteries for Generating Stations and Substations.
[4] ANSI/IEEE Std 450-1975, Recommended Practice for Maintenance, Testing, and Replacement of Large Lead Storage Batteries for Generating Stations and Substations.
[5] IEEE Std 323-1974, Qualifying Class 1E Equipment for Nuclear Power Generating Stations.

[6] HOXIE, E.A. Some Discharge Characteristics of Lead-Acid Batteries. *AIEE Transactions (Applications and Industry)*, vol 73, pp 17–22, 1954.

4. Defining Loads

4.1 General Considerations. The duty cycle imposed on the battery by any of the conditions described herein will depend on the dc system design and the requirements of the installation. The dc power requirements that the battery must supply occur when:

(1) Load on the dc system exceeds the maximum output of the battery charger

(2) Output of the battery charger is interrupted

(3) Auxiliary ac power is lost (may result in a greater dc power demand than (2) above)

The most severe of these conditions should be used to determine the battery size for the installation. A diagram of the duty cycle (Fig 1), showing the battery loads in amperes and the lengths of time for which they must be supplied, is normally plotted for this condition. The total time span of the duty cycle is determined by the requirements of the installation and need not exceed the time required to reduce the battery load to zero. This may be accomplished by restoration of ac power, restoration of battery charger output, or termination of battery loads.

4.2 Load Classifications. The individual dc loads supplied by the battery during the duty cycle may be classified as continuous or noncontinuous. Noncontinuous loads lasting one minute or less are designated "momentary loads" and should be given special consideration (see 4.2.3).

4.2.1 Continuous loads are energized throughout the duty cycle. These loads are those normally carried by the battery charger and those initiated at the inception of the duty cycle. Typical continuous loads are:

(1) Lighting
(2) Continuously operating motors
(3) Inverters
(4) Indicating lights
(5) Continuously energized coils
(6) Annunciator loads

4.2.2 Noncontinuous loads are energized during only a portion of the duty cycle. These loads may come on at any time within the duty cycle and may: be on for a set length of time; be removed automatically or by operator action; or continue to the end of the duty cycle. Typical noncontinuous loads are:

(1) Emergency pump motors
(2) Critical ventilation system motors
(3) Communication system power supplies
(4) Fire protection systems

4.2.3 Momentary loads can occur repeatedly during the duty cycle but are of short duration, not exceeding 1 min at any occurrence. Although momentary loads may exist for only a fraction of a second, each is considered to last for a full minute because the instantaneous battery voltage drop for a given momentary load is essentially the same as the voltage drop after 1 min. When several momentary loads occur within the same 1 min period and a discrete sequence cannot be established, the load shall be assumed to be the sum of all momentary loads occurring within that minute. If a discrete sequence can be established, the load for the 1 min period shall be assumed to be the maximum current at any instant. Typical momentary loads are:

(1) Switchgear operations
(2) Motor-driven valve operations
(3) Isolating switch operations
(4) Field flashing of generators
(5) Motor starting currents
(6) Inrush currents

4.2.4 The above lists of typical loads are not a full catalog of the dc loads at any one installation. The designer should review each system carefully to be sure he has included all possible loads.

4.3 Duty Cycle Diagram. A duty cycle diagram showing total current at any time during the cycle is an aid in the analysis of the duty cycle. To prepare such a diagram, all loads expected during the cycle are tabulated along with their anticipated inception and shutdown times.

4.3.1 Loads whose inception and shutdown times are known are plotted on the diagram as they would occur. If the inception time is known, but the shutdown time is indefinite, it shall be assumed that the load will continue through the remainder of the duty cycle.

4.3.2 Loads which occur at random shall be shown at the most critical time of the duty cycle in order to simulate the worst case load on the battery. These may be noncontinuous or momentary loads as described in 4.2.2 and 4.2.3. To determine the most critical time, it is necessary to size the battery without the random load(s) and to identify the portion of the duty cycle that controls battery size. Then the random load(s) shall be superimposed on the end of that controlling section as shown in Fig 1 (see 6.3.4).

4.3.3 Fig 1 is a diagram of a duty cycle made up of the following hypothetical loads.

L_1 40 amperes for 3 hours — continuous load

L_2 280 amperes for the 1st minute — momentary load, actually 5 seconds starting current to load L_3

L_3 60 amperes from the first minute through the 120th minute — noncontinuous load

L_4 100 amperes from the 30th minute through the 120th minute — noncontinuous load

L_5 80 amperes from the 30th minute through the 60th minute — noncontinuous load

L_6 80 amperes for the last minute — momentary load, actually a known sequence of:

      40 amperes for the first 5 seconds
      80 amperes for the first 10 seconds

      30 amperes for the next 20 seconds

L_7 100 amperes for 1 minute — random load. Actually this consists of four 25 amperes momentary loads that can occur at any time within the duty cycle. Therefore, the assumption is that they all occur simultaneously.

5. Cell Selection

This section summarizes some factors, other than capacity, that should be considered in selecting a cell design for a particular application.

5.1 Cell Designs. All lead storage cells used in applications covered by this Recommended Practice are categorized by the differences in positve plate design. The negative plates used in these cells are all of the "pasted plate" design.

5.1.1 The term "Planté" designates a positive plate made up largely or entirely of pure lead with the pure lead surface area increased by mechanical means. The plate may be a single sheet of pure lead or may have a lead-antimony frame supporting pure lead inserts.

5.1.2 The term "pasted plate" (or "Faure") designates a positive plate consisting of a lead-alloy grid pasted with lead compounds. This is

**Fig 1
Diagram of a Duty Cycle**

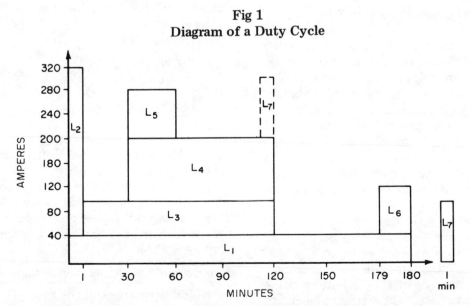

NOTE: This example is worked out in detail in the Appendix. There it will be found that the first 120 minutes is the controlling portion of the duty cycle. Therefore, the random load is located on the duty cycle so that the random load ends at the end of the 120th minute. This is indicated by the dashed lines.

the most widely used positive plate design. The grid alloy is either lead-antimony or lead-calcium.

5.1.3 The term "tubular" designates a positive plate made up of the following major components:

(1) Lead-antimony alloy grid consisting of parallel vertical spines

(2) Porous tubes which go over the spines

(3) Powdered compounds of lead which are used to fill the tubes

5.2 Selection Factors. The following factors should be considered in the selection of the cell:

(1) Physical characteristics, such as size and weight of the cells, container material, vent caps, intercell connectors, and terminals

(2) Planned life of the installation and expected life of the cell design

(3) Frequency and depth of discharge

(4) Ambient temperature. Note that high ambient temperatures result in reduced battery life. See 4.1.1.(3) of ANSI/IEEE Std 484-1975.

(5) Maintenance requirements of the various cell designs

(6) Seismic characteristics of the cell design

The battery manufacturers can provide detailed information on the many different cell designs.

6. Determining Battery Size

Three basic factors govern the size (number of cells and rated capacity) of the battery: the maximum system voltage; the minimum system voltage; and the duty cycle. Since a battery is usually composed of a number of identical cells connected in series, the voltage of the battery is the voltage of a cell times the number of cells in a series. The capacity of a battery is the same as the capacity of a single cell, which depends upon the size and number of plates.

If cells of sufficiently large capacity are not available, then two or more strings, of equal numbers of series-connected cells, may be connected in parallel to obtain the necessary capacity. The capacity of such a battery is the sum of the capacities of the strings. The conditions under which a battery is discharged can

change the effective capacity of that battery. For example:

(1) Low temperatures reduce the available capacity of the battery below its rated capacity

(2) The capacity decreases as the discharge rate increases

(3) The minimum specified cell voltage at any time during the battery discharge cycle limits the ampere-hours that the battery can supply

6.1 Number of Cells. The maximum allowable system voltage determines the number of cells in the battery consistent with the permissible battery float and equalizing voltages. It has been common practice to use 12, 24, 60, or 120 cells for system voltages of 24, 48, 125, or 250 V. In some cases, it may be desirable to vary from this practice to more closely match the battery to system voltage limitations.

6.1.1 Maximum System Voltage as Limiting Factor. When the battery voltage is not allowed to exceed a given maximum system voltage, the number of cells will be limited by the cell voltage required for satisfactory charging. That is,

$$\frac{\text{maximum allowable battery voltage}}{\text{cell voltage required for charging}} =$$

$$\text{number of cells}$$

Example. Assume 2.33 Volts per cell (VPC) required for charging and that the maximum allowable system voltage is (1) 140 V, or (2) 135 V. Then:

$$(1) \frac{140 \text{ V}}{2.33 \text{ VPC}} = 60.09 \text{ cells; use 60 cells}$$

$$(2) \frac{135 \text{ V}}{2.33 \text{ VPC}} = 57.94 \text{ cells; use 58 cells}$$

6.1.2 Minimum System Voltage as Limiting Factor. For a minimum battery voltage, determined by the minimum system voltage, the use of the largest possible number of cells allows the lowest end-of-discharge cell voltage and, therefore, the smallest size cell for the duty cycle. That is,

$$\frac{\text{minimum allowable battery voltage}}{\text{end-of-discharge cell voltage}} =$$

$$\text{number of cells}$$

Example. Assume that minimum allowable battery voltage is 105 V and that the desired end-of-discharge cell voltage is (1) 1.75 VPC, or (2) 1.81 VPC. Then:

(1) $\dfrac{105 \text{ V}}{1.75 \text{ VPC}} = 60$ cells

(2) $\dfrac{105 \text{ V}}{1.81 \text{ VPC}} = 58.01$ cells; use 58 cells

6.1.3 Float Voltage as Limiting Factor. In order to eliminate the need for frequent equalizing charges (refer to ANSI/IEEE Std 450-1975, Recommended Practice for Maintenance, Testing, and Replacement of Large Lead Storage Batteries for Generating Stations and Substations), it may be desirable to establish a float voltage at the high end of the manufacturer's recommended range. The float voltage must, however, be consistent with the maximum system voltage (see 6.1.1). This higher float voltage may then reduce the number of cells and may increase the cell size required for a given battery duty cycle.

6.1.4 Charging Rate as Limiting Factor. The time available to recharge the battery can affect both the number of cells and the cell size. The time required for a recharge decreases as the charging voltage per cell increases, assuming that the charging equipment can supply the high current necessary early in the recharge cycle. If the maximum charging voltage is limited, it is necessary to select the number of cells that can be recharged in the time avail-

able. This, in turn, may require using a larger cell than would otherwise have been necessary.

6.1.5 Rounding Off. If the calculations shown in 6.1.1 and 6.1.2 indicate a need for a fractional cell, round that result off to the nearest whole number of cells. Then adjust the end-of-discharge, float, and charge voltages accordingly.

6.2 Additional Considerations. Before proceeding to calculate the cell capacity required for a particular installation, the designer should consider factors that will influence cell size but that are not included in the general equation.

6.2.1 Temperature Correction Factor. The available capacity of a cell is affected by its operating temperature. The standard temperature for stating cell capacity is 25°C (77°F). If the lowest expected electrolyte temperature is below standard, select a cell large enough to have the required capacity available at the lowest expected temperature. If the lowest expected electrolyte temperature is above 25°C (77°F), it is a conservative practice to select a cell size to match the required capacity at the standard temperature and to recognize the resulting increase in available capacity as part of the overall design margin. Table 1 lists cell size correction factors for various temperatures. For unlisted temperatures within the range of Table 1, interpolate between adjacent values and round to two decimal places.

Table 1
Cell Size Correction Factors for Temperature

Electrolyte Temperature (°F)	(°C)	Cell Size Correction Factor	Electrolyte Temperature (°F)	(°C)	Cell Size Correction Factor
25	− 3.9	1.52	80	26.7	.98
30	− 1.1	1.43	85	29.4	.96
35	1.7	1.35	90	32.2	.94
40	4.4	1.30	95	35.0	.93
45	7.2	1.25	100	37.8	.91
50	10.0	1.19	105	40.6	.89
55	12.8	1.15	110	43.3	.88
60	15.6	1.11	115	46.1	.87
65	18.3	1.08	120	48.9	.86
70	21.1	1.04	125	51.7	.85
77	25.0	1.00			

NOTES:
(1) These correction factors were developed from manufacturers' published data.
(2) Table 1 is applicable regardless of the capacity rating factor used (see 6.3.3) and applies to all discharge rates.

6.2.2 Design Margin. It is prudent design practice to provide a capacity margin to allow for unforeseen additions to the dc system and less-than-optimum operating conditions of the battery due to improper maintenance, recent discharge, or ambient temperatures lower than anticipated, or both. A method of providing this design margin is to add 10–15 percent to the cell size determined by calculations. If the various loads are expected to grow at different rates, it may be more accurate to apply the expected growth rate to each load for a given time and to develop a duty cycle from the results.

The cell size calculated for a specific application will seldom match a commercially available cell exactly, and it is normal procedure to select the next higher capacity cell. The additional capacity obtained can be considered part of the design margin.

Note that the "margins" required by 6.3.1.5 and 6.3.3 of IEEE Std 323-1974, Qualifying Class 1E Equipment for Nuclear Power Generating Stations, are to be applied during "qualification" and are not related to "design margin."

6.2.3 Compensation for Age. IEEE Std 450-1975 recommends that a battery be replaced when its actual capacity drops to 80 percent of its rated capacity; therefore, the battery's rated capacity should be at least 125 percent of the load expected at the end of its service life.

6.2.4 Initial Capacity. Batteries may have less than rated capacity when delivered. Unless 100 percent capacity upon delivery is specified, initial capacity can be in the range of 90–95 percent of rated. This will rise to rated capacity in normal service after several charge-discharge cycles or after several years of float operation.

NOTE: If the designer has provided adequate compensation for aging (see 6.2.3), there is no need for the battery to have full rated capacity upon delivery, because the capacity normally available from a new battery will be above the duty cycle requirement.

6.3 Cell Size. This section describes and explains a proven method of calculating the cell capacity necessary for satisfactory performance on a given duty cycle. The Appendix demonstrates the application of this method to a specific duty cycle, using an optional preprinted work sheet to simplify the calculations. Section 6.4 provides instructions for the proper use of the work sheet.

6.3.1 Equation 1 (see 6.3.2) requires the use of a capacity rating factor C_T (6.3.3) that is based on the discharge characteristics of a particular plate type and size. Thus, the initial calculation must be based on a trial selection of positive plate type and capacity. Depending on the results of this initial calculation, it may be desirable to repeat the calculation for other types or sizes of plates to obtain the optimum cell type and size for the particular application. Use the capacity from the first calculation as a guide for selecting additional types to size.

6.3.2 The cell selected for a specific duty cycle must have enough capacity to carry the combined loads during the duty cycle. To determine the required cell size, it is necessary to calculate, from an analysis of each section of the duty cycle (see Fig 2), the maximum capacity required by the combined load demands (current versus time) of the various sections. The first section analyzed is the first period of the duty cycle. Using the capacity rating factor (see 6.3.3) for the given cell type, a cell size is calculated that will supply the required current for the duration of the first period. For the second section, the capacity is calculated assuming that the current A_1 required for the first period continued through the second period; this capacity is then adjusted for the change in current $(A_2 - A_1)$ during the second period. In the same manner, the capacity is calculated for each subsequent section of the duty cycle. This iterative process is continued until all sections of the duty cycle have been considered. The calculation of the capacity F_S required by each section S, where S can be any integer from 1 to N, can be expressed mathematically as follows:

$$F_S = \sum_{P=1}^{P=S} \frac{A_P - A_{(P-1)}}{C_T} \qquad \text{(Eq 1)}$$

NOTE: F_S will be expressed in ampere-hours or number of positive plates, depending upon which C_T is used (see 6.3.3).

The maximum capacity (max F_S) calculated determines the cell size that can be expressed by the following general equation:

$$\text{cell size} = \max_{S=1}^{S=N} F_S \qquad \text{(Eq 2)}$$

where

S = section of the duty cycle being analyzed. Section S contains the first S periods of the duty cycle (for example, section S_5 contains periods 1 through 5). See Fig 2 for a graphical representation of "section."

N = number of periods in the duty cycle

P = period being analyzed

A_P = amperes required for period P

T = time in minutes from the beginning of period P through the end of section S

C_T = capacity rating factor (see 6.3.3) for a given cell type, at the T minute discharge rate, at 25°C (77°F), to a definite end-of-discharge voltage

NOTE: If the current for period $P + 1$ is greater than the current for period P, then section $S = P + 1$ will require a larger cell than section $S = P$. Consequently, the calculations for section $S = P$ can be omitted.

6.3.3 There are two terms for expressing the capacity rating factor C_T of a given cell type in cell sizing calculations. One term R_T, is the number of amperes that each positive plate can supply for T minutes, at 25°C (77°F), to a definite end-of-discharge voltage. Therefore, $C_T = R_T$ and, combining equation 1 and equation 2:

cell size (positive plates) =

$$\max_{S=1}^{S=N} F_S = \max_{S=1}^{S=N} \sum_{P=1}^{P=S} \frac{A_P - A_{(P-1)}}{R_T} \quad \text{(Eq 3)}$$

The other term K_T, is the ratio of rated ampere-hour capacity (at a standard time rate, at 25°C [77°F] and to a standard end-of-discharge voltage) of a cell, to the amperes that can be supplied by that cell for T minutes at 25°C (77°F) and to a given end-of-discharge voltage. Therefore $C_T = 1/K_T$ and equation 3 can be rewritten:

cell size (ampere-hours) =

$$\max_{S=1}^{S=N} F_S = \max_{S=1}^{S=N} \sum_{P=1}^{P=S} [A_P - A_{(P-1)}] K_T$$

(Eq 4)

NOTE: R_T is not equal to $1/K_T$ because of the different units applied to each factor. However, R_T is proportional to $1/K_T$. The values may be obtained from battery manufacturers for each positive plate design and various end-of-discharge voltages.

6.3.4 When equipment loads that occur at random are included as part of the battery duty cycle, it is necessary to calculate the cell size required for the duty cycle without the random load(s) and then add to this the cell size required for the random load(s) only.

6.3.5 When used with the factor R_T (amperes per positive plate), the general equation expresses the cell size as the number of positive plates. In the manufacturers' literature, the cell size will be listed as the total number of positive and negative plates. The conversion from number of positive plates to the total number of plates is:

total number of plates =
1 + (2 × number of positive plates)   (Eq 5)

Fig 2
Generalized Duty Cycle Diagram

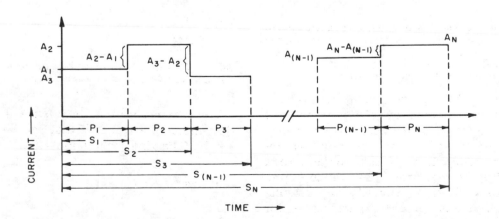

6.4 Cell Sizing Work Sheet. A work sheet, Fig 3, has been designed and may be used to simplify the manual application of the procedure described in 6.3. Examples of its use will be found in the Appendix. Following are instructions for proper use of the work sheet.

6.4.1 Fill in necessary information in the heading of the chart. The temperature and voltage recorded are those used in the calculations. The voltage used is the minimum allowable battery voltage divided by the number of cells in the battery.

6.4.2 Fill in the amperes and the minutes in columns (2) and (4) as indicated by the section heading notations.

6.4.3 Calculate and record the changes in amperes as indicated in column (3). Record whether the changes are positive or negative.

6.4.4 Calculate and record the times from the start of each period to the end of the section as indicated in column (5).

6.4.5 Record in column (6) the capacity factors (R_T or K_T, from the manufacturer's literature) for each discharge time calculated in column (5).

6.4.6 Calculate and record the cell size for each period as indicated in column (7). Note the separate subcolumns for positive and negative values.

6.4.7 Calculate and record in column (7) the algebraic subtotals and totals for each section as indicated.

6.4.8 Record the maximum section size (the largest total from column 7) on line (8), the random section size on line (9), and the uncorrected size (US) on lines (10) and (11).

6.4.9 Select the correction factor from Table 1 for the temperature shown in the main heading and record it on line (12).

6.4.10 Enter the design margin on line (13) and the aging factor on line (14). Combine lines (11), (12), (13), and (14) as indicated and record the result on line (15).

6.4.11 When line (15) is in terms of ampere-hours and does not match the capacity of a commercially available cell, the next larger cell is required. When line (15) shows a fractional number of positive plates, use the next larger integer. Show the result on line (16).

6.4.12 From the value on line (16), 6.3.5, and the manufacturer's literature, determine the commercial designation of the required cell and record it on line (17).

Project: _____ Date: _____ Page: _____

Lowest Expected Electrolyte Temp: °F	Minimum Cell Voltage:	Cell Mfg:	Cell Type:	Sized By:

(1) Period	(2) Load (amperes)	(3) Change in Load (amperes)	(4) Duration of Period (minutes)	(5) Time to End of Section (minutes)	(6) Capacity at T Min Rate (6A) Amps/Pos (R_T) or (6B) K Factor (K_T)	(7) Required Section Size (3) ÷ (6A) = Positive Plates or (3) × (6B) = Rated Amp Hrs	
						Pos Values	Neg Values

Section 1 — First Period Only — If A2 is greater than A1, go to Section 2.

1	A1=	A1—0=	M1=	T=M1=			* * *
					Sec 1 Total		* * *

Section 2 — First Two Periods Only — If A3 is greater than A2, go to Section 3.

1	A1=	A1—0=	M1=	T=M1+M2=			
2	A2=	A2—A1=	M2=	T=M2=			
					Sec 2 Sub Tot Total		* * *

Section 3 — First Three Periods Only — If A4 is greater than A3, go to Section 4.

1	A1=	A1—0=	M1=	T=M1+M2+M3=			
2	A2=	A2—A1=	M2=	T=M2+M3=			
3	A3=	A3—A2=	M3=	T=M3=			
					Sec 3 Sub Tot Total		* * *

Section 4 — First Four Periods Only — If A5 is greater than A4, go to Section 5.

1	A1=	A1—0=	M1=	T=M1+...M4=			
2	A2=	A2—A1=	M2=	T=M2+M3+M4=			
3	A3=	A3—A2=	M3=	T=M3+M4=			
4	A4=	A4—A3=	M4=	T=M4=			
					Sec 4 Sub Tot Total		* * *

Section 5 — First Five Periods Only — If A6 is greater than A5, go to Section 6.

1	A1=	A1—0=	M1=	T=M1+...M5=			
2	A2=	A2—A1=	M2=	T=M2+...M5=			
3	A3=	A3—A2=	M3=	T=M3+M4+M5=			
4	A4=	A4—A3=	M4=	T=M4+M5=			
5	A5=	A5—A4=	M5=	T=M5=			
					Sec 5 Sub Tot Total		* * *

Section 6 — First Six Periods Only — If A7 is greater than A6, go to Section 7.

1	A1=	A1—0=	M1=	T=M1+...M6=			
2	A2=	A2—A1=	M2=	T=M2+...M6=			
3	A3=	A3—A2=	M3=	T=M3+...M6=			
4	A4=	A4—A3=	M4=	T=M4+M5+M6=			
5	A5=	A5—A4=	M5=	T=M5+M6=			
6	A6=	A6—A5=	M6=	T=M6=			
					Sec 6 Sub Tot Total		* * *

Section 7 — First Seven Periods Only — If A8 is greater than A7, go to Section 8.

1	A1=	A1—0=	M1=	T=M1+...M7=			
2	A2=	A2—A1=	M2=	T=M2+...M7=			
3	A3=	A3—A2=	M3=	T=M3+...M7=			
4	A4=	A4—A3=	M4=	T=M4+...M7=			
5	A5=	A5—A4=	M5=	T=M5+M6+M7=			
6	A6=	A6—A5=	M6=	T=M6+M7=			
7	A7=	A7—A6=	M7=	T=M7=			
					Sec 7 Sub Tot Total		* * *

Random Equipment Load Only (if needed)

R	AR=	AR—0=	MR=	T=MR=			* * *

Maximum Section Size (8) _____ + Random Section Size (9) _____ = Uncorrected Size — (US) (10) _____

US (11) _____ × Temp Corr (12) _____ × Design Marg (13) 1. ____ × Aging Factor (14) 1. ____ = (15) _____.

When the cell size (15) is greater than a standard cell size, the next larger cell is required.

(A) — Positive Plates.

Required cell size (16) _____

(B) — Ampere Hours. Therefore cell (17) _____ is required.

Fig 3
Cell Sizing Work Sheet

Appendix

(This Appendix is not a part of IEEE Std 485-1978, Recommended Practice for Sizing Large Lead Storage Batteries for Generating Stations and Substations.)

In all the following examples, the duty cycle used is that of 4.3.3 and the lowest expected electrolyte temperature is 18.3°C (65°F). Section A1 provides several examples of calculations selecting the number of cells to be used in the battery and shows how the number of cells affects the required cell capacity. Section A2 shows how the cell sizing work sheet can be used to calculate the required cell size.

A1 Required Number of Cells

Example 1. The dc system voltage limits are 105 V to 140 V; the battery must be recharged at 2.33 V per cell (VPC) to reduce recharge time, and the battery and charger must remain directly connected to the dc system at all times.

Number of cells = 140V/2.33 VPC = 60.1, therefore 60 cells

End-of-discharge voltage = 105V/60 cells
        = 1.75 VPC

Corrected cell size = 1010.4 ampere hours at the 8 hour rate

This example is worked out in detail in Fig A1.

Example 2. Same dc system voltage limits as in Example 1, but the battery and charger can be isolated from the dc system during equalizing and recharging, and it is desired to float the battery at 2.25 VPC.

Number of cells = 140V/2.25 VPC = 62.2, therefore 62 cells

End-of-discharge voltage = 105V/62 cells
        = 1.69 VPC

A cell sizing work sheet is not provided for this example, but calculations show that the corrected cell size = 944 ampere hours at the 8 hour rate. The reduction in the end-of-discharge voltage results in about a 7 percent reduction in corrected cell capacity with only a 3 percent increase in the number of cells.

Example 3. Same conditions as Example 1, except the dc system voltage limits are now 105 V to 135 V.

Number of cells = 135V/2.33 VPC = 57.9, therefore 58 cells

End-of-discharge voltage = 105V/58 cells
        = 1.81 VPC.

A Cell Sizing Work Sheet is not provided for this example, but calculations show that the corrected cell size = 1186 ampere hours at the 8 hour rate. In comparison to Example 1, the increase in the end-of-discharge voltage results in a 17 percent increase in the corrected cell size with only a 3 percent reduction in the number of cells required.

A2 Required Cell Capacity

From the battery duty cycle diagram, Fig A2, we can construct Table A1, which will be of value in filling out the cell sizing work sheet.

Fig A3 is a hypothetical composite rating curve for the XYZ cell manufactured by the ABC Company. The graph gives values for both types of capacity rating factors for discharges started at 25°C (77°F) and terminated when the average cell voltage reaches 1.81, 1.75, or 1.69 V. Fig A4 shows the way in which the cell sizing work sheet and the R_T rating factor would be used to size the XYZ cell for the Fig 1 duty cycle. Fig A1 shows the application of the K_T rating factor to the same problem.

Table A1
Sample Cell Sizing Data

Period	Loads	Total Amperes	Duration (min)
1	$L_1 + L_2$	320	1
2	$L_1 + L_3$	100	29
3	$L_1 + L_3 + L_4 + L_5$	280	30
4	$L_1 + L_3 + L_4$	200	60
5	L_1	40	59
6	$L_1 + L_6$	120	1
R	L_7	100	1

Project: Example Using K_T Capacity Factor   Date: 9/17/77   Page: 1

Lowest Expected Electrolyte Temp: 65°F   Minimum Cell Voltage: 1.75   Cell Mfg: ABC Co   Cell Type: XYZ   Sized By: J.W.A.

(1) Period	(2) Load (amperes)	(3) Change in Load (amperes)	(4) Duration of Period (minutes)	(5) Time to End of Section (minutes)	(6) Capacity At T Min Rate (6A) Amps/Pos (R_T) or (6B) K Factor (K_T)	(7) Required Section Size (3) ÷ (6A) = Positive Plates or (3) × (6B) = Rated Amp Hrs — Pos Values	Neg Values
Section 1 — First Period Only — If A2 is greater than A1, go to Section 2.							
1	A1=320	A1—0=320	M1=1	T=M1=1	0.77	246.4	* * *
					Sec 1 Total	246.4	* * *
Section 2 — First Two Periods Only — If A3 is greater than A2, go to Section 3.							
1	A1=	A1—0=	M1=	T=M1+M2=			
2	A2=	A2—A1=	M2=	T=M2=			
					Sec 2 Sub Tot Total		* * *
Section 3 — First Three Periods Only — If A4 is greater than A3, go to Section 4.							
1	A1=320	A1—0=320	M1=1	T=M1+M2+M3=60	2.00	640.0	
2	A2=100	A2—A1=—220	M2=29	T=M2+M3=59	2.00		— 440.0
3	A3=280	A3—A2=180	M3=30	T=M3=30	1.44	259.2	
					Sec 3 Sub Tot	899.2	— 440.0
					Total	459.2	* * *
Section 4 — First Four Periods Only — If A5 is greater than A4, go to Section 5.							
1	A1=320	A1—0=320	M1=1	T=M1+...M4=120	2.91	931.2	
2	A2=100	A2—A1=—220	M2=29	T=M2+M3+M4=119	2.91		— 640.2
3	A3=280	A3—A2=180	M3=30	T=M3+M4=90	2.46	442.8	
4	A4=200	A4—A3=—80	M4=60	T=M4=60	2.00		— 160.0
					Sec 4 Sub Tot	1374.0	— 800.2
					Total	573.8	* * *
Section 5 — First Five Periods Only — If A6 is greater than A5, go to Section 6.							
1	A1=	A1—0=	M1=	T=M1+...M5=			
2	A2=	A2—A1=	M2=	T=M2+...M5=			
3	A3=	A3—A2=	M3=	T=M3+M4+M5=			
4	A4=	A4—A3=	M4=	T=M4+M5=			
5	A5=	A5—A4=	M5=	T=M5=			
					Sec 5 Sub Tot Total		* * *
Section 6 — First Six Periods Only — If A7 is greater than A6, go to Section 7.							
1	A1=320	A1—0=320	M1=1	T=M1+...M6=180	3.72	1190.4	
2	A2=100	A2—A1=—220	M2=29	T=M2+...M6=179	3.72		— 818.4
3	A3=280	A3—A2=180	M3=30	T=M3+...M6=150	3.33	599.4	
4	A4=200	A4—A3=—80	M4=60	T=M4+M5+M6=120	2.91		— 232.8
5	A5=40	A5—A4=—160	M5=59	T=M5+M6=60	2.00		— 320.0
6	A6=120	A6—A5=80	M6=1	T=M6=1	0.77	61.6	
					Sec 6 Sub Tot	1851.4	—1371.2
					Total	480.2	* * *
Section 7 — First Seven Periods Only — If A8 is greater than A7, go to Section 8.							
1	A1=	A1—0=	M1=	T=M1+...M7=			
2	A2=	A2—A1=	M2=	T=M2+...M7=			
3	A3=	A3—A2=	M3=	T=M3+...M7=			
4	A4=	A4—A3=	M4=	T=M4+...M7=			
5	A5=	A5—A4=	M5=	T=M5+M6+M7=			
6	A6=	A6—A5=	M6=	T=M6+M7=			
7	A7=	A7—A6=	M7=	T=M7=			
					Sec 7 Sub Tot Total		* * *
Random Equipment Load Only (if needed)							
R	AR=100	AR—0=100	MR=1	T=MR=1	0.77	77.0	* * *

Maximum Section Size (8) <u>573.8</u> + Random Section Size (9) <u>77.0</u> = Uncorrected Size — (US)(10) <u>650.8</u>.

US(11) <u>650.8</u> × Temp. Corr. (12) <u>1.08</u> × Design Margin (13) <u>1.15</u> × Aging Factor (14) <u>1.25</u> = (15) <u>1010.4</u>.

When the cell size (15) is greater than standard cell size, the next larger cell is required.

Required cell size (16) 1040 ^(A) — Positive plates.

<u>(B)</u> — Ampere hours. Therefore cell (17) XYZ-27 is required.

Fig A1
Sample Work Sheet Using K_T Capacity Factor

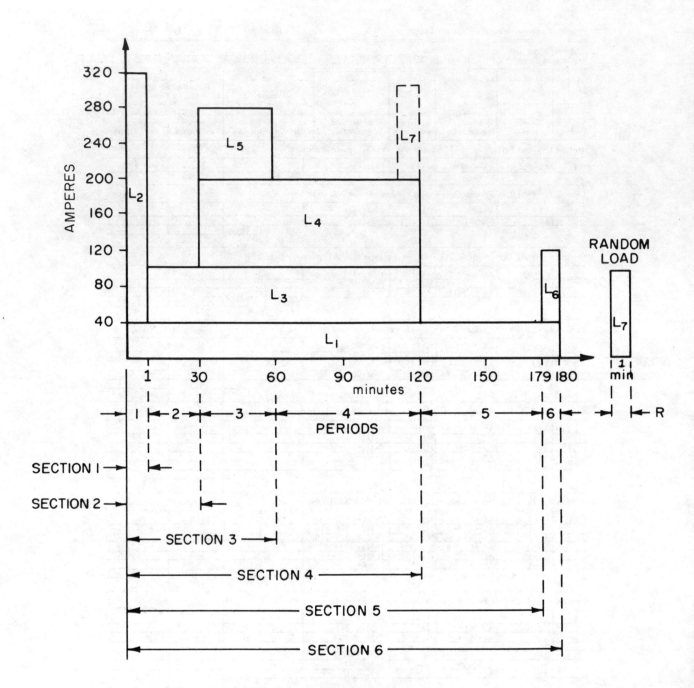

Fig A2
Battery Duty Cycle Diagram

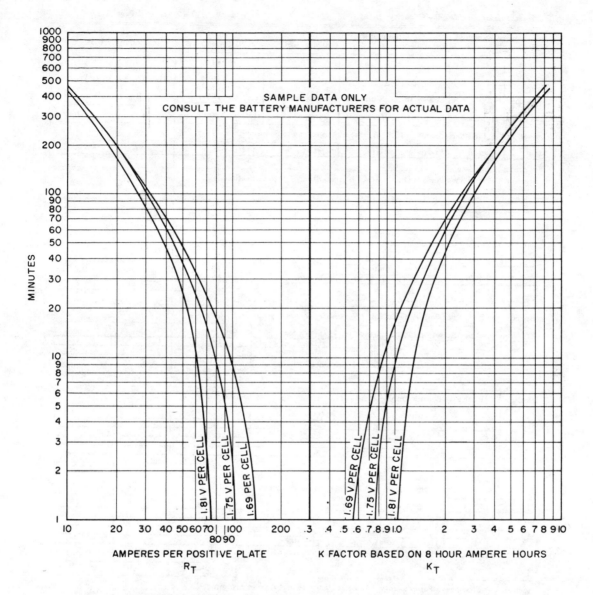

Available	Number of Plates	11	13	15	17	19	21	23	25	27
Sizes	8 h (Ah) (1.75 VPC)	400	480	560	640	720	800	880	960	1040

Fig A3
Hypothetical Composite
Rating Curve for
XYZ Cell Manufactured by ABC Company

Project: Example Using R_T Capacity Factor　　　　　　Date: 9/17/77　　　Page: 1

Lowest Expected Electrolyte Temp: 65°F	Minimum Cell Voltage: 1.75	Cell Mfg: ABC Co.		Cell Type: XYZ		Sized by: J.W.A.
(1)	(2)	(3)	(4)	(5)	(6)	(7)
Period	Load (amperes)	Change in Load (amperes)	Duration of Period (minutes)	Time to End of Section (minutes)	Capacity At T Min Rate (6A) Amps/Pos (R_T) or (6B) K Factor (K_T)	Required Section Size (3) ÷ (6A) = Positive Plates or (3) × (6B) = Rated Amp. Hrs.
						Pos Values — Neg Values

Section 1 — First Period Only — If A2 is greater than A1, go to Section 2.

1	A1=320	A1—0=320	M1=1	T=M1=1	104	3.08	* * *
					Sec 1 Total	3.08	* * *

Section 2 — First Two Periods Only — If A3 is greater than A2, go to Section 3.

1	A1=	A1—0=	M1=	T=M1+M2=			
2	A2=	A2—A1=	M2=	T=M2=			
					Sec 2 Sub Tot Total		* * *

Section 3 — First Three Periods Only — If A4 is greater than A3, go to Section 4.

1	A1=320	A1—0=320	M1=1	T=M1+M2+M3=60	40.0	8.00	
2	A2=100	A2—A1=—220	M2=29	T=M2+M3=59	40.0		— 5.50
3	A3=280	A3—A2=180	M3=30	T=M3=30	55.4	3.25	
					Sec 3 Sub Tot	11.25	— 5.50
					Total	5.75	* * *

Section 4 — First Four Periods Only — If A5 is greater than A4, go to Section 5.

1	A1=320	A1—0=320	M1=1	T=M1+...M4=120	27.5	11.64	
2	A2=100	A2—A1=—220	M2=29	T=M2+M3+M4=119	27.5		— 8.00
3	A3=280	A3—A2=180	M3=30	T=M3+M4=90	32.5	5.54	
4	A4=200	A4—A3=—80	M4=60	T=M4=60	40.0		— 2.00
					Sec 4 Sub Tot	17.18	—10.00
					Total	7.18	* * *

Section 5 — First Five Periods Only — If A6 is greater than A5, go to Section 6.

1	A1=	A1—0=	M1=	T=M1+...M5=			
2	A2=	A2—A1=	M2=	T=M2+...M5=			
3	A3=	A3—A2=	M3=	T=M3+M4+M5=			
4	A4=	A4—A3=	M4=	T=M4+M5=			
5	A5=	A5—A4=	M5=	T=M5=			
					Sec. 5 Sub Tot Total		* * *

Section 6 — First Six Periods Only — If A7 is greater than A6, go to Section 7.

1	A1=320	A1—0=320	M1=1	T=M1+...M6=180	21.5	14.88	
2	A2=100	A2—A1—220	M2=29	T=M2+...M6=179	21.5		—10.23
3	A3=280	A3—A2=180	M3=30	T=M3+...M6=150	24.0	7.50	
4	A4=200	A4—A3=—80	M4=60	T=M4+M5+M6=120	27.5		— 2.91
5	A5=40	A5—A4=—160	M5=59	T=M5+M6=60	40.0		— 4.00
6	A6=120	A6—A5=80	M6=1	T=M6=1	104.0	0.77	
					Sec 6 Sub Tot	23.15	—17.14
					Total	6.01	* * *

Section 7 — First Seven Periods Only — If A8 is greater than A7, go to Section 8.

1	A1=	A1—0=	M1=	T=M1+...M7=			
2	A2=	A2—A1=	M2=	T=M2+...M7=			
3	A3=	A3—A2=	M3=	T=M3+...M7=			
4	A4=	A4—A3=	M4=	T=M4+...M7=			
5	A5=	A5—A4=	M5=	T=M5+M6+M7=			
6	A6=	A6—A5=	M6=	T=M6+M7=			
7	A7=	A7—A6=	M7=	T=M7=			
					Sec 7 Sub Tot Total		* * *

Random Equipment Load Only (if needed)

R	AR=100	AR—0=100	MR=1	T=MR=1	104.0	0.96	* * *

Maximum Section Size (8) <u>7.18</u> + Random Section Size (9) <u>0.96</u> = Uncorrected Size — (US)(10) <u>8.14.</u>

US(11) <u>8.14</u> × Temp. Corr. (12) <u>1.08</u> × Design Margin (13) <u>1.15</u> × Aging Factor (14) <u>1.25</u> = (15) <u>12.64.</u>

When the cell size (15) is greater than standard cell size, the next larger cell is required.

Required cell size (16) <u>13</u> (A) — <u>Positive plates.</u>

　　　　　　　　　(B) — Ampere hours. Therefore cell (17) <u>XYZ-27</u> is required.

Fig A4
Sample Work Sheet Using R_T Capacity Factor

ANSI/IEEE Std 494-1975
(ANSI N41.28-1976)

An American National Standard

IEEE Standard Method for Identification of Documents Related to Class 1E Equipment and Systems for Nuclear Power Generating Stations

494

Sponsor

**Power Generation Committee of the
IEEE Power Engineering Society**

**Approved January 12, 1976
American National Standards Institute**

© Copyright 1974 by

The Institute of Electrical and Electronics Engineers, Inc.

Foreword

(This foreword is not a part of IEEE Std 494-1974, Standard Method for Identification of Documents Related to Class 1E Equipment and Systems for Nuclear Power Generating Stations.)

The Institute of Electrical and Electronics Engineers has generated this document to provide criteria for the uniform identification of documents relating to Class 1E equipment and systems for nuclear power generating stations. It supplements IEEE Std 279-1971, Criteria for Protection Systems for Nuclear Power Generating Stations (ANSI N42.7-1972), which describes the basic need for document identification. IEEE Std 308-1974, Standard Criteria for Class 1E Power Systems for Nuclear Power Generating Stations, also describes this need and includes the basic definition for Class 1E.

It is the responsibility of the applicant to define those documents considered significant.

It is hoped that other disciplines will give consideration to the adoption of the term *nuclear safety related* as it applies to their documentation.

This standard was prepared by the Adhoc Working Group on Document Identification of the Power Generation Committee. Members of this group were:

J. R. Hall, *Chairman*

I. Ahmed
J. E. Arden
R. W. Cantrell
C. L. Cobler
N. C. Farr

R. K. Jones
T. M. McMahon
R. E. Penn
W. G. Schwartz
Bill Williams

This standard was balloted by the Nuclear Power Engineering and the Power Generation Committees of the IEEE Power Engineering Society.

Comments, suggestions, and requests for interpretation should be addressed to:

Secretary
IEEE Standards Board
The Institute of Electrical and Electronics Engineers, Inc
345 East 47 Street
New York, NY 10017

Contents

IEEE Standard Method for Identification of Documents Related to Class 1E Equipment and Systems for Nuclear Power Generating Stations

1. Scope

This document establishes criteria for the uniform identification of documents significant to design, construction, testing, operation, and maintenance of Class 1E equipment and systems for nuclear power generating stations. It includes recommendations for identification of specific parts of these documents. Criteria are also established for identification on documents of redundant portions of Class 1E equipment and systems.

The quality assurance program, as applied to Class 1E equipment procurement, should provide records sufficient to meet the needs and requirements of IEEE Std 279-1971, Criteria for Protection Systems for Nuclear Power Generating Stations (ANSI N42.7-1972), and IEEE Std 308-1974, Standard Criteria for Class 1E Power Systems for Nuclear Power Generating Stations. Additional identification on procurement documents is, therefore, not required.

2. Purpose

The purpose of this document is to provide a method for identification of documents relating to Class 1E equipment and systems such that users of the documents will be able to recognize those systems which must meet the requirements set forth in IEEE Std 279-1971 (ANSI N42.7-1972) and IEEE Std 308-1974.

3. Definitions

Class 1E. The safety classification of the electric equipment and systems that are essential to emergency reactor shutdown, containment isolation, reactor core cooling, and containment and reactor heat removal, or otherwise are essential in preventing significant release of radioactive material to the environment.

documents. Drawings and other records significant to the design, construction, testing, maintenance, and operation of Class 1E equipment and systems for nuclear power generating stations.

NOTE: Within the context of this standard, documents include:
(1) Drawings such as instrument diagrams, functional control diagrams, one line diagrams, schematic diagrams, equipment arrangements, cable and tray lists, wiring diagrams
(2) Instrument data sheets
(3) Design specifications
(4) Instruction manuals
(5) Test specifications, procedures, and reports
(6) Device lists
Not to be included as documents within the context of this standard are: project schedules, financial reports, meeting minutes, correspondence such as letters and memoranda, and equipment procurement documentation covered by quality assurance programs.

nuclear safety related. That term used to call attention to safety classifications incorporated in the body of the document so marked.

NOTE: As used in this document, the term calls attention to the safety classification Class 1E.

redundant equipment or systems. A piece of equipment or a system that duplicates the essential function of another piece of equipment or a system to the extent that either may perform the required function regardless of the state of operation or failure of the other.

significant. Demonstrated to be important by the safety analysis of the station.

4. Requirements

4.1 Document Identification
4.1.1 Any document, as defined in Section 3, for Class 1E equipment and systems, in

whole or in part, shall be identified with the term *nuclear safety related*.

4.1.2 The term *nuclear safety related* shall be placed conspicuously on the title or first page. Drawings for Class 1E equipment and systems shall have these words located in or immediately adjacent to the title block.

4.1.3 The term *nuclear safety related* shall be displayed in a character size equal to or larger than the largest size used in words in the body of the document.

4.2 Additional Identification of Specific Parts of Nuclear Safety Related Documents

4.2.1 Except as provided in Section 4.2.2, documents that are only in part nuclear safety related shall have the nuclear safety related or the non-nuclear safety related information identified to differentiate the two.

4.2.2 Documents in the form of written descriptions in which the nuclear safety related portions are evident by the text do not require identification other than that specified in Section 4.1.2.

4.3 Additional Identification of Redundant Class 1E Equipment and Systems on Nuclear Safety Related Documentation

4.3.1 Information that pertains to a specific equipment or system requiring separation from a redundant equipment or system shall be identified as pertaining only to that equipment or system. This identification shall include the equipment or system designation.

4.3.2 Documents pertaining entirely to a single redundant equipment or system shall be identified as pertaining to that equipment or system by placing the proper designation in or adjacent to the title of the document.

4.3.3 In addition to the information contained in Section 4.3.1, which pertains to equipment or systems shared by more than one unit, the information itself shall be identified as being shared by those units.

4.3.4 Designations that are adopted for identification of specific equipment or systems shall be shown on the document immediately adjacent to portions pertaining to the specific equipment or systems.

4.4 Symbols and Abbreviations

4.4.1 Symbols and abbreviations may be used to satisfy the requirements of Sections 4.2 and 4.3. If symbols or abbreviations are used, the document shall contain an explanation of each symbol or abbreviation or shall contain a reference concerning where this information can be found.

5. References

IEEE 279-1971, Criteria for Protection Systems for Nuclear Power Generating Stations (ANSI N42.7-1972).

IEEE Std 308-1974, Standard Criteria for Class 1E Power Systems for Nuclear Power Generating Stations

Appendix
Examples

(This appendix is not a part of IEEE Std 494-1974, Standard Method for Identification of Documents Related to Class 1E Equipment and Systems for Nuclear Power Generating Stations.)

Figs A1 through A5 are examples of identification methods that meet the intent of this document. They are presented for illustrative purposes only; any suitable method of identification of nuclear safety related information may be used if that method meets the criteria outlined in the text. The parenthesis-enclosed numbers refer to those text paragraphs that pertain to the identification used in the illustrations and are not to be used on the documents.

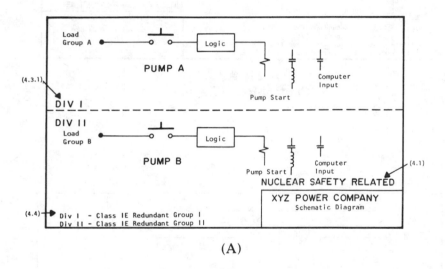

(A)

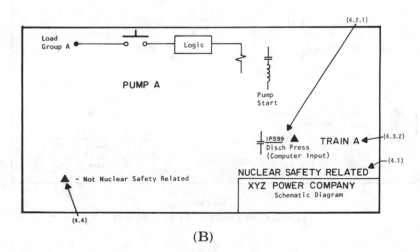

(B)

Fig A1
Schematic Diagram

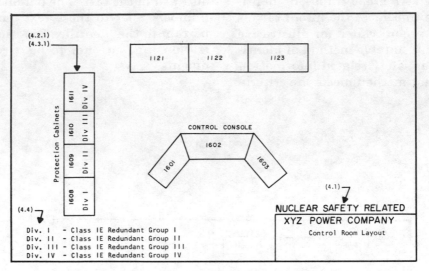

Fig A2
Control Room Layout

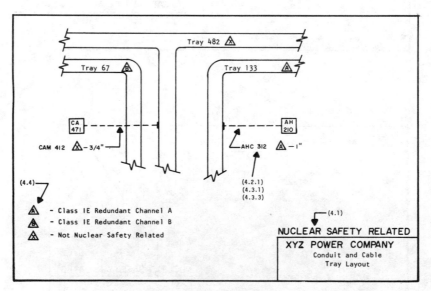

Fig A3
Conduit and Cable Tray Layout

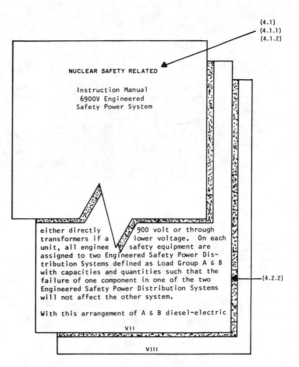

(4.1)
(4.1.1)
(4.1.2)

NUCLEAR SAFETY RELATED

Instruction Manual
6900V Engineered
Safety Power System

(4.2.2)

either directly   900 volt or through
transformers if a   lower voltage. On each
unit, all enginee   safety equipment are
assigned to two Engineered Safety Power Distribution Systems defined as Load Group A & B
with capacities and quantities such that the
failure of one component in one of the two
Engineered Safety Power Distribution Systems
will not affect the other system.

With this arrangement of A & B diesel-electric

VII

VIII

Fig A4
Instruction Manual

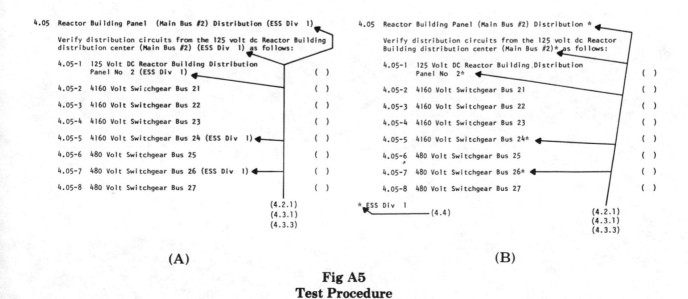

4.05 Reactor Building Panel (Main Bus #2) Distribution (ESS Div 1)

Verify distribution circuits from the 125 volt dc Reactor Building
distribution center (Main Bus #2) (ESS Div 1) as follows:

4.05-1   125 Volt DC Reactor Building Distribution
         Panel No 2 (ESS Div 1)   ( )

4.05-2   4160 Volt Switchgear Bus 21   ( )

4.05-3   4160 Volt Switchgear Bus 22   ( )

4.05-4   4160 Volt Switchgear Bus 23   ( )

4.05-5   4160 Volt Switchgear Bus 24 (ESS Div 1)   ( )

4.05-6   480 Volt Switchgear Bus 25   ( )

4.05-7   480 Volt Switchgear Bus 26 (ESS Div 1)   ( )

4.05-8   480 Volt Switchgear Bus 27   ( )

(4.2.1)
(4.3.1)
(4.3.3)

(A)

4.05 Reactor Building Panel (Main Bus #2) Distribution *

Verify distribution circuits from the 125 volt dc Reactor
Building distribution center (Main Bus #2)* as follows:

4.05-1   125 Volt DC Reactor Building Distribution
         Panel No 2*   ( )

4.05-2   4160 Volt Switchgear Bus 21   ( )

4.05-3   4160 Volt Switchgear Bus 22   ( )

4.05-4   4160 Volt Switchgear Bus 23   ( )

4.05-5   4160 Volt Switchgear Bus 24*   ( )

4.05-6   480 Volt Switchgear Bus 25   ( )

4.05-7   480 Volt Switchgear Bus 26*   ( )

4.05-8   480 Volt Switchgear Bus 27   ( )

* ESS Div 1   (4.4)

(4.2.1)
(4.3.1)
(4.3.3)

(B)

Fig A5
Test Procedure

ANSI/IEEE Std 497-1977
(ANSI N41.26)

Draft American National Standard

IEEE Trial-Use Standard Criteria for Post Accident Monitoring Instrumentation for Nuclear Power Generating Stations

497

Sponsor

**Nuclear Power Engineering Committee
of the
IEEE Power Engineering Society**

Foreword

(This foreword is not a part of IEEE Std 497-1977, Trial-Use Criteria for Post Accident Monitoring Instrumentation for Nuclear Power Generating Stations.)

The protection of nuclear power plants following a design basis event is provided automatically by the protection systems as defined by IEEE Std 279-1971 (ANSI N42.7-1972), Criteria for Protection Systems for Nuclear Power Generating Stations, and by required specified manual actions. Following the automatic initiation of the protection systems, the operator has the responsibility of bringing the nuclear power plant to and maintaining it in a safe condition. The operator requires specific information to assist him in fulfilling this role. The scope of this document is limited to the design criteria for the monitoring display instrumentation required by the operator during the post accident period.

The variables used in post accident monitoring should be chosen such that the operator will be allowed sufficient time to evaluate the trends, make reasoned judgments, and complete required manual actions. Although not required by this standard, an operator action sequence plan may be prepared to identify specific operator actions under specified conditions. This operator action sequence plan may be developed at the time of actual system design to determine which instruments are required. It could also be used to determine if sufficient time exists to check out and replace malfunctioning equipment, or to make available outside resources without adverse effects on plant safety or operating personnel.

The presence and need for the operator in performing safety functions during the post accident plant conditions impose different requirements than those used during normal plant operations. In many cases, redundant safety systems are employed. As a result of the accident, the reactor is usually shut down. The accident may cause system responses to be different than during normal conditions. The post accident plant characteristics and conditions that permit a manual operation (time to evaluate and take action) will be reflected in requirements for instrument reliability and response, removal for test, calibration and repair, in the failure analysis, etc.

The criteria in this document will also address certain information display channels pertinent to generating station safety (Section 4.20, IEEE Std 279-1971). The amount of instrumentation to be provided should be that minimum required by the operator during the post accident period. It does not include instrumentation used only for monitoring individual equipment operation (maintenance, degradation) or that which might be desirable to complement the post accident monitoring instrumentation.

It is the position of IEEE/NPEC that post accident monitoring instrumentation does not necessarily fit the definition of Class 1E and that IEEE Std 497-1977 shall delineate the requirements and stand on its own.

Subcommittee 6 recognizes the need for additional work in the following areas:

(1) Methods for the development of the specific design basis information required for post accident monitoring instrumentation.

(2) Guidance in the classification of post accident monitoring instrumentation.

Members of Working Group SC6.1 participating in the generation of these criteria at the time of final NPEC approval were:

N. C. Farr, *Chairman*

W. Bauer	A. Laird
R. C. Carruth	F. Paulitz
R. S. Darke	G. D. Quale
A. Hintze	C. E. Rossi
A. J. Kraft	D. Sokolsky

R. G. Walker

Members of the subcommittee participating in the generation of these revised criteria at the time of final NPEC approval were:

L. Stanley, *Chairman*

G. M. McHugh, Jr., *Vice Chairman*　　　　　　**P. M. Duggan,** *Secretary*

T. M. Bates, Jr.　　　　　　A. Laird
J. T. Beard　　　　　　J. I. Martone
C. J. Crane　　　　　　F. E. McLane
R. P. Daigle　　　　　　E. S. Patterson
R. H. Dalry　　　　　　C. E. Rossi
R. S. Darke　　　　　　H. K. Stolt
N. C. Farr　　　　　　D. Tondi

C. S. Walker

Members of the Nuclear Power Engineering Committee (NPEC) participating in the approval of these revised criteria were:

T. J. Martin, *Chairman*　　　　　　**A. J. Simmons,** *Vice Chairman*
L. M. Johnson, *Secretary*　　　　　　**R. E. Allen,** *Standards Coordinator*

J. F. Bates	R. I. Hayford	D. G. Pitcher
J. T. Bauer	T. A. Ippolito	H. V. Redgate
F. Baxter	I. Jacobs	B. M. Rice
J. T. Beard	A. Kaplan	J. C. Russ
R. G. Benham	R. F. Karlicek	W. F. Sailer
J. T. Boettger	A. Laird	J. H. Smith
K. J. Brockwell	G. M. McHugh, Jr.	A. J. Spurgin
D. E. Brosnan	T. J. McGrath	L. Stanley
F. W. Chandler	W. C. McKay	W. Steigelmann
C. M. Chiappetta	J. I. Martone	H. K. Stolt
N. C. Farr	W. P. Nowicki	D. F. Sullivan
J. M. Gallagher	W. E. O'Neal	P. Szabados
J. B. Gardner	E. S. Patterson	H. A. Thomas
	J. R. Penland	

Contents

IEEE Trial-Use Standard Criteria for Post Accident Monitoring Instrumentation for Nuclear Power Generating Stations

1. Scope

This standard applies to the design of instrumentation to monitor and display required post accident conditions within the nuclear power generating station.

Instrumentation addressed by this document includes that which enables the operator to: (1) identify the accident to the degree necessary for him to perform his role; (2) assess whether or not safety systems are accomplishing the required safety functions (for example, cooling the core, controlling containment pressure, etc); (3) determine when conditions exist that require specified manual actions and monitor the results of those actions; and (4) follow the course of the accident to determine whether or not conditions are evolving within prescribed limits.

This document does not include criteria for instrumentation used only to monitor status, maintenance, or degradation of individual items of safety system equipment during the post accident period.

2. Purpose

The purpose of this standard is to provide the minimum design criteria for permanently installed instrumentation used to provide the operator with information that may be necessary to perform his role in bringing the plant to and maintaining it in a safe condition following the accidents identified in the Design Basis.

3. Definitions

auxiliary supporting features. Installed systems or components which provide services such as cooling, illumination, and energy supply and which are required by the Post Accident Monitoring Instrumentation to perform its functions.

class 1E. The Safety Classification of the electric equipment and systems that are essential to emergency reactor shutdown, containment isolation, reactor core cooling, and containment and reactor heat removal, or are otherwise essential in preventing significant release of radioactive material to the environment.

components. Discrete items from which a system is assembled (for example, resistors, capacitors, wires, connectors, transistors, integrated circuits, switches, motors, relays, or solenoids).

information display channel. An arrangement of electrical and mechanical components or modules or both from measured process variable to display device as required to sense and display conditions within the generating station.

information display channel failure. A situation where the display disagrees, in a substantive manner, with the conditions or status of the plant. The display may fail to respond to a plant change, may improperly indicate a change when none has occurred, or may fail to provide any meaningful information.

margin. The difference between the most severe specified service conditions and the conditions used in design to account for uncertainties in defining satisfactory performance requirements under accident and post accident conditions.

modules. Any assembly of interconnected components which constitutes an identifiable device, instrument, or piece of equipment. A module can be disconnected, removed as a unit, and replaced with a spare. It has definable performance characteristics which permit it to be tested as a unit. A module could be a card or other subassembly of a larger device, provided it meets the requirements of this definition.

tolerable out-of-service time. The time an information display channel is allowed to be unavailable for use as a post accident monitoring display.

4. Design Basis

A specific design basis for the post accident monitoring instrumentation shall be established for each nuclear power generation station. The design basis information thus provided shall be available, as needed, for making judgments on the adequacy of design of the post accident monitoring instrumentation. The methods for development of the specific design basis information are not within the scope of this document.

The design basis shall document, as a minimum:

4.1 The generating station postulated accidents for which post accident monitoring instrumentation is required.

4.2 The safety systems that are required to mitigate the consequences of the postulated accidents referred to in 4.1.

4.3 The required operator actions and the conditions under which these actions are required during the post accident period.

4.4 The generating station variables to be used by the operator to: (a) identify the accidents mentioned in Section 4.1 above to the degree necessary for the operator to perform his role; (b) assess the accomplishment of the safety functions performed by the systems mentioned in Section 4.2 above; (c) guide the operator in accomplishing the required actions referred to in Section 4.3 above; and (d) follow the course of the accident to determine whether or not conditions are evolving within safe limits.

NOTE: Where practical, the same variable should be used for more than one of the above functions.

4.5 The portion of the post accident monitoring instrumentation that is Class 1E.

4.6 The events or conditions or both which determine the time period during which the monitoring of each variable referred to in 4.4 is required.

4.7 The time after the postulated accidents when each variable referred to in Section 4.4 is first required to be monitored and the time interval during which it is required to be monitored.

4.8 The minimum number and location of the sensor(s) required for any variable referred to in Section 4.4 that have a spatial dependence.

4.9 The locations at which the information must be available to the operator and the types of information (for example: discrete state, current value of a continuous variable, long term trend) which must be presented.

4.10 The range of transient and steady-state conditions of both the energy supply and the environment (for example: voltage, frequency, electromagnetic interference, temperature, humidity, pressure, vibration, and radiation) for which provisions must be incorporated to ensure adequate performance when required.

4.11 The malfunctions, accidents, or other unusual events (for example: fire, explosion, missiles, lightning, flood, earthquake, wind) which could physically damage components or could cause environmental changes leading to degradation of the performance of this instrumentation and which the design must withstand.

4.12 The maximum and minimum values and the maximum rate of change of each variable which must be accommodated by the post accident monitoring instrumentation and the maximum error within which the information must be conveyed to the operator for all of the applicable conditions listed in 4.10 and 4.11 above.

5. General Requirements

5.1 Functional Requirements.

5.1.1 *Failure Criteria.* No single failure within either the post accident monitoring instrumentation or its auxiliary supporting features, concurrent with the failures that are a condition or a result of a specific accident, shall prevent the operator from obtaining the information necessary for him to perform his role in bringing the plant to and maintaining it in a safe condition following that accident.

Where a single information display channel failure results in the information ambiguity (that is, the redundant displays disagree) which could lead the operator to defeat or fail to accomplish a required safety function, additional information shall be provided to allow the operator to deduce the actual conditions that are required for him to perform his role.

This may be accomplished by providing an additional independent channel of instrumentation on the same variable (additional redundancy), or by providing an independent channel which monitors a different variable which bears a known relationship to the redundant channels (diversity) or by providing the capability for the operator to perturb the measured variable and determine which channel has failed by observation of the response on each instrumentation channel.

Within each redundant portion of a safety system, redundant instrumentation display channels are not required if the failure of a single redundant portion of that safety system will not preclude accomplishing the system's safety function.

A single information display channel with a clearly identifiable failure mode is adequate where the mean time to replace or repair is less than the tolerable out-of-service time. Considerations used in determining the tolerable out-of-service time include: (1) the time until the unavailable information is required for a safety related operator action, (2) the time that equivalent information can be obtained by operator analysis of other information display channels, and (3) the time during which backup operator actions are available to eliminate the need for the unavailable information.

5.1.2 *Margin*. The characteristics of the information display channel (for example, range, accuracy, response time) shall include sufficient margin to account for the uncertainties in the analysis of the measured variable and the errors in the instrumentation itself (for example: temperature and voltage effects, drift).

NOTE: Margins for uncertainties in the predicted environmental and power supply conditions are accounted for in 5.2.2.

5.1.3 *Integrity*. Information display channels shall meet the performance requirements of the design basis including accuracy, response, the range under extreme conditions listed in Sections 4.10 and 4.11 during the time period which the particular information display channel is required. If justified by the design basis, different performance requirements may be applicable to different extremes of conditions.

5.1.4 *Channel Independence*. The redundant or diverse information display channels that are required to meet the failure criteria of Subsection 5.1.1 shall be independent and physically separated to accomplish decoupling of the effects of unsafe environmental factors, electrical transients, and physical accidents consequences documented in the design basis and to reduce the likelihood of interaction between channels during maintenance operations or in the event of channel malfunctions. This requirement does not preclude the association of these information display channels with safety system channels, provided this association does not compromise the performance of the safety system channel or the information display channel.[1]

5.1.5 *Trend Analysis*. Where the trend of a variable is essential for the operator to determine the required action, a means shall be provided to display that trend. The display channel shall have a response consistent with the transient characteristic of the variable.

5.1.6 *Derivation of Information Display Channels*. To the extent feasible and practical, information display channels shall directly measure the desired variable.

5.1.7 *System Repair*. The post accident monitoring system shall be designed to facilitate timely recognition, location, replacement, and repair or adjustment of malfunctioning equipment.

5.1.8 *Identification*. In order to provide assurance that the requirements given in this document can be applied during the design, construction, maintenance, and operation of the plant, post accident monitoring equipment shall be identified distinctively for each redundant portion of the post accident

---

[1] For guidance in the separation of redundant or diverse information display channels, refer to IEEE Std 384-1974 (ANSI N14.14), Trial-Use Standard Criteria for Separation of Class 1E Equipment and Circuits.

monitoring system. In the installed equipment, components and modules mounted in assemblies that are clearly identified as being in the post accident monitoring system do not themselves require identification.

5.2 Information Display Channel Requirements.

5.2.1 *Quality of Components and Modules.* Post accident monitoring instrumentation shall be of a quality that is consistent with minimum maintenance requirements and low failure rates. Information display channel equipment shall be designed, manufactured, inspected, installed, operated, and maintained in accordance with an approved quality assurance program.

5.2.2 *Equipment Qualification.* Information display channel equipment shall be qualified to substantiate that it will be capable of meeting, on a continuing basis, the performance requirements specified in the design basis.[2]

5.2.3 *Interaction Between Information Display Channels and Other Systems.*

5.2.3.1 *Classification of Equipment.* Information display channels that are used for both post accident monitoring display and for other operations and which are not part of a protection system[3] shall be classified as post accident monitoring equipment, and shall meet, as a minimum, the requirements of this document.

5.2.3.2 *Isolation.* The transmission of signals from post accident monitoring equipment to any system not meeting the minimum design requirements thereof, shall be through isolation devices which shall be classified as part of the post accident equipment and shall meet all the requirements of this document. No credible failure at the output of an isolation device shall prevent the associated post accident monitoring channel from meeting the performance requirements specified in the design basis.

Examples of credible failures include short circuit, open circuits, grounds, and the application of maximum credible ac or dc potential. A failure of an isolation device is evaluated in the same manner as a failure of other equipment in the post accident monitoring instrumentation.

5.2.4 *Capability for Test and Calibration.*

5.2.4.1 *Channel Availability Test.* Capability shall be provided for testing, with a high degree of confidence, the operational availability of each information display channel during plant operation. This may be accomplished in various ways, for example:

(1) by observing the effect of perturbing the monitored variable

(2) by observing the effect of introducing and varying, as appropriate, a substitute input to the sensor of the same nature as the measured variable

(3) by cross-checking between channels that bear a known relationship to each other.

5.2.4.2 *Calibration.* Capability shall be provided for calibration[4] of each information display channel during normal plant power or shutdown operation or both as determined by the required interval between calibrations.

During the post accident period, means shall be provided for validating the required information. This may be accomplished in various ways. For example:

(1) Recalibration

(2) Specifying a calibration interval to insure that the period during which the channel is needed will fall within the equipment's qualified calibration interval

(3) Selection of equipment that does not require periodic calibration. (For example: thermocouples.)

5.2.5 *Access to Calibration Adjustments and Test Points.* The design shall permit the administrative control of access to information display channel calibration adjustments and test points.

5.3 Display Requirements.

5.3.1 *Minimizing Displays.* To the extent feasible and practical, the same information display channel shall be used for normal operation and post accident monitoring.

---

[2] For guidance in the qualification of Class 1E post accident monitoring instrumentation equipment, refer to IEEE Std 323-1974, Standard for Qualifying Class 1E Equipment for Nuclear Power Generating Stations.

[3] For guidance on information display channels that are also part of the protective system, refer to IEEE Std 279-1971 (ANSI N42.7-1972), Criteria for Protection Systems for Nuclear Power Generating Stations.

[4] For guidance in the calibration of Class 1E information display channels refer to IEEE Std 498-1975, Standard Supplementary Requirements for the Calibration and Control of Measuring and Test Equipment Used in the Construction and Maintenance of Nuclear Power Generating Stations.

5.3.2 *Location and Identification.* Post accident monitoring displays shall be located accessible to the operator during the post accident period and shall be distinguishable from other displays. Post accident monitoring displays which enable the operator to determine when conditions exist that require specified manual actions, or monitoring the results of those actions, shall be located in the vicinity of the control stations used to effect the actions.

IEEE Standard Supplementary Requirements for the Calibration and Control of Measuring and Test Equipment Used in the Construction and Maintenance of Nuclear Power Generating Stations

498

Sponsor

**Nuclear Power Engineering Committee of the
IEEE Power Engineering Society**

Foreword

(This foreword is not part of IEEE Std 498-1975, Supplementary Requirements for the Calibration and Control of Measuring and Test Equipment Used in the Construction and Maintenance of Nuclear Power Generating Stations.)

This standard sets forth the quality assurance requirements for the calibration and control of measuring and test equipment used during the construction and maintenance of nuclear power generating stations.

This standard is intended to be used in conjunction with American National Standard Quality Assurance Program Requirements for Nuclear Power Plants, ANSI N45.2-1971.

This standard was prepared by the Institute of Electrical and Electronics Engineers in response to a request by American National Standards Committee N45 (Reactor Plants and Their Maintenance).

Committee N45 has been chartered to promote the development of standards for the location, design, construction, and maintenance of nuclear reactors and plants embodying nuclear reactors, including equipment, methods, and components specifically for this purpose.

In October of 1972 the IEEE Nuclear Power Engineering Committee (NPEC) established Subcommittee 8 (SC-8) to develop standards on quality assurance. A working group (SC-8.1) from this subcommittee was reorganized in April, 1973, to develop a standard to provide Supplementary Requirements for the Calibration and Control of Measuring and Test Equipment Used in the Construction and Maintenance of Nuclear Power Generating Stations.

This working group was composed of representatives of key segments of the nuclear industry, including utilities, reactor suppliers, construction contractors, component manufacturers, consultants, and the United States Atomic Energy Commission.

The initial draft of this standard was completed in August, 1973. Since then it has been revised to reflect the comments received from committee members of the IEEE, members of Subcommittee N45.2 on Quality Assurance Standards, and selected individuals from the nuclear industry.

The standard contained herein was developed from this activity.

In April of 1970, Committee N45 established Subcommittee N45.3.0 (now known as Subcommittee N45.2) to guide the preparation of nuclear quality assurance standards. This subcommittee is responsible for establishing guidelines and policy to govern the scope and content of the various standards; monitoring the status of standards in process, recommending preparation of additional standards, and giving final approval for standards prior to their submittal to Committee N45 for balloting.

Working with Subcommittee N45.2 and concurrently with the development of this standard, other N45.2 work groups are developing a series of standards that set forth both general and detailed technical provisions for certain activities to assure quality during the design, construction, and maintenance of nuclear power generating stations. Suggestions for improvement gained in the use of this standard will be welcomed. They should be addressed to:

Secretary
IEEE Standards Board
The Institute of Electrical and
Electronic Engineers, Inc
345 East 47th Street
New York, NY 10017

The working group (SC-8.1) that prepared this standard had the following membership:

S. G. Caslake, *Chairman*

B. J. Cochran	T. Hanson
J. C. Crews	T. E. Pinkham
A. C. D'Hoostelaere	T. W. Rudin
E. Funk	R. Santosuosso
M. Glotzer	R. A. Sessoms

When this standard was ballotted and approved by SC-8.2 membership was as follows:

D. R. Stone, *Chairman*

W. E. Airey	P. N. Gupta	F. N. Paplawsky
H. O. Denzer	R. B. Hubbard	R. S. Rick
E. P. Fogarty	R. E. Mills	Sivaram Timmaraju
	W. E. O'Neal	

This standard was also ballotted and approved by D. C. McClintok of the Power Generation Committee.

When NPEC balloted and approved this standard, membership was as follows:

T. J. Martin, *Chairman*   **A. J. Simmons,** *Vice Chairman*
L. M. Johnson, *Secretary*   **J. T. Boettger,** *Coordinator*

R. E. Allen	J. M. Gallagher	J. H. Smith
J. F. Bates	R. I. Hayford	A. J. Spurgin
R. G. Benham	T. A. Ippolito	L. Stanley
K. J. Brockwell	I. Jacobs	H. S. Staten
D. F. Brosnan	A. Kaplan	C. E. Stine
O. K. Brown	M. I. Olken	H. K. Stolt
S. G. Caslake	W. E. O'Neal	D. F. Sullivan
F. W. Chandler	E. S. Patterson	P. Szabados
E. F. Chelotti	D. G. Pitcher	W. A. Szelistowski
C. M. Chiappetta	H. V. Redgate	L. D. Test
D. H. Clark	J. C. Russ	H. A. Thomas
R. J. Cooney	W. F. Sailer	C. J. Wylie

Contents

IEEE Standard Supplementary Requirements for the Calibration and Control of Measuring and Test Equipment Used in the Construction and Maintenance of Nuclear Power Generating Stations

1. Introduction

1.1 Scope. This standard sets forth the requirements for a calibration program to control and verify the accuracy of M&TE (measuring and test equipment) which is used to assure that important parts of nuclear power generating stations are in conformance with prescribed technical requirements, and that data provided by testing, inspection, or maintenance are valid. These important parts include those structures, systems, and components whose satisfactory performance is required; for the plant to operate safely, to prevent accidents that could cause undue risk to the health and safety of the public, or to mitigate the consequences of such accidents if they were to occur. This standard is intended to be used in conjunction with American National Standard Quality Assurance Program Requirements for Nuclear Power Plants, ANSI N45.2-1971.

1.2 Applicability. The requirements of this standard apply to the M&TE used by any individual or organization that participates in installation, inspection, testing, or maintenance of those parts of a nuclear power generating station discussed in Section 1.1. The extent to which the individual requirements of this standard apply will depend upon the nature and scope of the work to be performed and the importance of the item or service involved.

1.3 Responsibility. It is the responsibility of the owner to provide for the establishment and execution of a calibration program for the plant consistent with the provisions of this standard. The work of establishing practices and procedures and providing the resources in terms of personnel, equipment, and services to implement the requirements of this standard may be delegated to other organizations; and such delegation shall be documented. In any case, the owner shall retain responsibility for overall program effectiveness.

1.4 Definitions. The following definitions are provided to assure a uniform understanding of selected terms as they are used in this standard.

accuracy. The quality of freedom from mistake or error.

calibration. Comparison of an item of M&TE with a reference standard or with an item of M&TE of equal or closer tolerance to detect and quantify inaccuracies and to report or eliminate those inaccuracies.

measuring and test equipment (M&TE). Devices or systems used to calibrate, measure, gauge, test, inspect, or control in order to acquire research, development, test, or operational data; to determine compliance with design, specifications, or other technical requirements. M&TE does not include per-

7

manently installed operating equipment or test equipment used for preliminary checks where accuracy is not required, for example, circuit checking multimeters.

precision. The quality of coherence or repeatability of measurement data.

reference standards. Standards (that is, primary, secondary, and working standards, where appropriate) used in a calibration program. These standards establish the basic accuracy limits for that program.

tolerance. The allowable deviation from a specified or true value.

1.5 Referenced Documents. Other documents that are required to be included as a part of this standard are either identified at the point of reference or described in Section 6 of this standard. The issue or edition of the referenced document that is required will be specified at the point of reference or in Section 6 of this standard.

2. General Requirements

A documented program shall be established and maintained for the calibration and control of M&TE and reference standards. It shall be designed to determine and assure the accuracy of M&TE and reference standards and shall provide for the prompt detection of inaccuracies and for timely, effective corrective action. This documented program shall include as a minimum the following general requirements.

2.1 Equipment Identification. A list of M&TE, reference standards, and their locations shall be prepared to specifically identify those items within the calibration program.

2.2 Calibration Procedures. Documented procedures for calibrating M&TE and reference standards shall be used. Procedures such as published standard practices, written instructions that accompany purchased equipment, or other acceptable instructions may be used.

Calibration procedures shall include the following minimum basic information:

(1) Identity of the item to be calibrated

(2) Calibration equipment and reference standards to be used

(3) Checks, tests, measurements, and acceptance tolerances

(4) Sequence of operations

(5) Special instructions when necessary

2.3 Records. Records shall be maintained for each individual piece of equipment to show that established schedules and procedures for the calibration of M&TE and reference standards have been followed. The records shall contain a history of calibration and other means of control showing calibration interval, date of last calibration, when next calibration is due, conformance or nonconformance to required tolerances prior to and following adjustments and any limitations on use.

Each record shall identify the equipment to which it applies, the procedure or instruction followed in performing the calibration, the calibration data, the identity of the standard used, the identity of the person performing the calibration, and the calibration date.

3. Elements of Control

The documented program shall include as a minimum the elements of control described in the following paragraphs:

3.1 Adequacy of Reference Standards. Reference standards used for calibrating M&TE shall have an accuracy level, acceptable calibration ranges, and precisions that are equal to or better than those required of M&TE. The accuracies of the M&TE and the reference standards should be chosen such that the equipment being calibrated can be calibrated and maintained within the required tolerances.

3.2 Environmental Controls. M&TE and reference standards shall be transported, stored, and calibrated in environments which will not adversely affect their accuracy. Environmental factors which shall be considered include, but shall not be limited to: temperature, humidity, vibration, radio frequency interference, electromagnetic interference, background radiation, dust, cleanness, and fumes. When inaccuracy of M&TE or reference standards, because of environmental effects, cannot be avoided, compensating corrections shall be determined and applied.

3.3 Intervals of Calibration. The program shall require that M&TE and reference standards be recalled for recalibration at prescribed intervals to verify the required accuracy. Such intervals may be in calendar time or relate to usage. Interval selection should consider experience, inherent stability, purpose of use, and accuracy required. Historical records which contain sufficient experience data for evaluating and adjusting calibration intervals shall be maintained.

3.4 Traceability. M&TE shall be calibrated utilizing reference standards whose calibration has a known valid relationship to nationally recognized standards or accepted values of natural physical constants. If no national standard exists, the basis for calibration shall be documented.

Reference standards used in the calibration program shall be identified on calibration data records and supported by certificates, reports, or data sheets attesting to the calibration date, calibration facility, environmental conditions, and data which shows conformance to accuracy requirements.

3.5 Labeling. M&TE and reference standards shall be labeled to indicate their control status. The label shall indicate when the next calibration is due. When size or functional characteristics of M&TE or reference standards prevent the application of a label, an identifying code shall be applied to reflect status. When neither labeling nor coding is practical, the procedures shall provide for monitoring of records to assure control. M&TE, whose use must be limited, shall be identified and controlled; for example, a multiscaled instrument which may be acceptable on one or more scales but limited on a specific scale or an instrument that is intended to be used for making preliminary checks.

3.6 Precalibration Checks. M&TE and reference standards, submitted for calibration, shall be checked and the results recorded before adjustments or repairs are made.

3.7 Nonconformance. M&TE and reference standards found to be out of calibration or which have not been properly maintained, or calibrated, or which have been subjected to possible damage, shall be identified as nonconforming and removed from service until such time as corrective measures have been taken. All equipment tested or calibrated by the item since the last calibration shall be identified and sufficient investigations performed to either reestablish the acceptability of the equipment or to confirm a nonconformance. The results of such investigations shall be documented.

3.8 Use of Measuring and Test Equipment and Reference Standards. Measuring and test equipment and reference standards shall be controlled to assure consistent results of acceptable accuracy. The following controls should be considered.

(1) Environmental and handling controls
(2) Training and qualification of personnel
(3) Checking calibration status before use
(4) Interim checks between calibrations
(5) Documenting and recalibrating possible damaged M&TE and reference standards
(6) Limiting use to authorized personnel

4. Audits

The calibration program, in its entirety, is subject to audit in accordance with American National Standard Requirements for Auditing of Quality Assurance Programs for Nuclear Power Plants, ANSI N45.2.12-1975.

5. Document Control

Equipment identification lists, procedures, calibration records, personnel qualification reports, and nonconformance reports shall be retained with other project records as required by codes, standards, specifications, or project procedures.

Collection, storage, and maintenance of these records shall be in accordance with American National Standard Requirements for the Collection, Storage and Maintenance of Quality Assurance Records for Nuclear Power Plants, ANSI N45.2.9-1974.

6. Revision of American National Standards Referred to in This Standard

When any of the following standards referred to in this document is superseded by a revision approved by the American National Standards Institute, the revision is not mandatory until it has been incorporated as a part of this standard.

American National Standard Quality Assurance Program Requirements for Nuclear Power Plants, ANSI N45.2-1971.

American National Standard Requirements for the Collection, Storage and Maintenance of Quality Assurance Records for Nuclear Power Plants, ANSI N45.2.9-1974.

American National Standard Requirements for Auditing of Quality Assurance Programs for Nuclear Power Plants, ANSI N45.2.12-1975.

IEEE Recommended Practice for the Design of Display and Control Facilities for Central Control Rooms of Nuclear Power Generating Stations

566

Sponsor

**Nuclear Power Engineering Committee
of the
IEEE Power Engineering Society**

Foreword

(This foreword is not part of IEEE Std 566-1977, Recommended Practice for the Design of Display and Control Facilities for Central Control Rooms of Nuclear Power Generating Stations.)

The nuclear power plant control room is the central location of the operator—power generation interface. It is the location where the plant operating personnel must make decisions and take action necessary to ensure the safe and efficient operation of the plant.

The assignment of functions to an operator; the type, layout, and accessibility of the necessary information and action devices; the design for the comfort and protection of the operator; and the degree to which automatic control is utilized must all be established recognizing the response capabilities of the operator, so as to optimize the use of his judgment as a valuable resource.

The selection of specific display and control equipment as well as the determination of information formats and control switchboard layouts has traditionally been and undoubtedly will continue to be largely the prerogative of the individual user. Also, the technical operating requirements for individual plant systems are determined and established largely by the individual system designers. Therefore, the successful integration of these requirements into a coordinated design that will meet operating and regulatory objectives requires that common guidelines be generated to guide all involved designers in making necessary selections and decisions. At this stage in the development of standards for the control room, it is difficult to apply mandatory standards to the design of design and control facilities. This document sets out guidelines to aid the designers in making decisions.

The committee recognizes that the method of dealing with functional classification of displays and controls is important. Appendix A documents one of several approaches which the designer may use to classify and group the various control room displays and controls. Investigations were made by the committee in an effort to quantify the relative importance of various types of controls and displays. The committee recommends that a systematic approach be used to determine the classification of the control and display facilities.

This work was performed under the auspices of the Power Generation Committee (PGC) and the Nuclear Power Engineering Committee (NPEC) by a combined working group of the Nuclear Power Plant—Control, protection, and Automation Subcommittee and of NPEC SC1.2.

The members of the working group at the time of development of this document were:

A. J. Spurgin, *Chairman*

O. M. Anderson	J. A. List	H. F. Reischel
C. L. Cobler	J. Owen	R. M. Reymers
W. A. Coley	R. W. Pack	D. C. Richardson
R. S. Darke	F. A. Palmer	R. A. Schmitter
J. R. Hall	D. S. Peikin	J. V. Stephens
D. A. Hansen	R. J. Reiman	M. D. Sulouff
W. J. Kerchner		V. D. Thomas
G. Lilly		W. E. Wilson

At the time it approved this standard, the Nuclear Power Engineering Committee had the following membership:

T. J. Martin, *Chairman*   **A. J. Simmons,** *Vice Chairman*

L. M. Johnson, *Secretary*

R. E. Allen	R. I. Hayford	J. R. Penland
J. F. Bates	T. A. Ippolito	D. G. Pitcher
J. T. Bauer	I. M. Jacobs	H. V. Redgate
F. D. Baxter	A. Kaplan	B. M. Rice
J. T. Beard	R. F. Karlicek	J. C. Russ
R. G. Benham	A. Laird	W. F. Sailer
J. T. Boettger	J. I. Martone	J. H. Smith
K. J. Brockwell	T. J. McGrath	A. J. Spurgin
D. F. Brosnan	G. M. McHugh	L. Stanley
F. W. Chandler	W. C. McKay	W. Steigelmann
C. M. Chiappetta	W. P. Nowicki	H. K. Stolt
N. C. Farr	W. E. O'Neal	D. F. Sullivan
J. M. Gallagher	E. S. Patterson	P. Szabados
J. B. Gardner		H. A. Thomas

Contents

IEEE Recommended Practice for the Design of Display and Control Facilities for Central Control Rooms of Nuclear Power Generating Stations

1. Scope

This document establishes guidelines to be used by power plant system designers in selecting information and control devices to be made available in the central control room, and in determining how and where they shall be made available so that they can most reliably and quickly be used by the operator. The guide addresses the functional requirements of the information systems, controls, and displays, but not the selection of specific devices or equipment. It does not apply to the physical design of the control room enclosure or structures mounted therein.

2. Purpose

To provide uniform guidelines for the functional selection, coordination, and organization of control and information systems in a nuclear power plant central control room.

3. References

The reference section is divided into two parts. The first contains the references mentioned in this document, and the second contains a set of related references to which reference is not made.

3.1 Specific to Document.

[1] IEEE Std 279-1971 (ANSI N42.7-1972), Criteria for Protection Systems for Nuclear Power Generating Stations.

[2] IEEE Std 308-1974, Standard Criteria for Class 1E Power Systems for Nuclear Power Generating Stations.

3.2 Other References.

[3] Code of Federal Regulations, Title 10, Part 50.

[4] General Design Criteria (Appendix A); Criterion 13, Instrumentation and Controls; Criterion 19, Control Room.

IEEE Standards and Guides:

[5] IEEE Std 336-1971 (ANSI N45.2.4-1972), Installation, Inspection, and Testing Requirements for Instrumentation and Electric Equipment During the Construction of Nuclear Power Generating Stations.

[6] IEEE Std 338-1975, Trial-Use Criteria for Periodic Testing of Nuclear Power Generating Station Class 1E Power and Protection Systems.

[7] IEEE Std 384-1974 (ANSI N14.14), Trial-Use Standard Criteria for Separation of Class 1E Equipment and Circuits.

[8] IEEE Std 420-1973, Trial-Use Guide for Class 1E Control Switchboards for Nuclear Power Generating Stations.

4. Definitions

4.1 accessibility. Relates to the accessibility of information to the operator on a "continuous," "sequenced," or "as called for" basis.

4.2 central control room. A continuously manned and protected enclosure from which actions are normally taken to operate the nuclear generating station under normal and abnormal conditions.

4.3 displays. Devices which convey information to the operator.

4.4 emergency operations area(s). Functional area(s) allocated for the displays used to assess the status of safety systems and the controls for manual operations required during emergency situations.

4.5 functional area(s). Location(s) designated within the control room to which displays and controls relating to specific function(s) are assigned.

4.6 information. Data describing the status and performance of the plant.

4.7 normal operations area. A functional area allocated for those displays and controls necessary for the tasks routinely performed during plant startup, shutdown, and power operation modes.

4.8 operating modes. The nuclear power plant modes as defined by the technical specifications for the plant.

4.9 operator. A person licensed to operate the plant.

4.10 supporting operations area(s). Functional area(s) allocated for controls and displays which support plant operation.

4.11 sensory saturation. The impairment of effective operator response to an event due to excessive amount of display information which must be evaluated prior to taking action.

5. Design Bases

5.1 General. The design bases for the control and display facilities in the control room should be established and documented, before beginning the detailed control room design, and updated as needed.

5.2 Contents. The design bases should include but not be limited to the following items:

   5.2.1 The operating modes for which the central control room display and control facilities should be designed.

   5.2.2 The number of operators and the responsibilities assigned to them under each operating mode.

   5.2.3 The functional areas into which the control room is to be organized. These may include the normal, emergency, and supporting operations areas.

NOTE: These functional areas need not be physically separate.

   5.2.4 The basis for grouping of display and control devices within any functional area. (See Section 6.)

   5.2.5 The limiting number of display devices which can be active at the same time, by type, established as a design goal for each functional area of the control room to avoid operator sensory saturation. (See Appendix B.)

   5.2.6 A listing and classification of the safety related display and control instrumentation and any post accident monitoring instrument for which specific requirements are already established by regulatory requirements, industry standards, or safety analysis reports. (See Ref [1], [2].)

   5.2.7 The requirements which are mandated by, or directed by, user company policies or contracts or both.

   5.2.8 The anthropometric relationship to be used for design of the control boards.

   5.2.9 The list of functions, the controls for which may be transferred from the central control room facilities to remote facilities.

   5.2.10 The sequence of events for the postulated design basis events.

   5.2.11 Data to be used for trend and historical record purposes.

6. Usage Analysis

The designer should establish and document a systematic method for the assignment of types and locations of the controls and displays. This method should include:

6.1 An identification of each function by its usage characteristics, including but not necessarily limited to:

   (1) priority and importance of information or action
   (2) plant systems
   (3) operating modes
   (4) frequency of use
   (5) response time
   (6) safety classification
   (7) the grouping of displays and controls in a functional area.

6.2 An agreed set of criteria for the determination of the device assignment and location based on an analysis of these characteristics.

6.3 Evaluation of criteria to ensure consistency with design bases.

(An example of one such systematic approach is described in Appendix A.)

7. Functional "Considerations"

7.1 General. The operator should be considered as one part of an integrated system that is necessary for the proper and efficient operation of a nuclear power plant.

7.2 Display Facilities. In support of the operator needs, the control room designer should arrange the display facilities so that the operator can readily observe the displays and analyze the status of any system.

7.2.1 *Accessibility.* As appropriate, the operator should have information available on a "dedicated," "intermittent − periodic," or "intermittent − as called for" basis. The need for information to be displayed and its accessibility to the operator depends on: (1) the consequence of the operator not taking corrective action, (2) the importance of the data to the operator in determining the plant status, (3) the degree of automation to be used in control system design, and (4) the use of such display techniques as "display by exception."

7.2.2 *Readability and Comprehension.* The display equipment should provide means to facilitate operator comprehension. These include consistent use of the following: (1) Physical differentiation of data which are presented, using such techniques as color coding, size, and shape. (2) Formats keyed to and consistent with the physical representation should be used, for example, a vertical bar indicator for level. (3) Graphic displays for: flow diagrams, one-line electric diagrams, bar charts, etc.

7.2.3 *Abnormal Conditions.* The operator should be alerted to abnormal or unsafe conditions or significant changes in the plant and its process systems or safety systems or both.

7.2.3.1 *Alarms.* The alarm function should be based (to the greatest extent possible) on a true abnormal condition (for example, low oil pressure on a shaft-driven oil pump on a condensate booster pump should be alarmed only when the booster pump is in service).

7.2.3.2 *System Modes.* Alarms should also be terminated or suppressed during modes of operation when they would be meaningless, due to changes in the operating mode (such as startup, power operation, shutdown, etc), so that information priority for the current mode of operation can be readily assessed.

7.2.3.3 *Limit Monitoring.* In addition to normal equipment protective limits, plant operational limits established by technical specifications and by plant administrative procedures shall be monitored by the operator. Provisions should be made to facilitate these requirements.

7.3 Control Facilities.

7.3.1 Control devices and their functionally associated displays should be located to facilitate operator action.

7.3.2 In determining whether control devices should be made available to the operator in the control room, the following factors should be considered: (1) the safety functions of the controlled equipment, (2) consequences of the operator not being able to take necessary action, (3) the degree of automation to be used for control, (4) the frequency of usage of the controls, and (5) the number of controls required to accomplish a given function.

7.3.3 Where the controls of equipment or devices which are part of safety systems can be transferred to points of control outside the control room, the mode of the active control should be indicated in the control room.

7.4 Device and Display Identification. Identification of control and display functions should be easily associated with the physical devices being monitored or controlled. Where alphanumeric identification systems are used, they should be supplementary to a functional identification.

7.5 Convention for Control Devices. A convention should be established to provide consistency in the operation of controls that perform similar functions, for example, control switches are to be turned clockwise to "close" (for circuit breakers).

7.6 Display and Control Facilities − Special. Special requirements such as safety surveillance, post accident monitoring, and remote shutdown should be considered in usage analysis described in Section 6.

7.6.1 *Safety System Status.* The operator

should be clearly informed of the status of the safety system by means of a display. This display should be used to enhance the normal plant administrative procedures.

7.6.2 *Redundant and Diverse Information.* Where a number of critical parameters require redundant or diverse displays as a means of checking the reasonability of information, the alternative information sources should be located to allow the operator to use both sources in arriving at a conclusion.

7.7 Area Arrangement. The normal operations area should be centrally arranged within the control room to provide the operator with surveillance and access capability to other operating areas within the control room. The emergency operations area should be readily accessible and visible from the normal operations area. This area should not be in a separate room or enclosure from the normal operations area.

7.8 Device Arrangement. Individual devices or groups of individual devices should be arranged to minimize operator motion including changes in direction of vision.

7.9 Equipment or System Status. Consideration should be given to provide indication when non-safety-related equipment is taken out of service for maintenance, calibration, or inspection, and when it is returned to service.

7.10 Communications. The methods provided for communication between the operator and various other personnel should not divert the operator from his principal duties.

7.11 Internal Security. Where display and alarm devices are provided within the central control room to alert the operator to unauthorized entry into vital areas, the devices should be clearly differentiated from any devices provided for plant functions by color, arrangement, or location.

Appendix A

(This appendix is not part of IEEE Std 566-1977, Recommended Practice for the Design of Display and Control Facilities for Central Control Rooms of Nuclear Power Generating Stations.)

Implementation of a Usage Analysis Method

Introduction

The prerequisites required for the performance of any usage analysis are:

(1) The determination of the plant display and control requirements to be included in the design basis (see Section 5).

(2) The determination of a set of characteristics which will subsequently be used to convert the plant control and display requirements into specific determinations of devices and their locations. Table 1 shows one possible set of characteristics. The columns show the characteristics and the rows show the various systems or subsystems. In deciding the appropriate characteristics to be recorded such questions as what is the system involved, what are its constituent parts, when does the operator need access to the systems controls, etc., have to be answered.

(A) The first step in the approach is, therefore, to identify the characteristics of the matrix and then fill in the matrix appropriately for each system and subsystem.

(B) The criteria for assignment or location of the display and control devices or both are then established to ensure that a consistent design results.

Typical criteria are:

(1) All control devices used frequently during startup, power, or hot standby modes of operation shall be located in the normal operations area.

(2) Functional controls shall be laid out on a system or subsystem basis.

(3) Safety system control devices shall be located in the emergency operations area.

(C) The selection and application of appropriate devices plus the specific layout of the control boards can proceed on a logical basis.

(D) A digital computer can be used advantageously to manipulate the developed data base to:

(1) store the characteristics of the system functions

(2) apply the criteria to the above characteristics, implementing the analysis by sorting techniques

(3) document any portion of the results.

Appendix B

(This appendix is not part of IEEE Std 566-1977, Recommended Practice for the Design of Display and Control Facilities for Central Control Rooms of Nuclear Power Generating Stations.)

Example of Operator Sensory Saturation

B.1 Steam Generator Tube Rupture on a Pressurized Water Reactor. In the event of a steam generator tube rupture the operator will be required to evaluate the information provided following the incident and take effective action to: (1) ensure that the plant is safely shut down and (2) minimize primary to secondary plant leakage. His response could be impaired by the number of displays to which he may be subjected. The alarms, indicators, and status lights

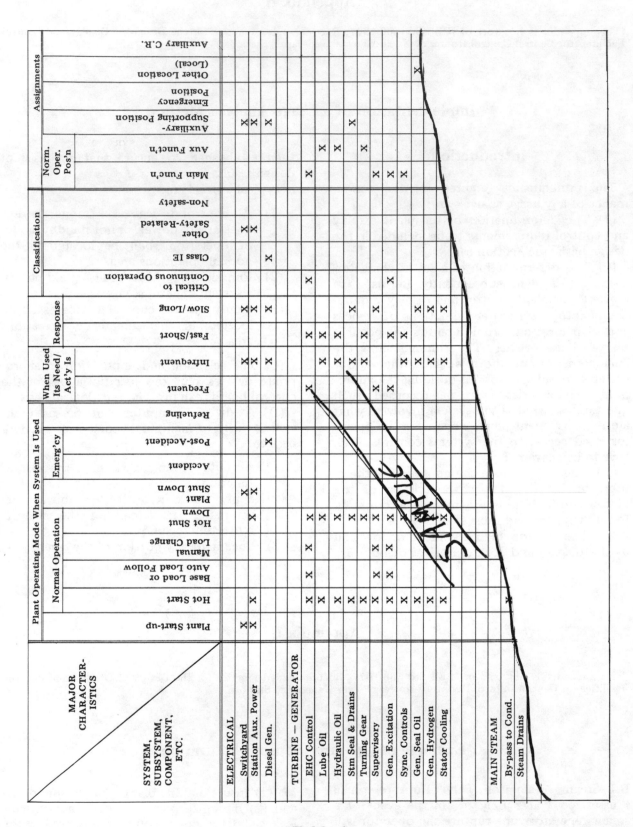

SYSTEM, SUBSYSTEM, COMPONENT, ETC.	Plant Start-up	Hot Start	Base Load or Auto Load Follow	Manual Load Change	Hot Shut Down	Plant Shut Down	Accident	Post-Accident	Refueling	Frequent	Infrequent	Fast/Short	Slow/Long	Critical to Continuous Operation	Class IE	Other Safety-Related	Non-safety	Main Func'n	Aux Func'n	Auxiliary-Supporting Position	Emergency Position	Other Location (Local)	Auxiliary C.R.
ELECTRICAL																							
Switchyard	X	X			X	X					X		X			X				X			
Station Aux. Power	X	X			X	X					X		X			X				X			
Diesel Gen.								X			X		X		X					X			
TURBINE — GENERATOR																							
EHC Control		X	X	X	X					X		X		X				X					
Lube Oil		X	X		X						X	X							X				
Hydraulic Oil		X	X		X						X	X							X				
Stm Seal & Drains		X	X		X						X	X	X							X			
Turning Gear		X			X						X		X						X				
Supervisory		X	X		X					X								X					
Gen. Excitation		X	X	X	X					X	X	X	X	X				X					
Sync. Controls		X									X	X	X					X					
Gen. Seal Oil		X									X		X					X					
Gen. Hydrogen		X									X		X										
Stator Cooling		X											X										
MAIN STEAM																							
By-pass to Cond.		X																				X	
Steam Drains																							

Table 1
System-Subsystem-Component Usage Analysis

which will confront him in the event of the incident are listed below. The symbols *, T, and S are used to indicate which information will be presented after the rupture and prior to the trip (*), after the trip (T), and following safety injection (S).

B.2 Alarms — (Annunciator Window Engravings).

* PSZR CONTROL HI/LO PRESSURE DEVIATION
* PSZR CONTROL LO LEVEL
* PSZR PROTECTION LO LEVEL
* PSZR PRESS LO/BACKUP HTRS ON
* PSZR LO LEVEL HTR CUTOFF AND LD ISOLATION
* VCT HI/LO LEVEL
* VCT HI/LO PRESS
* CENT CHG PUMP 1 AUTO START
* CENT CHG PUMP 2 AUTO START
* PSZR PROT LO PRESS.
* SG HI—HI/LO—LO LEVEL
* FEEDWATER ISOL
* SG ACTUAL LEVEL — SETPOINT HI/LO DEVIATION
* SG FLOW MISMATCH FS > FW
* CONDENSER 1 CONDUCTIVITY HI
* CONDENSER 2 CONDUCTIVITY HI
* HI TURB BLDG RADIATION
* HI RAD AIR EJECTOR

T ROD BOTTOM
T TWO OR MORE RODS AT BOTTOM
T ROD BOTTOM — ROD DROP AUTO WITHDRAWL STOP
T TURBINE STM STOP VALVE CLOSED
T REACTOR TRIP FROM TURBINE TRIP
T REACTOR TRIP PSZR LO PRESS
T REACTOR TRIP BKR A/B TRIPPED
T SG HI HI LEVEL TURBINE TRIP
T TURBINE TRIP REACTOR TRIP
T GENERATOR MOTORING
T FAST CLOS INTERCEPT VALVE ON
T UNIT TRANS 1 TRIP
T UNIT TRANS 2 TRIP
T GENERATOR TRIP
T TRIP OF 4kV BKRS TO UT
T CLOSURE OF 4kV BKRS TO ST
T TRIP OF 480V BKRS TO UT
T CLOSURE OF 480V BKRS TO ST
T TURBINE EXH LO VAC

S RHR PUMP 1 AUTO START
S RHR PUMP 2 AUTO START
S RHR PUMP 1 DISCH HI PRESS
S RHR PUMP 2 DISCH HI PRESS
S REACTOR TRIP PSZR SAF INJ
S CONDENSATE PUMP 1 TRIPPED
S CONDENSATE PUMP 2 TRIPPED
S DIESEL GEN 1 START
S DIESEL GEN 2 START
S AUX FEED PUMP START
S PHASE A CONT ISOL

S VENT ISOL
S CCW PMP 1 AUTO START
S CCW PMP 2 AUTO START
S REACTOR TRIP PSZR SAF INJ (SI)

B.3 Analog Indicators.

* INCR CHG FLOW
* DECR PSZR VAPR TEMP
* DECR PSZR LIQ TEMP
* DECR PSZR LEV (4)
* DECR PSZR PRESS (4)
* INCR CHG PMP AMPS
* DECR WIDE RANGE PRESS (RECORDER)
* INCR SG LEV (3)
* DECR SG FW FLOW (4)
* DECR SG WIDE RANGE LEVEL (R)
* INCR CONDENSATE CONDUCT (R)
* DECR COND VAC (R)
* DECR FW VALVE POS
* DECR RCS TEMP

T DECR NEUT FLUX (4)
T DECR IMPULSE CHAMBER PRESS (2)
T INCR/DECR SG HDR PRESS
T DECR FW TEMP (2)
T DECR COND FLOW (R)
T INCR SG REL VALVE FLOW
T INCR/DECR STEAM LINE PRESS (12)
T DECR ELEC OUTPUT
T DECR TUR—GEN SPEED
T PWR DIST SYSM STATUS (UNIT/STARTUP TRANS — AMPS, WATTS, VARS VOLTS ~30 INDICATORS)

S INCR SI FLOW
S INCR SI DISCH PRESS
S DECR RWST LEVEL
S AUX FEED PUMP FLOW (2)
S CCW PMP AMPS
S SI PMP AMPS
S AUX FW PMP AMPS (2)
S COND PMP AMPS (3)
S COND BOOSTER PMP AMPS (3)
S D/GEN STATUS (~60 INDICATIONS OF PWR, VOLTS, ETC)
S CONT FAN COOLER AMPS

B.4 Breakers/Valves Position Indicators (~200).

* CHG PMP
* PSZR HEATERS ON
* PSZR HEATERS OFF
* LETDOWN ISOL VALVE
* SG REL VALVE

S FAN COOLERS
S CONT ISOL VALVES
S SI VALVES
S FW ISOL VALVES
S AUX FW PUMP
S COND SYST PMP
S CONTRL ROOM VENT
S CCW PMP
S PWR DIST SYST BKR POS

IEEE Standard Requirements for Reliability Analysis in the Design and Operation of Safety Systems for Nuclear Power Generating Stations

577

Sponsor

Nuclear Power Engineering Committee of the IEEE Power Engineering Society

Foreword

(This foreword is not a part of IEEE Std 577-1976, Requirements Reliability Analysis in the Design and Operation of Safety Systems for Nuclear Power Generating Stations).

This standard has been prepared to standardize the application of reliability techniques in the design and operation of nuclear-power generating stations. It is directed towards those systems in the nuclear station which perform protective functions and fall within the scope of IEEE Std 279-1971, "Criteria for Protection Systems for Nuclear Power Generating Stations," and IEEE Std 308-1974, "Criteria for Class 1E Power Systems for Nuclear Power Generating Stations." However, the requirements of this standard may be applied to other systems within a nuclear-power generating station if appropriate. This standard may also be used as a guide to establish periodic testing programs.

IEEE Std 352-1975, "Guide for General Principles of Reliability Analysis of Nuclear Power Generating Station Protection Systems," supplements this standard by providing guidance in the application of reliability techniques.

IEEE Std 338-1975, "IEEE Criteria for the Periodic Testing of Nuclear Power Generating Station Class 1E Power and Protection Systems," requires that programs be established for periodic testing which are based, in part, upon the minimum acceptable analyses described in this standard.

Reliability analysis is a method which can be used to demonstrate compliance with reliability requirements stated in regulations and other standards. The subcommittee feels that when reliability analysis is used for this purpose, this standard describes an acceptable response to the requirements. The requirement that a reliability analysis be performed does not originate with this standard.

It is recognized that this standard leads current practice in the nuclear-power industry and is somewhat innovative in an attempt to fill a gap in nuclear standards. Because of this aspect, this standard should not be imposed as a requirement without consideration of the time necessary for organizations to gain experience with the analyses required in this standard.

This standard was prepared by the Reliability Subcommittee (SC-5) of the Nuclear Power Engineering Committee of the IEEE Power Engineering Society.

At the time it approved the document, Subcommittee SC-5 had the following membership:

I. M. Jacobs, *Chairman*

A. W. Barchas	R. L. Olson
L. E. Booth	E. S. Patterson
M. F. Chamow	J. R. Penland
D. E. Cole	H. Rizenstein
M. L. Faught	F. Rosa
W. Gangloff	G. Schoenbaum
B. Logan	W. L. Strong
A. J. McElroy	B. D. Sullivan
R. Miles	B. M. Tashjian
G. Stiehl	R. A. Waller

J. J. Wroblewski

The membership of the working group was as follows:

J. J. Wroblewski, *Chairman*

M. F. Chamow	J. R. Penland
W. C. Gangloff	W. L. Strong
R. Miles	B. D. Sullivan

B. M. Tashjian

The Liaison Representative for Probability Methods Subcommittee of the IEEE Power System Engineering Committee, at the time this document was approved by SC-5 was:

P. A. Albrecht

3

Contents

An American National Standard

IEEE Standard Requirements for Reliability Analysis in the Design and Operation of Safety Systems for Nuclear Power Generating Stations

1. Purpose

The purpose of this standard is to provide uniform, minimum acceptable requirements for the performance of reliability analyses for safety-related systems found in nuclear-power generating stations, but not to define the need for an analysis. The need for reliability analysis has been identified in other standards which expand the requirements of regulations (e g, IEEE Std 379-1972 (ANSI N41.2-1972), "Guide for the Application of the Single-Failure Criterion to Nuclear Power Generating Station Protection Systems," which describes the application of the single-failure criterion).

IEEE Std 352-1975, "Guide for General Principles of Reliability Analysis of Nuclear Power Generating Station Protection Systems," provides guidance in the application and use of reliability techniques referred to in this standard.

2. Scope

This standard sets forth the minimum acceptable requirements for the performance of reliability analyses when used to address the reliability considerations discussed in the standards listed below for safety-related systems.

This standard applies to all systems, or portions of systems, for which reliability considerations are discussed in the following IEEE Standards:

[1] IEEE Std 279-1971 (ANSI N42.7-1972), Criteria for Protection Systems for Nuclear Power Generating Stations.
[2] IEEE Std 308-1974, Standard Criteria for Class 1E Power Systems for Nuclear Power Generating Stations.
[3] IEEE Std 338-1975, IEEE Criteria for the Periodic Testing of Nuclear Power Generating Station Class 1E Power and Protection Systems.

[4] IEEE Std 379-1972 (ANSI N41.2-1972), Guide for the Application of the Single-Failure Criterion to Nuclear Power Generating Station Protection Systems.

NOTE: Revisions to the above standards may, or may not, result in changes to requirements for reliability analyses to be performed. These references are cited because they represent current requirements.

The methods of this standard may also be applied to other systems including the interactions, if any, between safety-related and non-safety- related system.

The requirements can be applied during the phases of design, fabrication, testing, maintenance, and repair of systems and components in nuclear-power plants. The timing of the analysis depends upon the purpose for which the analysis is performed. This standard applies to the plant owner and other organizations responsible for the activities stated above.

3. Definitions

The definitions listed below establish the meaning of words in the context of their use in this standard. Other definitions can be found in IEEE Std 352-1975, Guide for General Principles Analysis of Nuclear Power Generating Station Protection Systems.

3.1 availability. The characteristic of an item expressed by the probability that it will be operational at a randomly selected future instant in time.

3.2 interaction. A direct or indirect effect of one device or system upon another.

3.3 reliability. The characteristic of an item expressed by the probability that it will perform a required mission under stated conditions for a stated mission time.

3.4 risk. The expected detriment per unit time to a person or a population from a given cause.

3.5 safety-related (electrical). Any Class 1E power or protection system device included in the scope of IEEE Std 279-1971 [1] or IEEE Std 308-1974 [2].

3.6 unavailability. The numerical complement of availability. Unavailability may occur as a result of the item being repaired (repair unavailability), tested (testing unavailability), or it may occur as a result of undetected malfunctions (unannounced unavailability).

3.7 unavailability margin. The favorable difference between the desired goal and the calculated or observed unavailability.

4. Requirements

4.1 General. The purpose of reliability analysis is to assist in assuring that the nuclear-plant safety-related systems within the scope of this standard will perform their required functions with an acceptable probability of success. The actions required to perform a reliability analysis and evaluate results of the analysis include the following actions:

    establish availability/reliability goals
    evaluate system designs
    establish testing intervals
    evaluate the demonstrated operational performance of installed equipment
    take any necessary corrective action

4.1.1 When required, qualitative analysis shall be performed in accordance with Section 4.2 to assess safety-related system conformance to applicable design criteria.

4.1.2 When required, quantitative analysis shall be performed in accordance with Sections 4.3 and 4.4 to establish initial periodic testing intervals for safety-related system equipment, and to provide a means for evaluating operational performance against requirements.

4.1.3 Wherever standardized designs are used for any portion of more than one nuclear station, the analyses performed for the standardized portion of the first design will fulfill the requirements for that portion of later stations provided that the initial analyses are verified to be applicable.

4.2 Qualitative Analysis.
4.2.1 A qualitative analysis when performed shall be documented in a manner suitable for review.

4.2.2 The minimum information to be considered for documentation of qualitative analysis to satisfy applicable criteria (e g, single failure, independence, channel integrity, etc) shall include the following:

Level of analysis. The basic level of the system at which the faults of interest are investigated including a list of components, modules, or devices included in the analysis.

Failure mode. All applicable, significant failure modes for each class of component. (The *IEEE Standard Failure Rate Data Manual* includes information on failure modes.)

System diagram. A logical arrangement of components basic to the system's primary function or operational mode for which the analysis is performed (e g, schematics, process diagrams, etc).

Boundary of analysis. The area of design included within the scope of the work and germane to the analysis.

Results. The output of the analysis which is normally part of a standard worksheet (e g, cause of failure, method of detection, effects of the failure).

4.2.3 The analysis must consider multiple failures attributable to a single cause and cascading-type failures. Analyses performed using the methods described in Section 4.5 of IEEE Std 352-1975 are acceptable to fulfill this requirement.

4.2.4 Expected environmental and initial conditions assumed in the analysis shall be stated.

4.2.5 Qualitative analyses shall be performed in a manner which considers design changes. As a minimum, an analysis shall exist which reflects the final design. Partial analyses may be performed to account for changes to critical portions of a design.

4.3 Quantitative Analysis.
4.3.1 Quantitative analyses when performed may consist of any of the methods described in Section 5 or Appendix A of IEEE Std 352-1975. Documentation of the analysis shall be suitable for review. The model should be capable of being expanded into a higher level system model as suggested in Appendix A to IEEE Std 352-1975.

4.3.2 A quantitative analysis is performed to calculate the predicted availability or reliability or both of the various safety-related systems in the plant. The use of reliability or availability or both shall be selected in terms of the functions of the system in the operational mode being analyzed. This analysis shall include pertinent system interactions and shall include sufficient detail to establish testing intervals consistent with the goals for the system. Appendix A to IEEE Std 352-1975 illustrates an acceptable method of analysis.

4.3.3 The quantitative analyses shall be used to determine if a design can meet a specified goal. Goals for the safety-related systems shall be determined by the organization responsible for the designs. Determination of the goals shall consider the following, as appropriate:

overall plant goals,
system performance requirements,
rate of demand on the system,
complexity of system design,
consequences of system failure,
testing limitations,
alternative risks,
owner's requirements,
regulatory requirements.

Examples of acceptable model formats include:

fault tree,
reliability block diagram,
truth tables (or other appropriate tabular model).

Appropriate calculational techniques for quantification of the reliability or availability or both of systems modeled as above include the concepts and methods of:

Boolean algebra,
conditional probability,
minimum cut sets (appropriate bounds must be specified),
Monte Carlo Simulation (calculational uncertainties should be evaluated),
Markov matrices.

Combinations of any of the above model formats and calculational methods may be supplemented or replaced by a simple comparison with similar systems which have been analyzed in detail. Any difference between the similar systems shall be defined; analyses of such differences shall be performed to demonstrate that the existing detailed analysis is applicable.

4.3.4 Quantitative analyses shall be performed in a manner which considers the effects of design changes. As a minimum, an analysis shall exist which reflects the final design. Partial analyses may be performed to account for changes to critical portions of a design.

4.3.5 All component failure data sources and assumptions used in the analysis shall be documented.

4.3.6 Failure data shall be obtained from credible sources.[1] Standard failure data shall be modified by application factors when it reflects experience in a significantly different operating environment from that to which it is being applied.

4.3.7 Failure rates which are based on judgment may be used, provided the basis for the judgment is described and documented in the analysis. Sensitivity analyses shall be performed to assess the effects of data uncertainties when judgment is used to specify failure rates.

4.3.8 Quantitative analysis is intended to be suitable to constitute one of the bases for the plant Technical Specifications minimum surveillance requirements and limiting conditions for operation. The testing intervals shall be determined in this manner to meet the requirement of Sections 4.7, 4.8, and 6.5 of IEEE Std 338-1975[3].

4.4 Evaluation.

IEEE Std 338-1975[3] requires that periodic testing programs be established to assure that Class 1E power and protection systems function with high availability. The requirements stated below amplify or complement those of IEEE Std 338-1975[3].

4.4.1 If operational data reveal that the goals are being achieved with wide margins, the testing interval may be lengthened, redundancy requirements may be reduced, or limiting conditions for operation may be relaxed.

4.4.2 If actual performance falls significantly short of the goal, actions must be taken to assure that the goals can be attained. These actions include investigation for systematic causes such as design deficiencies or maintainability problems, shortening the test interval, requiring more stringent limiting conditions for operation, or reassessment of the goal.

4.4.3 The requirements of IEEE Std 338-1975[3] complemented by the methods of Section 7.3, IEEE 352-1975 shall be adhered to for changes in test intervals or operating limitations.

NOTE: The plant owner is urged to file failure reports which can be disseminated to the nuclear industry to increase the body of failure data and to participate in the Edison Electric Institute/Nuclear Plant Reliability Data System.

[1] The *IEEE Reliability Data Manual*, when published, may be a credible data source for most electrical components.

Draft American National Standard

IEEE Trial-Use Standard
Criteria for Safety Systems for
Nuclear Power Generating Stations

603

Sponsor

**Nuclear Power Engineering Committee
of the
IEEE Power Engineering Society**

Corrected edition October 25, 1977

Foreword

(This foreword is not a part of IEEE Std 603-1977, Trial-Use Standard Criteria for Safety Systems for Nuclear Power Generating Stations.)

This standard establishes functional and design criteria for nuclear power generating station "safety systems," that is, that collection of systems required to minimize the probability and magnitude of release of radioactive material to the environment by maintaining plant conditions within the allowable limits established for each design basis event. These criteria are established to provide a means of promoting safe practices for design and evaluation of safety system performance and reliability. The criteria herein represent an acceptable basis for ascertaining the adequacy of safety system performance and reliability. Adherence to these criteria does not necessarily fully establish the adequacy of safety system functional performance and reliability. However, omission of any of these criteria will, in most instances, be an indication of safety system inadequacy.

The safety system encompasses both the protection system (a sense and command aspect) and protective action system (execute aspect). The safety system also encompasses the reactor trip system, engineered safety features, and auxiliary supporting features. Fig 1 has been provided to clarify the interrelationships among protection and protective action systems as well as the reactor trip system, engineered safety features, and auxiliary supporting features.

Application

The safety system criteria established herein are to be applied to those systems required to protect the public health and safety by functioning to mitigate the consequences of design basis events. However, this standard does not necessarily apply to all of the safety related systems, structures, and equipment required for complete plant safety. Although the scope is limited to those systems which perform to mitigate the consequences of design basis events, that is, safety systems, many of the underlying principles may be applicable to equipment provided for safe shutdown, post-accident monitoring display instrumentation, preventive interlock features, or any other systems, structures, or equipment related to safety, or all of the preceding.

To determine which systems are subject to these criteria, an analysis of the overall plant response to postulated design basis events must be performed. A detailed presentation of analytical techniques that can be used in such an analysis is contained in Ref [C4] of Appendix C. Good engineering judgment must be exercised in this assessment to assure that adequate margin exists in the design while not imposing unduly restrictive criteria such that the goal of protecting the public health and safety is replaced by one of protecting every plant component.

Interdisciplinary Approach

The safety system criteria herein are established using a "systems approach" to the design of the power, instrumentation, and control portion of the safety system, as opposed to a specific engineering discipline approach (that is, electrical, mechanical, or civil). It must be recognized that the protective functions cannot be accomplished without mechanical as well as electric equipment and circuitry. Therefore, the design by other than electrical engineering disciplines, primarily mechanical and nuclear engineering, is also affected by these criteria. In order for the safety system to meet the requirements of this standard and the supportive standards, the aggregate design of the safety system (without regard to discipline) may be constrained. Such constraints are implicit interface requirements imposed upon the individual constituent parts to enable the entire safety system to meet these requirements.

While this standard does take a systems approach to the design of the power, instrumentation, and control portion of the safety system, it does not attempt to establish new and different criteria for mechanical equipment or components. Such an attempt by a user is a misapplication of the standard. Nor does this standard attempt to establish the system level requirements which may be required by mechanical or civil equipment; for example, inservice inspection of piping is intentionally excluded. This standard is to provide criteria for the safety system without conflicting with existing standards. This standard is not intended to duplicate or conflict with component design requirements such as the ASME Boiler and Pressure Vessel Code. Rather, this standard is to complement and interface with such documents. This standard and others (for example, Refs [C1] and [C4] in

Appendix C) establish "system level" criteria while other codes and standards establish such detailed design requirements as deemed necessary to assure the functional adequacy of the individual constituent parts of the safety system.

Evolution

This standard evolved by expanding the scope of IEEE Std 279-1971, Criteria for Protection Systems for Nuclear Power Generating Stations (ANSI N42.7-1972). Work was initiated in 1971 to establish criteria for the actuator system (now termed the protective action system). It was decided, after many draft revisions, to assign a new number to the expanded scope, and due to the codification of IEEE Std 279-1971 in Federal Regulation 10CFR50.55a(h), to temporarily retain the existing number 279 for the protection system scope. Following the trial-use period, this standard will be issued as a full standard replacing IEEE Std 279-1971. Therefore, this standard represents the third Subcommittee 6 publication of a systems level criteria document: first, IEEE Std 279-1968 as a trial-use protection system standard; second, IEEE Std 279-1971 as a full protection system standard; and now IEEE Std 603-1977 as a trial-use safety system standard. The significant modifications accomplished with this standard over IEEE Std 279-1971 are as follows:

(1) The scope of this standard has been expanded from the protection system (IEEE Std 279-1971) to the safety system (IEEE Trial-Use Std 603-1977). The safety system encompasses both the protection and protective action systems. The protection system scope has been clarified as the system providing the "sense and command" aspects of the protective function. In contrast, the protective action system has been established as the system providing the "execute" aspects of the protective function. Fig 1 has been added to clarify the interrelationships among protection, protective action, and safety systems as well as the reactor trip system, engineered safety features, and auxiliary supporting features.

(2) The requirements in the former Section 4 of IEEE Std 279-1971 were the primary bases for establishing the requirements contained in Sections 4, 5, and 6 of this standard. Section 4 of this standard contains the functional and design requirements applicable to both the protection and protective action systems. Those additional requirements unique to the sense and command aspects are contained in Section 5 of this standard, while those requirements unique to the execute aspect are contained in Section 6. A conscious attempt was made to establish most of the requirements in Section 4 — stating these requirements broadly enough to apply to both the protection and protective action systems, yet with sufficient specificity to be clearly applied by the user. When the characteristics of the protection or protective action system were considered to be such that a requirement could not be generalized in Section 4 without losing this objective, separate Section 5 or Section 6 requirements, or both, were established containing the desired degree of specificity.

(3) The safety system scope, initially conceived as being limited to the reactor trip system and engineered safety features, has been expanded to include auxiliary supporting features.

(4) To assist in the application of IEEE Trial-Use Std 603-1977, the foreword has been expanded and Appendix A contains an illustrative example.

(5) A close alignment of IEEE Trial-Use Std 603-1977 criteria with related criteria has been established by:

(a) Expanded and revised definitions, developed to encourage consistent terminology throughout IEEE, the American Nuclear Society, and other technical society standards

(b) Specific references to supplemental standards listed in Appendixes B and C.

(6) Section 3 has been reorganized so that the design basis considerations unique to automatically initiated protective actions are grouped in 3.4; those unique to manually initiated protection actions are grouped in 3.5; and those common to both automatically and manually initiated protective actions follow.

Relationship to Class 1E

The term "Class 1E" has been intentionally avoided in the title, scope, and requirements of this standard for the following reasons:

(1) The safety system is limited in scope to those systems required to minimize the probability and magnitude of release of radioactive material to the environment. The scope of Class 1E may encompass more than just the safety system. However, it should be clear from the definition of Class

1E that the electric portions of the safety systems are definitely Class 1E.

(2) The systems approach taken by this standard is interdisciplinary. Use of Class 1E would limit its application to "electric" power, instrumentation, and control portions. These criteria apply to safety systems pneumatically, hydraulically, or electrically powered, instrumented, or controlled, or all of the preceding.

(3) Use of Class 1E in combinations with the term safety system could incorrectly imply that non-Class 1E electric safety systems also exist.

Hence through its avoidance of Class 1E, it is the intent of Subcommittee 6 to broaden the inter-disciplinary application of this standard and yet restrict its application to features required to mitigate the consequences of design basis events.

Relationship to IEEE Std 308-1974

The relationship between this trial-use standard and IEEE Std 308-1974, Criteria for Class 1E Power Systems for Nuclear Power Generating Stations, deserves a special explanation. The major role of the Class 1E Power System is to provide electric power to the reactor trip system, engineered safety features, and auxiliary supporting features and therefore the Class 1E Power System itself is an auxiliary supporting feature.

In its role as an auxiliary supporting feature, the Class 1E Power System is covered by both IEEE Std 308-1974 and IEEE Trial-Use Std 603-1977. (Auxiliary supporting feature coverage is indicated on the lower half of Fig 1 of this standard.) The Class 1E Power System is unique in that it extends throughout the plant having far more complex interfaces than other auxiliary supporting features. Most other auxiliary supporting features are limited to one area or a single process in the plant and are basically mechanical systems. Characteristic of the complex interfaces of the Class 1E Power System is the fact that it is an auxiliary supporting feature for other auxiliary supporting features; other auxiliary supporting features are auxiliary supporting features for it; and it may provide support for nonsafety system equipment, as well as providing the means for the execution of the safety system protective actions.

In its role in the protective action system, some Class 1E Power System equipment, switchgear, circuit breakers, power cabling, and loads (primarily motors) are not only part of the Class 1E Power System but are integral parts of the engineered safety features as well. This can be seen in the upper right-hand portion of Fig 1 of this standard.

Consequently, the Class 1E Power System is subject to the requirements of IEEE Std 308-1974 and of IEEE Trial-Use Std 603-1977, and neither can be subjugated by the other.

Relationship to Other Standards

This standard establishes functional and design criteria that are general in nature. It requires supportive standards containing both general and detailed criteria to comprise a minimal set of requirements for the safety system. The American National Standards Institute standards (in part) Refs [C1], [C4], [C8], [C9], and [C10] in Appendix C also contain functional and design criteria for safety systems.

Other IEEE standards prepared in support of the IEEE Std 279-1971 criteria are referenced throughout this standard to aid the user in applying the criteria.

Future Tasks

Subcommittee 6 recognizes the need for standards which are applicable to other overall specific safety related systems, structures, and equipment. Hence Subcommittee 6 acknowledges the need to undertake the additional work described below:

(1) Development of criteria for post-accident monitoring display instrumentation Ref [B11].

(2) Development of criteria for preventive interlocks (Project P659).

(3) Continued consideration of design requirements in IEEE Trial-Use Std 603-1977 pertinent to:

    (a) Diversity and redundancy

    (b) Reliability goals

    (c) Interlocks

(d) Interface requirements

(e) Shared systems

(f) Design basis events.

(4) Pursuit of joint societal sponsorship of this standard.

Acknowledgements

Members of NPEC's Subcommittee 6 participating in the generation of these revised criteria at the time of final NPEC approval were:

G. M. McHugh, Jr, *Chairman*

H. K. Stolt, *Vice Chairman*

R. S. Darke, *Secretary*

T. M. Bates, Jr.	R. P. Daigle	E. S. Patterson
J. T. Beard	P. M. Duggan	C. E. Rossi
W. W. Bowers	N. C. Farr	W. F. Schmauss
G. W. Brastad	A. Laird	L. Stanley
R. C. Carruth	T. Matsumoto	J. E. Thomas
C. J. Crane	J. I. Martone	D. Tondi
B. Curtis	F. E. McLane	C. S. Walker
	D. C. Nau	

Liaison was provided by:

R. Brockman — liaison from SC-4

A. R. Kasper — liaison from ANS 50

L. K. Holland — liaison from ANS 4

E. A. Kollitides — liaison from SC-4

A. Nathan — liaison from SC-4

The subcommittee wishes to emphasize its acknowledgement to the following members:

L. Stanley — for this three years of service as Subcommittee Chairman

R. P. Daigle — for his two years of service as Working Group Chairman.

The subcommittee also wishes to acknowledge the contributions of the following former members;

R. H. Dalry		R. L. McIntyre
S. J. Ditto		R. D. Pollard
	A. J. Spurgin	

Members of the Nuclear Power Engineering Committee (NPEC) participating in the approval of these revised criteria were:

T. J. Martin, *Chairman*

A. J. Simmons, *Vice Chairman*

R. E. Allen, *Standards Coordinator*

L. M. Johnson, *Secretary*

J. F. Bates	N. C. Farr	G. M. McHugh, Jr	W. F. Sailer
J. T. Bauer	J. M. Gallagher	W. C. McKay	J. H. Smith
F. D. Baxter	J. B. Gardner	W. P. Nowicki	A. J. Spurgin
J. T. Beard	R. I. Hayford	W. E. O'Neal	L. Stanley
R. G. Benham	T. A. Ippolito	E. S. Patterson	W. Steigelmann
J. T. Boettger	I. M. Jacobs	J. R. Penland	H. K. Stolt
D. F. Brosnan	R. F. Karlicek	D. G. Pitcher	D. F. Sullivan
F. W. Chandler	A. Laird	H. V. Redgate	P. Szabados
C. M. Chiappetta	J. I. Martone	B. M. Rice	H. A. Thomas
	T. J. McGrath	J. C. Russ	

The IEEE will maintain this standard current with the state of the technology. Comments are invited on this standard as well as suggestions for additional material that should be included. These and requests for interpretations should be addressed to:

Secretary
IEEE Standards Committee
The Institute of Electrical and Electronics Engineers, Inc
345 East 47th Street
New York, New York 10017

Contents

IEEE Trial-Use Standard
Criteria for Safety Systems for
Nuclear Power Generating Stations

1. Scope

The criteria contained in this standard establish minimum functional and design requirements for safety systems for nuclear power generating stations. The safety system is the collection of systems required to minimize the probability and magnitude of release of radioactive material to the environment by maintaining plant conditions within the allowable limits established for each design basis event. Safety system functional and design criteria are also contained in other standards.[1]

1.1 Illustration. Fig 1 illustrates the scope of the safety system.

1.2 Application. The safety system criteria established herein are to be applied to those systems required to protect the public health and safety by functioning to mitigate the consequences of design basis events. However, this standard does not necessarily apply to all of the safety related systems, structures, and equipment required for complete plant safety.

The criteria herein are directed to the power, instrumentation, and control portions of the safety system. To satisfy these criteria, interface requirements may be imposed on the other portions of the safety system.

2. Definitions

The definitions in this section establish the meanings of words in the context of their use in this standard.[2]

actuated equipment. The assembly of prime movers and driven equipment used to accomplish a protective action.

NOTE: Examples of prime movers are: turbines, motors, and solenoids. Examples of driven equipment are: control rods, pumps, and valves.

actuation device. A component or assembly of components that directly controls the motive power (electricity, compressed air, hydraulic fluid, etc) for actuated equipment.[2]

NOTE: Examples of actuation devices are: circuit breakers, relays, and pilot valves.

administrative controls. Written rules, orders, instructions, procedures, policies, practices, and designations of authority and responsibility for the operation and maintenance of a nuclear generating station.

auxiliary supporting features. Systems or components which provide services (such as cooling, lubrication, and energy supply) which are required for the safety system to accomplish its protective functions.

channel. An arrangement of components and modules as required to generate a single protective action signal when required by a generating station condition. A channel loses its identity where single protective action signals are combined.[2]

class 1E. The safety classification of the electric equipment and systems that are essential to emergency reactor shutdown, containment isolation, reactor core cooling, and containment and reactor heat removal, or are otherwise essential in preventing significant release of radioactive material to the environment.

components. Discrete items from which a system is assembled.

NOTE: Examples of components are: wires, transistors, switches, motors, relays, solenoids, pipes, fittings, pumps, tanks, or valves.

design basis events. Postulated abnormal events used in the design to establish the acceptable performance requirements of the structures, systems, and components.

---

[1] Refer to Refs [B1], [B2], [C1], [C2], [C3], [C4], [C8], [C9], and [C10] in Appendixes B and C.

[2] Certain definitions represent recent refinements of the definitions in Ref [B6] in Appendix B and are expected to be reflected in its next revision.

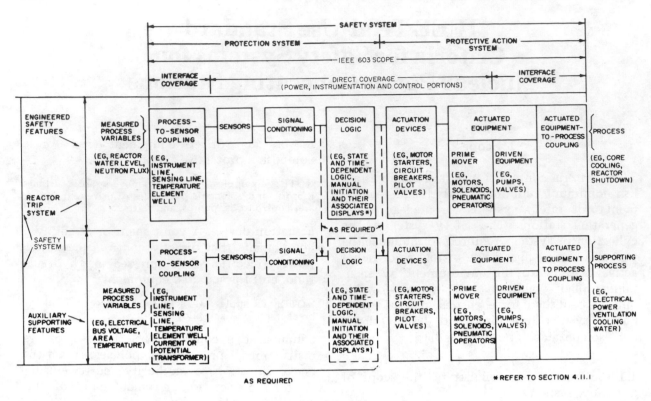

Fig 1
Safety System

detectable failures. Failures that will be identified through periodic testing or will be revealed by alarm or anomalous indication. Component failures which are detected at the channel or system level are detectable failures.[2]

maintenance bypass. Removal of the capability of a channel, component, or piece of equipment to perform a protective action due to a requirement for replacement, repair, test, or calibration.

NOTE: A maintenance bypass is not the same as an operating bypass. A maintenance bypass may remove a protective action, but it will not result in the loss of a protective function.

module. Any assembly of interconnected components which constitutes an identifiable device, instrument, or piece of equipment. A module can be disconnected, removed as a unit, and replaced with a spare. It has definable performance characteristics which permit it to be tested as a unit. A module could be a card, a drawout circuit breaker, or other subassembly of a larger device, provided it meets the requirements of this definition.

operating bypass. Inhibition of the capability to accomplish a protective function that could otherwise occur in response to a particular set of generating station conditions.

NOTE: An operating bypass is not the same as a maintenance bypass. Different modes of plant operation may necessitate an automatic or manual bypass of a protective function. Operating bypasses are used to permit mode changes (for example, prevention of initiation of safety injection during the cold shutdown mode).

protection system. The electrical and mechanical devices (from measured process variables to protective action system input terminals) involved in generating those signals associated with the protective functions. These signals include those that initiate reactor trip, engineered safety features (for example, containment isolation, core spray, safety injection, pressure reduction, and air cleaning), and auxiliary supporting features.

protective action. The initiation of a signal or operation of equipment within the protection system or protective action system for the purpose of accomplishing a protective function in response to a generating station condition having reached a limit specified in the design basis:

(1) *Protection System:* Protective action at

the channel level is the initiation of a signal by a single channel when the sensed variable(s) reaches a specified limit.

(2) *Protective Action System:* Protective action at the system level is the operation of sufficient actuated equipment, including the appropriate auxiliary supporting features, to accomplish a protective function. Examples of protective actions at the system level are: rapid insertion of control rods, closing of containment isolation valves, operation of safety injection, and core spray.

protective action system. The electrical and mechanical equipment (from the protection system output to and including the actuated equipment-to-process coupling) that performs a protective action when it receives a signal from the protection system.

NOTES: 1. Examples of protective action systems are: control rods, and their trip mechanisms; isolation valves, their operators, and their contractors; and emergency service water pumps and associated valves, their motors, and circuit breakers.
2. In some instances protective actions may be performed by protective action system equipment that responds directly to the process conditions (for example, check valves, self-actuating relief valves).

protective function. The completion of those protective actions at the system level required to maintain plant conditions within the allowable limits established for a design basis event (for example, reduce power, isolate containment, or cool the core).

redundant equipment or system. A piece of equipment or a system that duplicates the essential function of another piece of equipment or system to the extent that either may perform the required function regardless of the state of operation or failure of the other.[2]

NOTE: The term redundant could include identical equipment, equipment diversity, or functional diversity.

safety system. The collection of systems required to minimize the probability and magnitude of release of radioactive material to the environment by maintaining plant conditions within the allowable limits established for each design basis event.

NOTE: The safety system is the aggregate of one or more protection systems, and one or more protective action systems. It includes the engineered safety features, the reactor trip system, and the auxiliary supporting features. See Fig 1.

sensor. That portion of a channel which responds to changes in a plant variable or condition, and converts the measured process variable into a safety system signal (for example, electric, pneumatic).

3. Design Basis

A specific basis[3] shall be established for the design of the safety system of each nuclear power generating station. The design basis shall also be available as needed to facilitate the determination of the adequacy of the safety system, including design changes.

The design basis shall document, as a minimum:

3.1 The design basis events applicable to each mode of operation of the generating station along with the allowable limits of plant conditions for each such event

3.2 The protective functions and corresponding protective actions at the system level for each design basis event

3.3 The permissive conditions for each operating bypass capability that is to be provided

3.4 Those protective actions at the system level, identified in 3.2, requiring automatic initiation, and shall document for each:

3.4.1 The variables or combinations of variables, or both, that are to be monitored to initiate each protective action; the ranges (normal, abnormal, and accident conditions); and the rates of change of these variables to be accommodated until proper completion of the protective actions is assured

3.4.2 The safety limits applicable to each measured variable or combination of variables (an example is illustrated by Fig 2)

3.4.3 The minimum performance requirements, including the following for the appropriate combinations of those conditions of 3.7 and 3.8: the limiting safety system setting for each variable or combination of variables (as illustrated by Fig 2); the increment allotted for inaccuracies, calibration uncertainties, and errors; and the overall response times of the safety system used in establishing the limiting safety system setting in the station safety

[3]Refer to Ref [C4] in Appendix C.

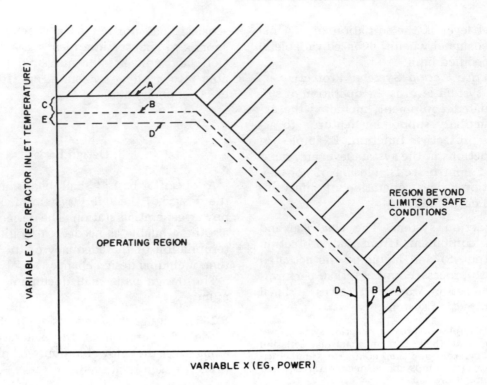

Fig 2
Example of Relationship Between Nominal Protection System
Setpoints and Limits of Safe Conditions

A   Process Safety Limit — Represents limits of safe conditions

B   Limiting Safety System Setting — Represents the least restrictive setpoints which may exist at any time

C   Allowance for calibration errors, instrument accuracy, and transient overshoot (this allowance may be a function of variable X or variable Y or both)

D   Safety System Setting — Nominal setpoints set to ensure that drift will not result in setpoints exceeding the limiting safety system settings of B

E   Allowance for instrument and setpoint drift (this allowance may be a function of variable X or variable Y or both)

analysis. The basis shall be provided to demonstrate that the assumed values used for instrumentation inaccuracy, calibration uncertainties and error, and time response are acceptable and reasonable.

3.4.4 The increment (as illustrated by Fig 2) between the limiting safety system setting and the nominal safety system setting to accommodate drift during the interval between calibration verification tests. The basis shall be provided to demonstrate that the assumed

value used for instrument drift is acceptable and reasonable.

3.5 Those protective actions, identified in 3.2, that may be initiated solely by manual means, and shall document for each:

3.5.1 The justification for permitting manual initiation

3.5.2 The variables to be monitored to facilitate the manual initiation of protective action

3.5.3 The minimum performance requirements including the following for the appropriate combinations of those conditions of 3.7 and 3.8:

3.5.3.1 System response times with appropriate interpretive information

3.5.3.2 System accuracies

3.5.4 The range of environmental conditions imposed upon the operator during normal, abnormal, and accident circumstances throughout which the manual operations must be performed

3.6 For those variables in 3.4.1 or 3.5.2 that have a spatial dependence (that is, where the variable varies as a function of position in a particular region), the minimum number and locations of sensors required for protective purposes

3.7 The acceptable range of transient and steady-state conditions of both motive power and the environment (for example, voltage, frequency, radiation, temperature, humidity, pressure, and vibration) during normal, abnormal, and accident circumstances throughout which the safety system must perform

3.8 The conditions having the potential for functional degradation of safety system performance and for which provisions must be incorporated to retain the capability for the protective functions

3.9 The methods to be used to determine that the reliability of the safety system design is appropriate for the safety system functions, including any qualitative or quantitative reliability goals which may be imposed on the system design

3.10 The critical points in time or the plant conditions, after the onset of a design basis event, including:

3.10.1 The point in time by which the protective action at the system level must be initiated

3.10.2 The point in time after which some protective actions may be manual

3.10.3 The plant conditions after which a deliberate operator intervention may prevent the completion of protective action at the system level.

3.10.4 The point in time, or plant conditions, which define the proper completion of the protective action at the system level

4. Safety System Functional and Design Requirements

The safety system shall, with precision and reliability, be capable of accomplishing the protective functions required for each design basis event.

4.1 Automatic Initiation. Means shall be provided to initiate automatically at the system level those protective actions identified in 3.4. The safety system design shall be such that the operator is not required to take any action prior to the point in time specified in 3.10 following the onset of each design basis event. At the option of the safety system designer, means may be provided to initiate automatically at the system level those protective actions of 3.5.

4.2 Manual Initiation

4.2.1 Means shall be provided in the control room to implement manual initiation at the system level of the automatically initiated protective actions. The means provided shall minimize the number of discrete operator manipulations and shall depend on the operation of a minimum of equipment.

4.2.2 Means shall be provided in the control room to implement manual initiation of the protective actions identified in 3.5 that have not been selected for automatic initiation under 4.1.

4.2.3 Means shall be provided to implement the manual actions necessary to maintain safe conditions after the protective actions are completed as specified in 3.10. The number of available qualified operators, the information provided to these operators, the actions required of these operators, and the quantity and location of associated displays and controls shall be appropriate for the time period within which the actions must be accomplished. Such displays and controls shall be located in areas that are accessible and in an environment suitable for the operator.

4.3 Single Failure Criterion. The safety system shall perform the protective actions required to accomplish a protective function in the presence of any single detectable failure within the safety system concurrent with all identifiable but nondetectable failures, all failures caused by the single failure, and all failures caused by the design basis event requiring the protective function.[4] The single failure criterion applies to the safety system whether initiation is by automatic or manual means.

The performance of a probabilistic assessment[5] of the safety system may be used to demonstrate that certain postulated failures need not be considered in the application of this criterion. A probabilistic assessment is intended to eliminate consideration of events and failures that are not credible; it shall not be used in lieu of the single failure criterion.

4.4 Completion of Protective Action. The safety system shall be designed so that, once initiated automatically or manually, the intended sequence of protective actions at the system level shall continue until completion. Deliberate operator action shall be required to return the safety system to normal. This requirement shall not preclude the use of equipment protective devices or the provision for those deliberate operator interventions which are identified in 3.10 of the design basis.

4.5 Quality. Components and modules shall be of a quality that is consistent with minimum maintenance requirements and low failure rates. Safety equipment shall be designed, manufactured, inspected, installed, tested, operated, and maintained in accordance with an acceptable quality assurance program.[6] For those systems for which either quantitative or qualitative reliability goals have been established, appropriate analysis of the design shall be performed in order to confirm that such goals have been achieved.[5]

4.6 Equipment Qualification. Safety system equipment shall be qualified by type test, previous operating experience, or analysis or any combination of these three methods to substantiate that it will be capable of meeting, on a continuing basis, the performance requirements as specified in the design basis.[7]

4.7 System Integrity. The safety system shall maintain the capability to accomplish its protective functions under the full range of applicable conditions enumerated in the design basis.

4.8 Independence[8]

4.8.1 *Between Redundant Portions of the Safety System.* Redundant portions of the safety system provided for a specific protective function shall be independent of and physically separated from each other to the degree necessary to retain the capability to accomplish the protective function during and following any design basis event.

4.8.2 *Between Safety System and Effects of Design Basis Event.* Equipment required to mitigate the consequences of a specific design basis event shall be independent of and physically separated from the effects of the design basis event to the degree necessary to retain the capability to meet the requirements of this standard.

4.8.3 *Between Safety System and Other Systems.* The safety system design shall be such that no credible failure in or consequential action by other systems shall prevent the safety system from meeting the minimum performance requirements of this standard in the presence of such failure specified in the design basis.

4.8.3.1 *Interconnected Equipment*

(1) *Classification:* Equipment that is shared and consequently is used for both protective actions and nonprotective actions shall be classified as part of the safety system. Isolation devices used to effect a safety system boundary shall be classified as part of the safety system.

(2) *Isolation:* The safety system shall be isolated from the effects of credible failures in, or actions by, other systems to the degree necessary to retain the capability to accomplish its protective functions. No credible failure on the nonsafety side of an isolation device shall prevent any portion of the safety system from meeting its minimum performance requirements. A failure in an isolation device is evaluated in the same manner as a failure of other equipment in the safety system.

4.8.3.2 *Equipment in Proximity*

(1) *Separation:* Equipment in other systems not shared with but in physical proximity to safety system equipment shall be physically separated from safety system equipment by

---

[4]Refer to Ref [B5] in Appendix B and Ref [C13] in Appendix C.
[5]Refer to Ref [B10] in Appendix B.
[6]Refer to Refs [C5] and [C6] in Appendix C.
[7]Refer to Ref [B3] in Appendix B.

[8]Refer to Ref [B7] in Appendix B and Refs [C2], [C8], [C11], and [C12] in Appendix C.

physical barriers or acceptable separation distance. This does not apply to associated or Class 1E circuits.

(2) *Barriers:* Physical barriers used to effect a safety system boundary shall meet the requirements of 4.5, 4.6, and 4.7 for the applicable conditions specified in 3.7 of the design basis.

4.8.3.3 *Effects of a Single Random Failure.* Where a single random failure can result in a generating station condition requiring protective action and can also prevent proper action of a portion of the safety system designed to protect against that condition, the remaining redundant portions shall be capable of providing the protective action at the system level even when degraded by any separate single failure.[4]

4.9 Capability for Test and Calibration. Capability for test and calibration of safety system equipment shall be provided while retaining the capability of the safety system to accomplish its protective functions. The capability for test and calibration of safety system equipment should be provided during power operation and should duplicate, as closely as practicable, performance of protective actions at the system level.[9] Where this capability cannot be provided during power operation without adversely affecting the safety or operability of the generating station then:

(1) Appropriate justification shall be provided (for example, demonstration that no practical design exists)

(2) Acceptable reliability of equipment operation shall be otherwise demonstrated

(3) The capability shall be provided while the generating station is shut down.

4.10 Operating Bypasses. Whenever the applicable permissive conditions are not met, the safety system shall automatically accomplish one of the following:

(1) Prevent the activation of an operating bypass

(2) Remove any active operating bypass

(3) Obtain or retain the permissive conditions for the operating bypass

(4) Initiate the protective function.

4.11 Information Displays

4.11.1 *Displays for Protective Actions Initiated Solely by Manual Means.* The display instrumentation provided for the manually initiated actions required for the safety system to accomplish its protective function shall be part of the safety system. The design shall minimize the possibility of anomalous indications which could be confusing to the operator.

4.11.2 *System Status Indication.* The display instrumentation provided for safety system status indication need not be part of the safety system. The display instrumentation shall provide accurate, complete, and timely information pertinent to safety system status. This information shall include indication and identification of protective actions at the channel level and the system level. The design shall minimize the possibility of anomalous indications which could be confusing to the operator.

4.11.3 *Indication of Bypasses.* If the protective actions of some part of the safety system have been bypassed or deliberately rendered inoperative for any purpose, continuing indication of this fact at the system level shall be provided in the control room.

4.11.4 *Location.* Information displays shall be located accessible to the operator. Information displays provided for manually initiated protective actions shall be visible from the location of the controls used to effect the actions.

4.12 Control of Access. The design shall permit the administrative control of access to safety system equipment. These administrative controls shall be supported by provisions within the safety system, by provisions in the generating station design, or by a combination thereof.

4.13 Repair. The safety system shall be designed to facilitate timely recognition, location, replacement, repair, and adjustment of malfunctioning equipment.

4.14 Identification. In order to provide assurance that the requirements given in this standard can be applied during the design, construction, maintenance, and operation of the plant, the following should apply:

(1) Safety system equipment shall be identified distinctively for each redundant portion of the safety system.[10]

(2) Components or modules mounted in equipment or assemblies that are clearly identified as being in a single redundant portion of

---

[9] Refer to Ref [B4] in Appendix B.

[10] Refer to Refs [B7] and [B8] in Appendix B.

the safety system do not themselves require identification.

(3) Identification of safety system equipment shall be distinguishable from any identifying markings placed on equipment for other purposes (for example, identification of fire protection equipment, phase identification of power cables).

(4) Identification of the safety system equipment shall not require the use of reference material.

(5) The associated documentation shall be identified distinctively.[11]

5. Protection System Functional and Design Requirements

In addition to the functional and design requirements in Section 4, the following shall apply to the protection system.

5.1 Interaction Between the Protection System and Other Systems.

5.1.1 Where a single credible event, including all direct and consequential results of that event, can cause a nonsafety system action that results in a condition requiring protective action and can concurrently prevent the protective action in those protection system channels designated to provide principal protection against the condition, one of the following shall be met:

(1) Alternate channels and equipment, not subject to failure resulting from the same single event, shall be provided to detect the event and limit the consequences of this event to a value specified by the design basis. In the selection of alternate channels, consideration should be given to:

(a) Channels that sense a set of variables different from the principal channels

(b) Channels that use equipment different from that of the principal channels to sense the same variable

(c) Channels that sense a set of variables different from those of the principal channels using equipment different from that of the principal channels.

Both the principal and alternate channels shall be a part of the protection system.

(2) Equipment not subject to failure caused

by the same single credible event shall be provided to detect the event and limit the consequences to a value specified by the design bases. Such equipment is considered a part of the protection system.

5.1.2 Provisions shall be included so that the requirements in 5.1.1 can still be met in conjunction with the requirements of 5.4 if a channel is in maintenance bypass. Acceptable provisions include reducing the required coincidence, defeating the nonsafety system signals taken from the redundant channels, or initiating a protective action from the bypassed channel.

5.2 Derivation of System Inputs. To the extent feasible and practical, protection system inputs shall be derived from signals that are direct measures of the desired variables as specified in the design basis.

5.3 Capability for Sensor Checks. Means shall be provided for checking, with a high degree of confidence, the operational availability of each protection system input sensor during reactor operation. This may be accomplished in various ways, for example:

(1) By perturbing the monitored variable

(2) Within the constraints of 4.10, by introducing and varying, as appropriate, a substitute input to the sensor of the same nature as the measured variable

(3) By cross checking between channels that bear a known relationship to each other and that have readouts available.

5.4 Maintenance Bypass. Capability of the safety system to accomplish its protective functions shall be retained while protection system equipment is in maintenance bypass. During such operation the protection system shall continue to meet the requirements of 4.3 and 5.1.

EXCEPTION: "One-out-of-two" portions of the protection systems are not required to meet 4.3 and 5.1 when one portion is rendered inoperable provided that acceptable reliability of equipment operation is otherwise demonstrated (that is, that the period allowed for removal from service for maintenance bypass is sufficiently short so as to have no significantly detrimental effect on overall protection system availability).

5.5 Multiple Set Points. Where it is necessary to change to a more restrictive set point to provide adequate protection for a particular mode of operation or set of operating conditions, the design shall provide positive means of as-

---

[11] Refer to Ref [B9] in Appendix B.

suring that the more restrictive set point is used. The devices used to prevent improper use of less restrictive set points shall be a part of the protection system.

6. Protective Action System Functional and Design Requirements[12]

In addition to the functional and design requirements in Section 4, the following shall apply to the protective action system.

6.1 Manual Initiation. If manual initiation of any actuated component in the protective action system is required to fulfill a design basis objective, the additional design features in the protective action system necessary to accomplish such manual initiation shall not defeat the requirements of 4.2 or 4.3.

6.2 Completion of Protective Action. The design of the protective action system shall be such that once initiated, a protective action at the system level will go to completion. This requirement shall not preclude the use of equipment protective devices or the provision for the deliberate operator interventions which are identified in 3.10 of the design basis. After the initial protective action has gone to completion, the protective action system may require manual control or automatic control (that is, cycling) of specific equipment to maintain completion of system level protective action.

6.3 Maintenance Bypass. Capability of the safety system to accomplish its protective functions shall be retained while protective action system equipment is in maintenance bypass. Portions of the protective action system with a degree of redundancy of one[13] shall be designed such that when a portion is placed in maintenance bypass (that is, reducing temporarily its degree of redundancy to zero[13]), the remaining portions provide acceptable reliability.

---

[12]Refer to Ref [B2] in Appendix B.

[13]Redundancy of one: 1 out 2, 2 out 3, 3 out of 4, etc; redundancy of zero: 1 out of 1, 2 out of 2, 3 out of 3, etc.

Appendixes

(These appendixes are not a part of IEEE Std 603-1977, Trial-Use Standard Criteria for Safety Systems for Nuclear Power Generating Stations.)

Appendix A
Illustration of Scope of Safety System

A1. Purpose

The purpose of this appendix is to illustrate the scope of the safety system by means of a specific, illustrative example. The following example is provided to illustrate the scope of IEEE Trial-Use Std 603-1977. This example is representative and does not imply suitability to any specific system or equipment application.

The example presented is a portion of a hypothetical safety system necessary to accomplish an individual protective function. Illustrated is the relationship between the ESF (engineered safety feature) and the ASF (auxiliary supporting feature) portions of the safety system.

Fig A1
Safety System Partial Example
PS Signals from the Protection System

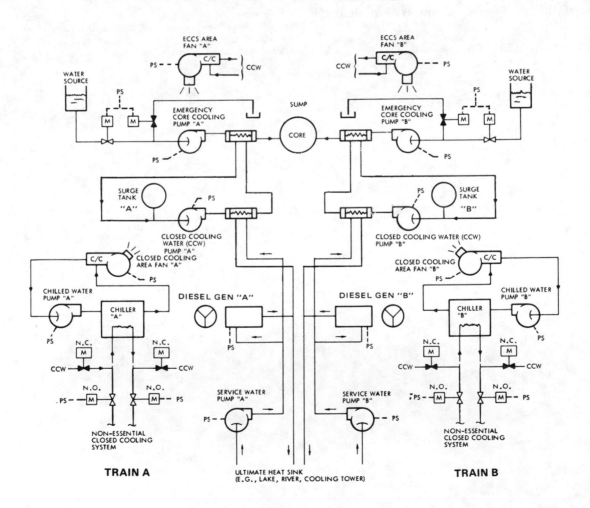

A2. Example

Fig A1 is a simplified diagram of a portion of a representative safety system. This safety system must accomplish a single protective function: cool the core following a particular design basis event.

A3. Description

The function is primarily accomplished by the ECCS (emergency core cooling system) which consists of two sets of independent redundant equipment, either of which is capable of accomplishing the protective function. The emergency core cooling pumps, heat exchangers, storage tanks, and associated valves, piping, instrumentation, and controls constitute the ESF portion of this safety system.

A3.1 Essential services such as electrical power, auxiliary cooling water, and equipment area ventilation are provided to support the ESF system under all design basis conditions for the periods of time necessary to establish and maintain safe conditions. The systems (including the associated equipment, piping, wiring, instrumentation, and controls) which must operate to provide these essential services are the ASF of the safety system. Some ASF systems directly support ESF systems equipment, and some provide indirect support.

A3.2 Table A1 lists the systems and system functions within the safety system of Fig A1 and identifies their type with respect to the terminology of IEEE Trial-Use Std 603-1977.

Table A1
Safety System Example Functional Breakdown
Protective Function: Cool the Core

System	Portion of Safety System	System Function
(1) Emergency core cooling system (ECCS)	ESF	Establish and maintain water flow to core
(2) Service water system	ASF	Provide cooling water to remove heat from ECCS system
	ASF	Provide cooling water to remove heat from closed cooling system
	ASF	Provide cooling water to remove heat from diesel generator heat exchanger
(3) Closed cooling system	ASF	Provide cooling water to remove heat from ECCS pump bearings
	ASF	Provide cooling water to cooling coils of ECCS equipment area air handling unit
	ASF	Provide cooling water for chilled water system heat exchanger
(4) ECCS equipment area ventilation system	ASF	Maintain ambient temperature of ECCS equipment area with specified limits
(5) Closed cooling system equipment area ventilation system	ASF	Maintain ambient temperature of closed cooling system equipment area within specified limits
(6) Chilled water system	ASF	Provide cooling water for cooling coils of closed cooling system equipment area air handling unit
(7) Class 1E power system	ASF	Provide electrical motive, control, and instrumentation power for item (1) (ESF)
	ASF	Provide electrical motive, control, and instrumentation power for items (2) through (7) (ASF)
(8) Ultimate heat sink	ASF	Provide source of water and heat sink for service water system

Appendix B
References

[B1] IEEE Std 279-1971, Criteria for Protection Systems for Nuclear Power Generating Stations (ANSI N42.7-1972).

[B2] IEEE Std 308-1974, Criteria for Class 1E Power Systems for Nuclear Power Generating Stations (ANSI N41.12-1975).

[B3] IEEE Std 323-1974, Qualifying Class 1E Equipment for Nuclear Power Generating Stations.

[B4] IEEE Std 338-1975, Trial-Use Criteria for the Periodic Testing of Nuclear Power Generating Station Class 1E Power and Protection Systems.

[B5] IEEE Std 379-1977, Guide for the Application of the Single-Failure Criterion to Nuclear Power Generating Station Protection Systems (ANSI N41.2 — Draft Standard).

[B6] IEEE Std 380-1975, Definition of Terms Used in IEEE Standards on Nuclear Power Generating Stations.

[B7] IEEE Std 384-1977, Standard Criteria for Independence of Class 1E Equipment and Circuits

[B8] IEEE Std 420-1973, Trial-Use Guide for Class 1E Control Switchboards for Nuclear Power Generation Stations (ANSI N41.17 — Draft Standard).

[B9] IEEE Std 494-1974, Standard Method for Identification of Documents Related to Class 1E Equipment and Systems for Nuclear Power Generating Stations (ANSI N41.28-1976).

[B10] IEEE Std 577-1976, Requirements for Reliability Analyses in the Design and Operation of Safety Systems for Nuclear Power Generating Stations.

[B11] IEEE Std 497-1977, Trial-Use Standard Criteria for Post Accident Monitoring Instrumentation for Nuclear Power Generating Stations (ANSI N41.26 — Draft Standard)

Appendix C
Bibliography

[C1] ANSI N18.2-1973, Nuclear Safety Criteria for the Design of Stationary Pressurized Water Reactor Plants.

[C2] ANSI N18.2a-1975, Supplement to Nuclear Safety Criteria for the Design of Stationary Pressurized Water Reactor Plants.

[C3] ANSI N18.7-1972, Administrative Controls for Nuclear Power Plants.

[C4] ANSI N18.8-1973, DRAFT STANDARD Criteria for Preparation of Design Bases for Systems that Perform Protection Functions in Nuclear Power Generating Stations (issued for trial use and comment).

[C5] ANSI N45.2-1971, Quality Assurance Program Requirements for Nuclear Power Plants.

[C6] ANSI N45.2.11-1974, Quality Assurance Requirements for the Design of Nuclear Power Plants.

[C7] ANSI N182-1976, Protection Criteria for Systems and Components Important to Safety.

[C8] ANSI N212-1974, DRAFT STANDARD Nuclear Safety Criteria for the Design of Stationary Boiling Water Reactor Plants (issued for trial use and comment).

[C9] ANSI N213-1975, Nuclear Safety Criteria for the Design of Stationary Gas-Cooled Reactor Plants (issued for trial use and comment).

[C10] ANSI N214-1976, General Safety Design Criteria for an LMFBR Nuclear Power Plant (issued for trial use and comment).

[C11] ANSI N176-1974, DRAFT STANDARD Design Basis for Protection of Nuclear Power Plants Against Effects of Postulated Pipe Rupture (issued for trial use and comment).

[C12] ANSI N177-1974, DRAFT STANDARD Plant Design Against Missiles (issued for trial use and comment).

[C13] ANSI N658-1976, Single Failure Criteria for PWR Fluid Systems (issued for trial use and comment).

IEEE Standard Cable Penetration Fire Stop Qualification Test

Sponsor

**Insulated Conductors Committee of the
IEEE Power Engineering Society**

634

© Copyright 1978 by

The Institute of Electrical and Electronics Engineers, Inc.

Foreword

(This Foreword is not a part of IEEE Std 634-1978, Cable Penetration Fire Stop Qualification Test.)

This standard provides qualification test procedures for type testing cable penetration fire stops when mounted in rated fire barriers.

In the course of construction of all types of buildings, cables in raceways penetrate barriers such as walls, floors, or floor-ceiling assemblies of that building. If these barriers are rated as fire resistive barriers, the penetrations should be as resistant to fire as required of the barriers. Thus, in order to test the penetration and rate it, the penetration should be mounted in a rated wall, floor, or floor ceiling assembly as it would be used in practice and the combination exposed to the same standard fire as used for the wall, floor, or floor-ceiling assembly.

Rating of a Fire Resistive Barrier, with No Penetrations

This rating is expressed in hours and represents the ability of that barrier to withstand, without failure, exposure to a standard fire for that length of time. A fire rating for a barrier may be arrived at by testing it according to the procedure outlined in ANSI A2.1-1972, Methods of Fire Tests of Building Construction and Materials (ASTM E119-1971) (ISO 834).

A barrier achieves its rating if, during the specified time, it contains the fire, and its surface unexposed to the fire does not heat up sufficiently to ignite cotton waste or the temperature does not exceed 250°F above ambient. In addition, following the fire, the barrier is required to withstand a specified standard fire hose test on the hot face.

Caution Re: ANSI A2.1-1972, Limitations

ANSI A2.1-1972 cautions that its results give only a *relative* measure of fire performance of comparable barriers (see 2.2), that it does not measure degree of control or limitation of smoke or products of combustion through the assembly (see 2.4.3), and does not consider the effect of conventional openings, that is, electrical receptacle outlets or plumbing pipe, etc (see 2.4.6).

Standard Fire in ANSI A2.1-1972

The standard fire is defined by a time-temperature relationship which must be produced by the test furnace. The seven defined points on this curve are given as follows:

$$1000°F \quad (538°C) \quad \text{at 5 min}$$
$$1300°F \quad (704°C) \quad \text{at 10 min}$$
$$1550°F \quad (843°C) \quad \text{at 30 min}$$
$$1700°F \quad (927°C) \quad \text{at 1 hr}$$
$$1850°F \quad (1,010°C) \quad \text{at 2 hr}$$
$$2000°F \quad (1,093°C) \quad \text{at 4 hr}$$
$$2300°F \quad (1,260°C) \quad \text{at 8 hr or more}$$

A more detailed description is given in ANSI A2.1-1972, Appendix A1 which lists intervening points and tabulates the integrated area under the time-temperature curve as a function of time.

The same standard fire is used on the cable penetration fire stop qualification test.

Fundamental Difference Between a Fire Test on a
Barrier Alone and a Penetration-Barrier Combination

The fire resistive barrier described above has a relatively low thermal conductivity so that it can maintain a 1300-1600°F temperature difference between the face exposed to the fire and the opposite face. A cable penetration has a metallic electrical conductor which has a very high thermal conductance. It may have many large copper conductors and steel trays or conduits or metal parts of the penetration, all of which pass through the barrier. On the cool side of the barrier, these metal parts are necessarily at a higher temperature than the wall adjacent to the penetration. The stop material filling the interstices between cables or between cables and the barrier should give comparable thermal conductance to the barrier itself, in addition to resisting the fire.

Thus the higher temperature rise of the metallic parts of the penetration presents a new and different problem and may make it impossible to use the same pass-fail criteria as for the barriers. An obvious failure occurs when sufficient heat is transmitted so that the insulation of the cable on the cold side bursts into flame. This is discussed further in 2.3.

Maximum Allowable Cable Penetration Fire Stop Face Temperature

If one examines the temperatures across the unexposed face of the cable penetration fire stop near the end of a 3 h test, the temperatures will vary widely depending on the distance from a cable or a raceway. The temperature of the unexposed face of the cable penetration fire stop material at a point away from the cable or the raceway will also depend on the thermal conductivity of the cable penetration fire stop material. The *maximum* temperature on that face is the important one. If this temperature is at the interface between the cable jacket and the cable penetration fire stop material, and if this temperature rises to the self-ignition temperature of the cable jacket or the stop material, a fire may result.

Thus, the test procedure finds the maximum temperature on the unexposed cable penetration fire stop face and compares it with a maximum allowable temperature. The maximum allowable temperature is defined as one at which the insulation systems expected to be used should not ignite.

The maximum allowable temperature is arrived at by an examination of the known ignition temperatures of insulating materials. Ignition temperature is measured by a procedure in ANSI K65.111-1971, Method of Test for the Ignition Properties of Plastics (ASTM D129-1968). This is described as a hot-air ignition furnace. The values obtained represent the lowest ambient air temperature that will cause ignition of the material under the conditions of test. Measured properties are "flash-ignition temperature" where an igniting source is present (small gas flame) and "self-ignition temperature," where ignition occurs spontaneously.

For ignition, there must be adequate temperature; the combustible gases released from the hot insulation must be mixed with the correct proportion of air.

The required temperature to cause ignition would be much higher than the ASTM value because the hot gases released are swept away by air drafts, and a higher temperature is needed to produce a higher rate of release of gases so that an ignitable gas-air ratio can be attained. Thus, there is a good factor of safety in the assigned maximum allowable temperature.

Typical values of the ignition temperatures as determined in ANSI K65.111-1971 are given below in degrees Fahrenheit:

Material	Flash-Ignition	Self-Ignition
Cotton	446-511	490
Newspaper	445	445
Pine shavings	406-507	500
Wool	401	——
Polyethylene	645	660
Polyvinyl chloride	735	850
Polytetrafluroethylene	——	986
Polyvinyl chloride-acetate	608-644	815-1035
Polystyrene	635-680	910-925
Nylon 66	750-790	788-806

The maximum allowable temperature selected for a cable penetration fire stop should be based on the self-ignition temperature of the outer cable covering the fire stop materials, or materials in contact with the cable penetration fire stop, whichever has the lower self-ignition temperature.. For cable penetration fire stops the self-ignition temperatures of the outer cable covering and fire stop materials are generally above 700°F.

The maximum allowable temperature is the actual measured temperature on the unexposed side and *not* temperature rise. This is because the ignition of a given material occurs at a specific temperature of degrees Fahrenehit.

What This Standard Does Not Do, and Problems Yet to Be Covered

Pressure Seals

A penetration fire stop and the fire barrier itself should, in some locations, function as a seal to maintain any existing pressure difference and should not pass through hot gases or smoke. It should maintain that ability for the duration of the rating test. While this problem is recognized, the present standard does not address it, nor does the ANSI A2.1-1972 test. This should be a future task.

If it is desired by the user of this standard, he can specify an added test, outside the scope of this standard and supplementing the information it provides, which would require a check of the ability of the penetration to maintain a differential pressure before, during, and after the fire test. There has been no standard method yet proposed and accepted for checking this seal during a fire test.

Ampacity Derating Due to Penetration Stops

It is recognized that the thermal insulating characteristics of a penetration fire stop may have an effect on the ampacity of the cables passing through the penetration. Design of the fire stop should address this effect. However, ampacity considerations are not a part of the qualification test and, consequently, are not within the scope of this standard.

Adequacy of Test Furnace

Furnaces used in these rating tests are sometimes operated at lower than atmospheric pressure, and thus hot gases or smoke would not tend to leak outward, but cold air would tend to flow inward toward the fire. This test may not represent a typical situation in a real fire and should be the subject of future investigations.

Test Limitation and Cautions

Just as in the case of the fire barrier in ANSI A2.1-1972, this test is run with a specific standard fire. This fire may or may not be as severe as fires actually experienced and hence may not predict the performance of the cable penetration fire stop barrier combination in actual service. It is the judgment of those experienced in the field that *relative* performance is accurately portrayed, and the relative values may be used as a basis for engineering judgment in a particular design situation.

The test, as already pointed out, gives no information on the necessity, if any, for ampacity derating of cables within the cable penetration, nor does it give any indication of the capability of the stop to maintain a pressure differential between the opposing faces of the barrier before, during, or after a fire test.

Furthermore, the user must consider the higher temperature of those components emerging from the face of the barrier not exposed to fire, for example, the conductors and metallic elements, such as the tray, conduit, or structural parts of the penetration. These higher temperatures must be considered by the designer who will perform a hazards analysis and will take steps necessary to counter these hazards if any are found.

Electric Penetration Assemblies in Containment Structures

Electric penetration assemblies in containment structures are not covered in this standard. For guidance in this area, refer to IEEE Std 317-1976, Electric Penetration Assemblies in Containment Structures for Nuclear Power Generating Stations.

Seismic, Radiation, Aging, and LOCA

Although it is recognized that seismic, radiation, aging, and LOCA conditions may be required to be considered and evaluated for nuclear power plants, these effects are not within the scope of this standard. For guidance in these areas, refer to IEEE Std 344-1975, Recommended Practices for Seismic Qualification of Class 1E Equipment for Nuclear Power Generating Stations, and IEEE Std 323-1974, Qualifying Class 1E Equipment for Nuclear Power Generating Stations.

This standard was prepared by Task Force 12-40 of the Insulated Conductors Committee and at the time of approval had the following membership:

Contents

IEEE Standard Cable Penetration Fire Stop Qualification Test

1. Scope

This standard provides direction for establishing type tests for qualifying the performance of cable penetration fire stops when mounted in rated fire barriers.

2. Purpose

The purpose of this standard is to establish type tests to assure that cable penetration fire stops meet the required fire rating.

2.1 General. The requirements presented include the principles and procedures for testing. These test requirements, when met, will confirm the adequacy of the cable penetration fire stop design under fire conditions tested.

2.2 Applicability. Cable penetration fire stops that meet the requirements outlined herein are intended for use in power-generating stations including nuclear-generating stations, as well as other applicable commercial and industrial installations. Among the categories of cables covered, but not limited to, are those used for power, control, and instrumentation services.

2.3 Method of Approach. When a cable penetration is used in a rated fire-resistive barrier, the fire stop should remain intact and prevent the spread of fire and restrict the passage of hot gases through that barrier for the required rated time. A fire barrier meeting the requirements of ANSI A2.1-1972, Methods of Fire Tests of Building Construction and Materials (ASTM E119-1971)[1] (ISO 834), must limit the flow of heat or gases through from the fire side as indicated by a relatively cool surface, one whose temperature will not ignite gases, cotton waste, or National Fire Prevention Association Class A materials which require a temperature of approximately 400°F (in ANSI A2.1-1972 this is expressed as 250°F above ambient). With a fire stop, however, there are always metallic conductors and perhaps structural portions of the penetration which present good thermal conduction paths through the fire stop and whose temperatures at the point of exit may exceed markedly the approximately 400°F expected of the unpenetrated wall. The temperature can be such that the insulation and jacket on the cable may ignite, indicating a failure of the stop. These higher temperatures of the metallic through-portions of the penetration must be considered and evaluated by the user/designer.

3. Definitions

These definitions establish the meanings of words in the context of their use in this standard.

cable penetration. An assembly or group of assemblies for electrical conductors to enter and continue through a fire-rated structural wall, floor, or floor-ceiling assembly.

cable penetration fire stop. Material, devices, or an assembly of parts providing cable penetrations through fire-rated walls, floors, or floor-ceiling assemblies, and maintaining their required fire rating.

fire rating. The term applied to cable penetration fire stops to indicate the endurance in time (hours and minutes) to the standard time-

---

[1] ANSI documents are available from American National Standards Institute, 1430 Broadway, New York, N.Y. 10018.

temperature curve in ANSI/ASTM E119-76, while satisfying the acceptance criteria specified in this standard.

fire resistive barrier. A wall, floor, or floor-ceiling assembly erected to prevent the spread of fire. (To be effective, fire barriers must have sufficient fire resistance to withstand the effects of the most severe fire that may be expected to occur in the area adjacent to the fire barrier and must provide a complete barrier to the spread of fire.)

fire resistive barrier rating. This is expressed in time (hours and minutes) and indicates that the wall, floor, or floor-ceiling assembly can withstand, without failure, exposure to a standard fire for that period of time. The test fire procedure and acceptance criteria are defined in American National Standard A2.1-1972.

module. An opening in a fire resistive barrier so located and spaced from adjacent modules (openings) that its respective cable penetration fire stop's performance will not affect the performance of cable penetration fire stops in any adjacent module. A module may take on any shape to permit the passage of cables from one or any number of raceways.

raceway. Any channel that is designed and used expressly for supporting or enclosing wires, cable, or bus bars. Raceways consist primarily of, but are not restricted to, cable trays and conduits.

unexposed side. The side of a fire-rated wall, floor-ceiling assembly, or floor which is opposite to the fire side. Also referred to as cold side.

4. References

The following standards were used as references in preparing this guide and may be useful in interpretation of its meaning:

[1] ANSI A2.1-1972, Methods of Fire Tests of Building Construction and Materials (ASTM E119-1971) (ISO 834).

[2] ASTM E84-1976a, Test for Surface Burning Characteristics of Building Materials.

[3] ASTM D2863-1976, Measuring of Test for Flammability of Plastics Using Oxygen Index Method.

[4] IEEE Std 317-1976, Electric Penetration Assemblies in Containment Structures for Nuclear Power Generating Stations.

[5] ANSI K65.111-1971, Methods of Test for the Ignition Properties of Plastics (ASTM D 1929-1968).

5. Test Description

5.1 General. This section describes the methods for testing cable penetration fire stops around cables penetrating a fire resistive barrier.

5.1.1 *Applicability.* These methods shall be applicable to assemblies or groups of cables and materials or components which comprise the fire stop that will be installed in a fire resistive barrier wall, floor, or floor-ceiling assembly. It is not the intent of this standard to test the wall, floor, floor-ceiling assembly or other structural members of the fire resistive barrier. Therefore, no simulated structural loading is required.

5.1.2 *Penetration Fire Stop Components — Excluding Cable.* Individual components of the fire stop system shall have a flame spread rating of 25 or less in accordance with Ref [2]. Components to which the test in Ref [2] are not applicable shall be tested in accordance with Ref [3] and shall have a minimum limiting oxygen index of 25.

5.1.3 *Method of Testing.* Qualification shall be by type testing of an actual full-sized cable penetration fire stop or module indicative of installed conditions.

5.1.4 *Test Experience.* Cable penetration fire stops or modules or both that have successfully functioned under test can be considered qualified for equal or less severe fire rating. Testing in the floor-ceiling position qualifies the cable penetration fire stops for either floor or wall penetration provided the cable penetration fire stop under test is constructed symmetrically so as to provide equal resistance to fire from either side. For unsymmetrical design, refer to 5.3.5.

5.2 Test Specimens

5.2.1 *General.* The type tests specified shall be for power, control, and instrumentation (including signal and communications) cables. The cable penetration fire stops shall be installed in modules or openings through fire-rated barriers, which may be lined with metallic components. Cables may penetrate these openings either directly without a raceway or within a metallic raceway depending on the intended installed configuration.

5.2.2 *Cable Selection and Raceway Fill.* The selection of the sizes, construction, and materials of the cables and cable penetration opening fill to be used in the test shall be representative of the cables used in the fire stop under actual installed conditions.

The cable sizes and cable penetration fire stop fill listed in Table 1 may be used. If these sizes, constructions, or fills are not indicative of the actual installed conditions, more suitable selections shall be used. It is not the intent that different construction types, that is, instrumentation and medium voltage power cable, be installed in the same test cable penetration unless this is indicative of actual conditions.

In order to assess the design of cable penetration fire stops by type testing, similar designs with maximum and minimum, or zero, percent cable fills shall be tested.

When large modules in the fire resistive barrier are used to permit several cable systems to pass through, intermediate percent fills as well as minimum and maximum should be tested in the openings. If these designs are successfully tested, then all designs within these extremities of fill also are qualified. For further guidance, refer to Appendix A2.

5.2.3 *Cable Penetration Fire Stop Opening Dimensions and Type.* The opening dimensions and type of cable penetration fire stops to be tested shall be representative of the type to be used. In order to facilitate the selection of test specimens where several variations of the same type penetration are used, the sizes and type of cable penetration fire stop openings listed in Table 2 may be used as a basis for selection.

If these sizes or types are not indicative of the actual installed condition, more suitable selections shall be used.

In order to assess the design of the cable penetration fire stop by type testing, the largest module or opening or both shall be tested and the cable selected in accordance with 5.2.2.

If the largest cross-sectional module design

is successfully tested, then all designs of the same type and size module or smaller modules are also qualified. Likewise, arrays of openings or modules which are successfully tested shall qualify similar arrays with the same or larger spacing.

The user of cable penetration fire stops and modules qualified by themselves shall demonstrate that the influence of adjacent cable penetration fire stops or modules or both does not compromise their qualification. For further guidance, refer to Appendix A2.

5.3 Fire Test Facility and Procedure

5.3.1 *Test Room.* The fire test shall be conducted in a suitable room or area as defined in American National Standard A2.1-1972, 10.1.

5.3.2 *System Test.* The cable penetration fire stop shall be tested as a complete system. The raceway mounting and anchoring to the fire stop assembly, the cable arrangement, including attachment to raceway and the raceway fill, shall be representative of the actual installed conditions.

5.3.3 *Cable Installation.* The cable within the penetration shall protrude 3 ft to 5 ft on the unexposed side and the ends capped. The cable on the side to which the flame is to be applied shall protrude a minimum of 1 ft. Vertical cables in floor penetration tests shall be supported on the unexposed side to simulate continuous cables in an actual installation.

5.3.4 *Raceway Installation.* If the penetration under test is to simulate an actual penetration in which the raceway passes through the fire barrier, the test raceway shall protrude 3 ft to 5 ft on the unexposed side and a minimum of 1 ft on the exposed side.

5.3.5 *Orientation.* Testing in the floor-ceiling position qualifies the cable penetration fire stop for either a floor or wall penetration. Cable penetration fire stops that are symmetrical with respect to design and location in the wall-floor need only be fire tested on one side. Cable penetration fire stop designs which are unsymmetrical in design or location may require testing on both sides for qualification. For example of unsymmetrical designs and location, refer to Appendix A2.

5.3.6 *Time-Temperature Curve.* The test penetration module shall be subjected to the standard time-temperature curve in ANSI A2.1-1972 (reproduced in Appendix A1) for the time necessary to obtain the required fire rating.

Table 1
Suggested Representative Cables and Cable Penetration Fire Stop Opening Fill for Type Tests

Cable Fire Stop Penetration Type Cable	Size Cable and Construction	Fraction of Total* Fill for Each Penetration Type
Medium voltage power (2 - 15 kV)	3/C No 6 AWG	$\frac{1}{3}$
	3/C No 2/0	$\frac{1}{3}$
	3/C No 4/0	$\frac{1}{3}$
Low voltage power	3/C No 6 AWG	$\frac{1}{3}$
	3/C No 2/0	$\frac{1}{3}$
	3/C No 4/0	$\frac{1}{3}$
Control and instrumentation	7/C No 12	$\frac{1}{2}$
	1 pr No 16 AWG shielded	$\frac{1}{2}$

*Total fill is the total quantity of cable to be installed in the test penetration. For example, this could be 40 percent of the cross-sectional area of the raceway penetration or raceway.

Table 2
Suggested Representative Penetration Opening Dimensions

Cable Fire Stop Penetration Type — Structural	Cross-Sectional Dimensions (Inches)	Slab Thickness (Inches)
Round — No metal sleeve; cables pass through without raceway	6 (diameter)	12 or 6
Round — No metal sleeve; cables pass through in metal raceway	6 (diameter)	12 or 6
Round — Metal sleeve; cables pass through without raceway	6 (diameter)	12 or 6
Round — Metal sleeve; cables pass through in raceway	6 (diameter)	12 or 6
Rectangular — No metal sleeve; cables pass through without raceway	8 × 42 or 48	12 or 6
Rectangular — No metal sleeve; cables pass through in metal raceway	8 × 42 or 48	12 or 6
Rectangular — Metal sleeve; cables pass through without raceway	8 × 42 or 48	12 or 6
Rectangular — Metal sleeve; cables pass through in raceway	8 × 42 or 48	12 or 6

5.3.7 *Exposed Side Test Instrumentation.* The temperature fixed by the curve shall be deemed to be the average temperature obtained from the readings of not less than three thermocouples symmetrically disposed and distributed to show the temperature for each cable penetration fire stop. Additional thermocouples shall be used, as necessary, for larger test specimens. The thermocouples shall be enclosed in sealed porcelain tubes ¾ in (19 mm) in outside diameter and ⅛ in (3 mm) in wall thickness, or, as an alternative in the case of base metal thermocouples, enclosed in sealed, standardweight, ½ in (13 mm), black wrought steel or black wrought iron pipe. The exposed length of the pyrometer tube and thermocouple in the flame area shall be not less than 12 in (305 mm). Other types of protecting tubes or pyrometers may be used that, under test conditions, give the same indications as the above standard. For cable penetrations through floors or floor-ceiling assemblies, the junction of the thermocouples shall be placed 12 in away from the exposed face of the test penetration at the beginning of the test and, during the test, shall not touch the sample as a result of its deflection. In the case of cable penetration through walls, the thermocouples shall be placed 6 in (152 mm) away from the exposed face of the test penetration at the beginning of the test and shall not touch the test penetration during the test, in the event of deflection.

5.3.8 *Exposed Side Temperature Reading Intervals.* The temperatures shall be read at intervals not exceeding 5 min during the first 2 h, and thereafter the intervals may be increased to not more than 10 min.

5.3.9 *Flame Source Accuracy.* The accuracy of the flame source control shall be such that the area under the time-temperature curve, obtained by averaging the results from the pyrometer readings, is within the following tolerances, or exceeds the corresponding area under the standard time-temperature curve in Appendix A1.

Fire Test Duration	Tolerance (%)
1 h or less	10
Over 1 h to 2 h	7.5
Over 2 h	5

5.3.10 *Unexposed Side Temperature.* Temperatures on the penetration cold side surfaces shall be measured with thermocouples. A minimum of three thermocouples shall be located on the surface of each fire stop under test. The maximum temperature on the face of the cable penetration fire stop shall be measured. As a minimum, temperature shall be measured at the cable jacket, cable penetration fire stop interface, the interface between the fire stop, and through metallic components, other than the insulated cable conductor, and on the surface of the fire stop material.

5.3.11 *Unexposed Side Temperature Reading Intervals.* Temperature readings shall be taken at intervals not exceeding 15 min until a reading exceeding 212°F (100°C) has been obtained at any one point. Thereafter, the readings may be taken more frequently at the discretion of the tester, but the intervals need not be less than 5 min.

5.3.12 *Hose Stream Test.* A hose stream test shall be conducted immediately following the end of the fire endurance test and removal, if necessary, of the test slab.

For power-generating stations including nuclear-generating stations, a 1½ in hose discharging through a nozzle approved, for use on fires in electrical equipment producing a long-range-narrow-angle (30-90° set at 30° included angle) high velocity spray only shall be used. The hose stream shall be applied to the exposed side. The water pressure shall be 75 p/in², calculated, at the base of the nozzle and minimum flow of 75 gal/min with a duration of application of 2½ min per 100 ft² of test slab. The nozzle distance shall be 10 ft from the center of the exposed surface of the test specimen.

For other applicable industrial and commercial establishments, the hose stream shall be applied to the exposed surface for a period calculated on a basis of 2½ min per 100 ft² of test slab. The stream shall be delivered through a 2½ in national standard playpipe equipped with 1⅛ in tip, nozzle pressure of 30 p/in² calculated, located 20 ft from the exposed face.

6. Evaluation of Test Results

Cable penetration fire stops which allow cables or fire stop materials on the unexposed side to ignite, or allow thermocouples on the unexposed side to exceed the temperature limits specified, or any visible flame on the unexposed side, within the specified fire rating time, or the hose stream to cause through-openings, fail the test.

6.1 Acceptance. The test can be considered acceptable and the cable penetration fire stop suitable for use in accordance with the fire rating, provided the following is met:

6.1.1 The cable penetration fire stop shall have withstood the fire endurance test as specified without passage of flame or gases hot enough to ignite the cable or other fire stop material on the unexposed side for a period equal to the required fire rating.

6.1.2 Transmission of heat through the cable penetration fire stop shall not raise the temperature on its unexposed surface above the self-ignition temperature as determined in ANSI K65.111-1971 of the outer cable covering, the cable penetration fire stop material, or material in contact with the cable penetration fire stop, when measured in accordance with 5.3.10 and 5.3.11. For power generating station, the maximum temperature is 700° F.

6.1.3 The fire stop shall have withstood the hose stream test without the hose stream causing an opening through the test specimen.

7. Documentation of Testing

Following the procedures outlined in this standard, provide data necessary to document satisfactory compliance. Type test data derived from tests shall be organized to present the results in an orderly manner so as to be easily understood and located.

Specifically, the following data shall be recorded:

(1) Manufacturer of cable

(2) Manufacturer's designation for cable and generic name of materials used

(3) Temperature, current, and voltage rating of cable

(4) Physical dimensions including conductor size insulation and jacket thickness

(5) Miscellaneous construction details including type of raceway, etc

(6) Manufacturer of fire stop materials or devices

(7) Manufacturer of fire stop designation and generic name of materials or devices or both

(8) Environmental conditions, such as air ambient, air currents

(9) Details of hose stream test

(10) Complete description of materials surrounding the fire stop, including test results of 5.1.2

(11) The temperature and time readings taken

The test equipment shall be described in detail, supplemented with record of fuel supply, photographs, dimensioned drawings, and written specifications with not less data than that necessary to reproduce accurately the same test.

The results, pass or fail, shall be recorded and supplemented with photographs and a statement of the conclusions drawn made by those conducting the test.

Engineering data and references to other publications which were used to make the test and select the equipment shall be included in the documentation.

Installation methods shall be described including any Quality Assurance data applicable to the specific materials and installation methods used.

Appendix

A1. Standard Time-Temperature Curve for Control of Fire Tests

Time (h:min)	Temperature (°F)	Area Above 68° F Base (°F-min)	(°F-h)	Temperature (°C)	Area Above 20° C Base (°C-min)	(°C-h)
0:00	68	00	0	20	00	0
0:05	1 000	2 330	39	538	1 290	22
0:10	1 300	7 740	129	704	4 300	72
0:15	1 399	14 150	236	760	7 860	131
0:25	1 510	28 050	468	821	15 590	260
0:30	1 550	35 360	589	843	19 650	328
0:35	1 584	42 860	714	862	23 810	397
0:40	1 613	50 510	842	878	28 060	468
0:45	1 638	58 300	971	892	32 390	540
0:50	1 661	66 200	1 103	905	36 780	613
0:55	1 681	74 220	1 237	916	41 230	687
1:00	1 700	82 330	1 372	927	45 740	762
1:05	1 718	90 540	1 509	937	50 300	838
1:10	1 735	98 830	1 647	946	54 910	915
1:15	1 750	107 200	1 787	955	59 560	993
1:20	1 765	115 650	1 928	963	64 250	1 071
1:25	1 779	124 180	2 070	971	68 990	1 150
1:30	1 792	132 760	2 213	978	73 760	1 229
1:35	1 804	141 420	2 357	985	78 560	1 309
1:40	1 815	150 120	2 502	991	83 400	1 390
1:45	1 826	158 890	2 648	996	88 280	1 471
1:50	1 835	167 700	2 795	1 001	93 170	1 553
1:55	1 843	176 550	2 942	1 006	98 080	1 635
2:00	1 850	185 440	3 091	1 010	103 020	1 717
2:10	1 862	203 330	3 389	1 017	112 960	1 882
2:20	1 875	221 330	3 689	1 024	122 960	2 049
2:30	1 888	239 470	3 991	1 031	133 040	2 217
2:40	1 900	257 720	4 295	1 038	143 180	2 386
2:50	1 912	276 110	4 602	1 045	153 390	2 556
3:00	1 925	294 610	4 910	1 052	163 670	2 728
3:10	1 938	313 250	5 221	1 059	174 030	2 900
3:20	1 950	332 000	5 533	1 066	184 450	3 074
3:30	1 962	350 890	5 848	1 072	194 940	3 249
3:40	1 975	369 890	6 165	1 079	205 500	3 425
3:50	1 988	389 030	6 484	1 086	216 130	3 602
4:00	2 000	408 280	6 805	1 093	226 820	3 780
4:10	2 012	427 670	7 128	1 100	237 590	3 960
4:20	2 025	447 180	7 453	1 107	248 430	4 140
4:30	2 038	466 810	7 780	1 114	259 340	4 322
4:40	2 050	486 560	8 110	1 121	270 310	4 505
4:50	2 062	506 450	8 441	1 128	281 360	4 689
5:00	2 075	526 450	8 774	1 135	292 470	4 874
5:10	2 088	546 580	9 110	1 142	303 660	5 061
5:20	2 100	566 840	9 447	1 149	314 910	5 248
5:30	2 112	587 220	9 787	1 156	326 240	5 437
5:40	2 125	607 730	10 129	1 163	337 630	5 627
5:50	2 138	628 360	10 473	1 170	349 090	5 818
6:00	2 150	649 120	10 819	1 177	360 620	6 010
6:10	2 162	670 000	11 167	1 184	372 230	6 204
6:20	2 175	691 010	11 517	1 191	383 900	6 398
6:30	2 188	712 140	11 869	1 198	395 640	6 594

A2. Cable Penetration Type Tests

A2.1 Typical Cross Sections

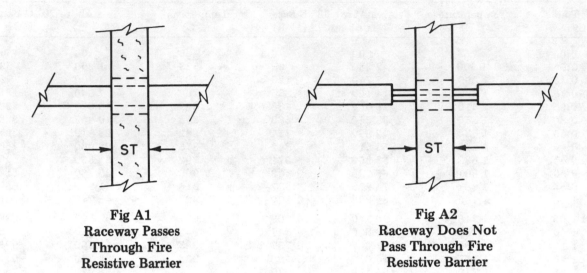

**Fig A1
Raceway Passes
Through Fire
Resistive Barrier**

**Fig A2
Raceway Does Not
Pass Through Fire
Resistive Barrier**

A2.2 Example of Single Type Test

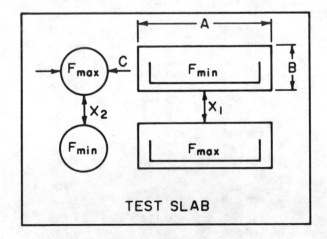

**Fig A3
Four Individual Modules Each
with One Opening**

A2.3 Multiopening Single Module Type Test Example

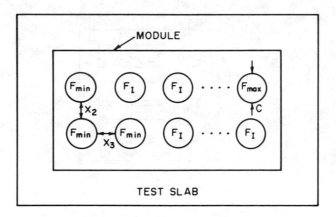

Fig A4
Typical Conduit or Sleeve Penetration

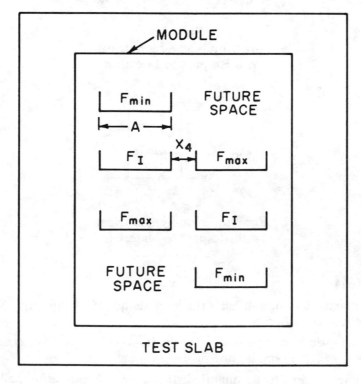

Fig A5
Typical Tray Opening Penetration

NOTE: If test facility will permit, both multiopening single modules shown above could be tested simultaneously.

A2.4 Example of Modules with Nonsymmetrical Fire Stops

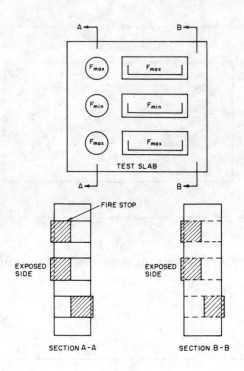

**Fig A6
Fire Stop Non-Symmetrical
with Respect to Location**

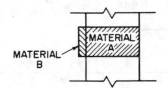

**Fig A7
Fire Stop Non-Symmetrical
with Respect to Materials**

A2.5 Symbol Definitions

ST	Slab thickness. If minimum slab thickness is qualified, all larger thicknesses of similar design are also qualified.
A, B, C	Largest dimensions of opening to be qualified. If largest A, B, C dimension is qualified, all smaller A, B, C of similar design are also qualified.
X_1, X_2, X_3	Minimum separation to be qualified. If X_1, X_2, X_3 is qualified, all larger X_1, X_2, X_3 are also qualified.
F_{max}	Maximum percent cable fill to be qualified.
F_{min}	Minimum percent cable fill used. If design is to be qualified for spares, then $F_{min} = 0$ percent.
F_I	Intermediate percent cable fill, usually taken as $(F_{max} + F_{min})/2$.

18

IEEE Std 645-1977
(Supplement to
IEEE Std 325-1971,
Reaff 1977)

IEEE Test Procedures for High-Purity Germanium Detectors for Ionizing Radiation

Sponsor

**Nuclear Instruments and Detectors Committee of the
IEEE Nuclear and Plasma Sciences Society**

645

© Copyright 1977 by

The Institute of Electrical and Electronics Engineers, Inc.

Foreword

(This foreword is not a part of IEEE Std 645-1977, Test Procedures for High-Purity Germanium Detectors for Ionizing Radiation.)

IEEE Std 325-1971 (Reaff 1977), Test Procedures for Germanium Gamma-Ray Detectors (ANSI N42.8-1972) (Reaff 1977), presents test procedures for germanium detectors for ionizing radiation. The germanium detectors that existed at the time of publication of IEEE Std 325-1971 (Reaff 1977) were made by the lithium drift process. Detectors made from high-purity germanium have since become available. This standard has therefore been prepared as a supplement to IEEE Std 325-1971 (Reaff 1977) to provide such additional test procedures as are required for high-purity germanium detectors.

This standard is not intended to imply that all tests and procedures described herein are mandatory for every application, but only that such tests as are carried out on high-purity germanium detectors should be performed in accordance with the procedures given in this standard.

This standard was prepared and balloted by the Nuclear Instruments and Detectors Committee of the IEEE Nuclear and Plasma Sciences Society. At the time of approval of this standard, the committee membership was as follows:

D. E. Persyk, *Chairman*

J. A. Coleman, *Vice Chairman* **Louis Costrell,** *Secretary*

R. F. Butenhoff	T. R. Kohler	J. H. Trainor
D. C. Cook	H. W. Kraner	S. Wagner
J. F. Detko	R. S. Larsen	F. J. Walter
F. S. Goulding	W. W. Managan	H. R. Wasson
	G. L. Miller	

Robert N. Hall served as project leader for the development of this standard.

At the time is approved this standard, American National Standards Committee N42 on Radiation Instrumentation had the following personnel:

Louis Costrell, *Chairman* **D. C. Cook,** *Recording Secretary*

Organization Represented	*Name of Representative*
American Chemical Society	(*Vacant*)
American Conference of Governmental Industrial Hygienists	Jesse Lieberman
American Industrial Hygiene Association	(*Vacant*)
American Nuclear Society	F. W. Manning
American Society of Mechanical Engineers	P. E. Greenwood
American Society of Safety Engineers	(*Vacant*)
Atomic Industrial Forum	(*Vacant*)
Health Physics Society	J. B. Horner Kuper
	Robert L. Butenhoff (*Alt*)
Institute of Electrical and Electronics Engineers	Louis Costrell
	D. C. Cook (*Alt*)
	J. Forster (*Alt*)
	P. J. Spurgin (*Alt*)
Instrument Society of America	M. T. Slind
	J. E. Kaveckis (*Alt*)
Manufacturing Chemists Association	(*Vacant*)
National Electrical Manufacturers Association	Theodore Hamburger
Oak Ridge National Laboratory	Frank W. Manning
	D. J. Knowles (*Alt*)
Scientific Apparatus Makers Association	Robert Breen
US Energy Research and Development Administration, Division of Biomedical and Environmental Research	Hodge R. Wasson
US Department of the Army, Materiel Command	Abraham E. Cohen
US Defense Civil Preparedness Agency	Carl R. Siebentritt, Jr
US Department of Commerce, National Bureau of Standards	Louis Costrell
US Naval Research Laboratory	D. C. Cook
Members-at-Large	J. G. Bellian
	O. W. Bilharz
	John M. Gallagher, Jr
	S. H. Hanauer
	W. C. Lipinski
	Voss A. Moore
	R. F. Shea
	E. J. Vallario

Contents

IEEE Test Procedures for High-Purity Germanium Detectors for Ionizing Radiation

1. Introduction

1.1 General. IEEE Std 325-1971 (Reaff 1977), Test Procedures for Germanium Gamma-Ray Detectors (ANSI N42.8-1972) (Reaff 1977), was issued for the purpose of establishing uniform testing procedures for germanium detectors for ionizing radiation [1]. The detectors which existed at that time were made by the lithium drift process and were used primarily for the analysis of gamma radiation. Detectors made from high purity germanium (HP Ge) have since become available [2], [3]. In addition, a strong interest has developed in the use of small germanium detectors for gamma and x-ray spectroscopy at photon energies up to a few hundred kiloelectron volts. The purpose of this document is to provide test procedures applicable to HP Ge detectors.

The principal new performance feature offered by HP Ge detectors is that of temperature cyclability. Since such detectors do not depend upon lithium compensation, they may be brought to room temperature on an occasional or repeated basis without degradation caused by lithium precipitation.

Another distinctive feature of HP Ge detectors is that they are generally made from material which contains a nonzero residual impurity concentration so that depletion starts at one contact and extends to the other only after a sufficient voltage V_D (the depletion voltage) has been applied. The operating voltage must exceed V_D for the full spectrometer efficiency of the detector to be realized. The performance of HP Ge detectors is closely similar to that of lithium drifted detectors, except that the latter do not generally display a well-defined depletion voltage.

The performance of a detector may be adversely affected by several factors which include the screening effects of surface charges induced by external electric fields and adsorbed impurities, nonuniformity of the residual impurity distribution, and structural defects (such as lineage) in the germanium. The use of a colli-mated beam of radiation to scan detectors is a useful tool for diagnosing various detector abnormalities. A brief description of this test method is included in this standard as a step toward achieving greater understanding and uniformity of the interpretation of these deficiencies.

Further information on the operation of semiconductor detectors is to be found in Refs [4] and [5].

1.2 Symbols

C_d	Detector capacitance
C_F	Coupling capacitor in filter box
e	Electron charge
E	Gamma-ray energy
ϵ	Average energy required to form one hole-electron pair
N_{imp}	Net residual impurity concentration
Q_γ	Charge produced by gamma-ray absorption
Q_P	Charge produced by test pulse
R_F	Filter resistor in filter box
V_D	Depletion or punch-through voltage
V_P	Amplitude of test pulse
w	Thickness of high-purity germanium region

1.3 References

[1] IEEE Std 325-1971 (Reaff 1977), Test Procedures for Germanium Gamma-Ray Detectors (ANSI N42.8-1972) (Reaff 1977).

[2] BAERTSCH, R. D., and HALL, R. N. Gamma Ray Detectors made from High Purity Germanium. *IEEE Transactions on Nuclear Science*, vol NS-17, pp 235—240, June 1970.

[3] PEHL, R. H., CORDI, R. C., and GOULD-ING, F. S. High Purity Germanium: Detector Fabrication and Performance. *IEEE Transactions on Nuclear Science*, vol NS-19, pp 265—269, Feb 1972.

[4] CERNY, J. ed. *Nuclear Spectroscopy and Reactions.* New York: Academic Press, vol 40-A, 1974. (See Chapter III, Part A, Semiconductor Radiation Detectors, F. S.

Goulding and R. H. Pehl, and III, Part D, Semiconductor Detector Spectrometer Electronics, F. S. Goulding and D. A. Landis.

[5] GOULDING, F. S. *Nuclear Instruments and Methods*, vol 43, p 1, 1966.

[6] GOODMAN, A. M. *Journal of Applied Physics*, vol 34, p 329, Feb 1963.

[7] PEHL, R. H., GOULDING, F. S., LANDIS, D. A., and LENZLINGER, M. *Nuclear Instruments and Methods*, vol 59, p 45, 1968.

[8] BAERTSCH, R. D. Surface Effects on P Type High Purity Germanium Detectors at 77°K. *IEEE Transactions on Nuclear Science*, vol NS-21, pp 347—359, Feb 1974.

[9] MALM, H. L., and DINGER, R. J. Charge Collection in Surface Channels on High-Purity Ge Detectors. *IEEE Transactions on Nuclear Science*, vol NS-23, pp 76—80, Feb 1976.

2. Temperature Cyclability

2.1 Discussion. Unlike Ge(Li) detectors, HP Ge detectors may be cycled between room temperature and their operating temperature without affecting the bulk properties of the germanium (unless the detector has been radiation damaged). Any changes that may take place can be attributed to surface effects. The causes of these surface effects include adsorption and desorption of gas molecules or ions, motion of charges over the surface of the semiconductor or nearby insulator surfaces, and chemical reactions at the surface such as the growth of an oxide layer. The extent to which such changes may affect the operation of the detector depends not only upon the way in which the detector and its associated system have been prepared and assembled, but also upon the manner in which the detector has been treated by the user. Because of the sensitivity of semiconductor surfaces to very small traces of contamination, the user is cautioned to avoid procedures which might induce such changes and to follow carefully the operating procedures recommended by the manufacturer. For example, a detector system containing sorption material such as zeolite should not be subjected to accelerated warm-up since this is likely to cause condensable vapors to be transferred from the sorption material to the detector, which is usually the last portion of the system to reach room temperature.

In conducting tests of cyclability, each temperature cycle shall be of such duration that the detector will be at room temperature for at least 12 hours, followed by cooling to the operating temperature for a period of time (typically 2 to 6 hours) sufficient to permit its spectrometer characteristics to stabilize. To assure stabilization, power shall be applied to the electronics for at least 1 hour prior to measurement of the spectrometer characteristics. The bias voltage shall also be applied for at least 1 hour prior to measurement. The bias voltage should be applied gradually and only when the detector is cold; otherwise damage to the detector or FET may result. The vacuum system is not to be opened for repumping during this series of test cycles.

2.2 Classes of Temperature Cyclability. Several classes of temperature cyclability are defined below, corresponding to current and anticipated usage. The manufacturer shall designate the class of temperature cyclability of the detector. He may select those parameters for which he wishes to make claims (for example, FWHM, FWTM, efficiency, peak-to-Compton) and shall specify that the values of these parameters shall still meet his warranted values after cycling (rather than that they change by less than a certain amount), throughout the duration of the warranty period.

2.2.1 *Cyclable.* A cyclable detector remains under vacuum as an integral part of the cryostat system. It may be cycled repeatedly between room temperature and liquid nitrogen temperature according to the claims of the manufacturer and may be shipped at room temperature at the option of the user. This class of detector is particularly useful for systems which are expected to undergo frequent warm-ups, such as portable systems, and those where a continuous supply of liquid nitrogen is not assured or convenient.

To be designated "cyclable" by the manufacturer each detector must be cycled, meaning that it has been subjected to a stated number of temperature cycles and that the values of the designated parameters have remained better than the warranted values during or after the series of cycles. The manufacturer shall state whether the warranted parameters were measured after each cycle or only after the stated number of cycles. Unless stated otherwise by the manufacturer, the claim of cyclability implies also indefinite room temperature storage capability.

2.2.2 *Hermetic.* A hermetic detector is supplied as a separate unit in its own sealed enclosure so that it can be mounted in the user's cryostat. It can be held at room temperature indefinitely (or as specified) and may be cycled repeatedly with performance to be as designated by the manufacturer.

2.2.3 *Room Air Passivated.* The surface of a room air passivated detector has been sufficiently protected that it can be stored in room air at ambient temperature without significant degradation. The user can place the detector in his cryostat, pump down, and then cool the detector for operation. It may then be warmed and returned to room atmosphere storage and the cycle repeated as designated by the manufacturer.

2.2.4 *Annealable.* Annealable detectors are cyclable detectors (see 2.2.1) which can be subjected to a specified temperature for a given time to reduce the effects of radiation damage. The cryostat, detector, and mount must be capable of accommodating the annealing schedule (such as 120°C for 24 hours). It should be specified that in the absence of radiation damage, the detector will be able to tolerate this annealing schedule without degrading its performance beyond specified limits.

3. Depletion Voltage

3.1 Discussion. The thickness of the depleted region of a high-purity detector increases with applied voltage until a value V_D is reached where the detector becomes fully depleted. Best performance is obtained at operating voltages significantly greater than V_D but below the range where the leakage current becomes excessive. This behavior is in contrast with that of an accurately compensated lithium drifted detector, where even a small voltage is sufficient to cause the internal field to extend from one electrode to the other.

Knowledge of the depletion voltage is not essential for establishing the optimum conditions for detector operation. The material in this section is provided for its tutorial value, and it is not mandatory to specify a value for the depletion voltage.

In a plane parallel HP Ge detector with a uniform impurity concentration, the thickness of the depleted region increases with the square root of the applied voltage below the depletion voltage. The depletion voltage is given by

$$V_D = 5.66 \times 10^{-8} w^2 N_{\text{imp}},$$

where V_D is the depletion voltage in volts, w is the thickness of the high purity germanium region in centimeters, and N_{imp} is the net concentration of electrically active impurities in atoms per cubic centimeter. Analogous considerations apply for coaxial and other geometries.

In an ideal detector, V_D would be constant and could be determined uniquely by any of several different methods of measurement. In actual detectors, apparent changes in V_D may be observed if the surface properties of the semiconductor change. Furthermore, different measurement methods may yield somewhat different values for V_D due to geometrical factors or to nonuniform impurity distributions within the semiconductor. Therefore, any statement of a value for V_D should be accompanied by a statement of the method used for its determination.

3.2 Measurement Based on Detector Capacitance. The recommended method for determining the depletion voltage is to measure the variation of detector capacitance with bias voltage. If both detector contacts are accessible or if one of them is grounded to the cyrostat, the measurement procedure using a capacitance bridge is relatively straightforward (see Ref [6]). On the other hand, if the detector is connected to an internal preamplifier, and if its other terminal is accessible, its capacity can be measured by means of the test circuit shown in Fig 1.

An electronic test pulse is fed through the detector (via the high-voltage filter box) which causes a charge Q_P,

$$Q_P = \frac{V_P C_d}{1 + C_d / C_F},$$

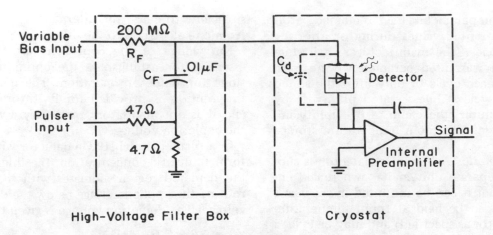

NOTE: C_F is a high-voltage low-noise capacitor, for example, a high-quality plastic type.

Fig 1
Test Circuit for Measuring Detector Capacitance

to flow into the preamplifier input. Here V_P is the amplitude of the test pulse actually applied to the detector, that is, 0.091 times the amplitude at the pulser input for the circuit components used in Fig 1. The capacitor C_F, whose role in the high-voltage filter box is to smooth rapid bias supply changes, serves also as a large series coupling capacitor for the test pulse. By making $C_F \gg C_d$ the correction term C_d/C_F is made negligible. The high-voltage filter resistor R_F has a large value so that it does not shunt the pulser signal.

The detector capacitance can be determined by observing the charge $Q_\gamma = eE/\epsilon$ which is created by a totally absorbed gamma ray of energy E, where ϵ is the average energy required to create an electron-hole pair in Ge, approximately 2.97 eV at 77 K [7], and e is the electron charge. This observation should be made at a sufficiently high detector bias that a negligible amount of charge is lost by trapping. With the pulse amplitude adjusted so that $Q_P = Q_\gamma$, we have

$$C_d = eE/\epsilon V_P .$$

The detector bias is varied and the value of V_P necessary to maintain a constant value of Q_P is measured. These data are used to construct a $C-V$ plot. The depletion voltage of the detector is the voltage above which the capacitance of the detector does not change.

The use of the calibration pulse Q_γ is not essential for the determination of V_D but is recommended because of the information which it affords. Furthermore, for a detector

with simple geometry, it can provide a check that total depletion throughout the whole detector has been achieved. If the transition from a voltage dependent capacitance to the expected geometrical value of capacitance is sharp, it indicates that the depleted region has reached the corresponding electrode surface simultaneously over its entire area. A gradual transition can be an indication of a nonuniform impurity distribution, voltage dependent surface condition, or complex detector geometry.

3.3 Measurement Based on Variation of Efficiency with Bias Voltage. The sensitive volume of a detector increases with bias voltage until the detector becomes fully depleted. A transition corresponding to V_D may be observed by measuring the full energy count rate as a function of bias with the detector exposed to penetrating gamma radiation. This method is subject to errors arising from charge-carrier trapping which are not present in the capacitance method described in 3.2, and the results should be interpreted with this in mind.

A radiation source should be selected which produces an isolated gamma ray with a $1/e$ absorption length in germanium that is at least as great as the thickest dimension of the detector measured parallel to the direction of the incident radiation. Recommended sources are the 1.33 MeV ^{60}Co gamma ray and the 2.61 MEV ^{208}Tl gamma ray (from ^{228}Th source). These gamma rays have $1/e$ lengths of 3.8 and 4.8 cm, respectively. The count rate should be measured at the output of a single channel analyzer, set to accept the entire full energy peak. It must be demonstrated that the source intensity is low enough to avoid significant pile-up errors.

3.4 Measurement Using Weakly Penetrating Radiation. A direct measurement of the depletion voltage is possible if the cryostat configuration is such that radiation can be directed upon the electrode toward which the depletion boundary moves as full depletion is approached. The count rate is measured as a function of bias voltage, using a source which produces radiation having a $1/e$ absorption length that is small compared with the detector thickness. Maximum sensitivity is achieved when the radiation is just sufficiently energetic to penetrate the container and detector dead layer. If the detector depletes uniformly over its area, the count rate will begin to increase rapidly just before full depletion is reached and remain constant at higher voltages.

This method has often been used with a collimated radiation source as discussed more fully in the following section.

4. Scanning

4.1 Discussion. The material in this section is presented for its tutorial value and does not necessarily imply a recommendation for routine use of such tests.

Because the internal detector field is determined by the distribution of a very few fixed charges (few active impurities), surface charges and relatively small variations in the internal

**Table 1
Radiation Sources Convenient for Scanning**

Source	Half life	Radiation (keV)	$1/e$ absorption length in Ge (mm)
^{109}Cd	453 d	22 } Ag x-rays 25 }	0.0587 0.0835
		88	2.59
^{241}Am	433 y	60	0.962
^{57}Co	270 d	14	0.019
		122	5.31

impurity distribution can significantly affect the internal and surface fields of high-purity detectors. The field lines along which charge is collected may intersect the surface of the detector, and surface channels may be operative as collection pathways. Therefore, it can be instructive to inspect the active volume of a detector as a function of bias using a collimated beam of radiation.

To illustrate this method, assume that a plane parallel detector is to be investigated, as shown in Fig. 2. In this example the detector is made from p-type germanium so that as the bias voltage is increased the boundary of the sensitive volume moves closer to the P^+ outer contact. The location of this boundary within the detector can be determined by scanning it with a collimated beam of gamma or x-rays having energies such that they are strongly absorbed by a few millimeters of Ge. Useful information may be obtained by moving the beam across the P^+ entrance window of the detector (transverse scan) or along its lateral surface (side scan).

4.2 Radiation Sources. Several radiation sources which are convenient for scanning are noted in Table 1. For all of these source energies, a

**Fig 2
Scanning Test Plane Parallel Detector**

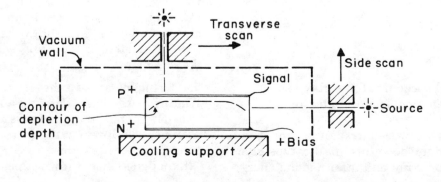

collimator consisting of a 0.5 mm hole through 3 mm of Pb can be used to produce a suitable scanning spot. A 10 mCi point source placed 3 cm behind the collimator hole will provide a beam of suitable intensity.

The maximum measured count rate which may be observed in the scans is that in the full energy peak of the source used. The count rate will be reduced below this value if the beam is attenuated by passage through an undepleted layer of semiconductor or if it is absorbed in a region where there is poor charge collection.

4.3 Transverse Scans.

A transverse scan is made by passing the beam through the contact toward which the depletion boundary moves with increasing bias voltage. Since the maximum count rate is approached only as this boundary moves within the $1/e$ absorption length of this contact, it can be used to determine whether full depletion is reached uniformly over the contact area.

A transverse scan can also be used to diagnose the surface conditions of the detector. As the beam is scanned across a contact, the count rate should begin to decrease when it comes within a beam diameter of the edges of a fully depleted detector. This decrease may take place earlier, however, if the field there is distorted by charges on the side surface of the detector.

4.4 Side Scans.

If the radiation being used is moderately penetrating (for example, 88 or 122 keV), the response to a side scan will be averaged over a distance of several millimeters beneath the surface of the detector and can be used to locate the boundary of the depleted region. On the other hand, if less penetrating radiation is used (22 or 25 keV) the count rate becomes very sensitive to surface conditions and can be used to reveal the presence of dead layers. For the interpretation of such scanning measurements, see Refs [8] and [9] and papers cited therein.

5. Test Procedures for X-Ray and Low-Energy Gamma Detectors

5.1 General.

The procedures for testing small planar germanium detectors of the type commonly used for x-ray and intermediate energy gamma-ray spectroscopy are similar to the general procedures described in IEEE Std 325-1971 (Reaff 1977). Therefore, only those parts of the test procedure which are particular to x-ray detectors are outlined here.

5.2 Detector Energy Resolution.

Although germanium x-ray detectors are normally used at energies higher than 20 keV it is accepted practice to specify their energy resolution at the 5.9 keV K_a line of Mn. The source used for this measurement is ^{55}Fe. The other energy resolution measurement normally specified is made at the 122 keV line of ^{57}Co.

5.3 Resolution and Peak Position Versus Count Rate Performance.

The count rate performance and the energy resolution performance of a germanium x-ray detector system set conflicting requirements on the value of the feedback resistor of charge-sensitive preamplifiers using resistive feedback. Best energy resolution, especially at low energies, is achieved by using values of the feedback resistor as high as possible (for example, practically infinitely high as in an optoelectronically coupled system). However, good performance at high count rates requires a low value of the feedback resistor.

Another conflict arises in the setting of the time constants of the main amplifier. Optimum energy resolution, particularly at low energies, usually requires a long shaping time. High count rate performance, on the other hand, can be achieved only by employing shorter time constants. The performance is determined by the actual pulse length whose relationship to the nominal time constant varies between manufacturers. The energy resolution of a germanium x-ray detector should be given at a specified count rate, for example, 1000 counts per second. The input count rate can be determined either by measuring the output count rate and the percent dead time with an MCA or by using a fast amplifier — discriminator — scaler system at the preamplifier output. Complete information on time constants and corresponding input and output count rates shall be stated in reporting energy resolution measurements.

5.4 Thickness of the Detector Entrance Window.

As these detectors are supplied in sealed systems, the thickness of the detector entrance window is difficult to measure directly.

Experience has shown that detectors with

thick dead layers on the front face display poor peak-to-background ratios. While the test as described herein does not provide a quantitative measurement, there is a definite correlation between these ratios and the detector dead layer thickness.

The test consists of placing the active side of an [55]Fe source adjacent to the center of the Be window of the detector system, adjusting gains so that 20 channels appear across the FWHM, and counting until 20 000 counts are accumulated in the peak channel at 5.9 keV. The number of counts in the 5 channels centered around 5.4 keV and 4.5 keV, respectively, are averaged to give values C_1 and C_2, respectively.

The peak-to-background ratios R_1 and R_2 are calculated from $R_1 = 20\,000/C_1$ and $R_2 = 20\,000/C_2$. These ratios may then be quoted to indicate a measure of the dead layer. Typical values are from 50 to 120, for the 5.9 keV/5.4 keV ratio, and 120 to 300 for the 5.9 keV/4.5 keV ratio. These values are functions, of course, of detector resolution and, therefore, also detector size. Values lower than this indicate a significant dead layer.

It should be noted that any deterioration in peak resolution from contributions due to system noise (ground loops, ac noise, etc) or improper equipment settings (pole zero misadjustment, less than optimum shaping times, etc), or the presence of other radiation sources in the vicinity, will lower the peak-to-background ratio and will give a false indication of a thick dead layer.

6. Test Procedures for High-Energy Gamma Detectors

The procedures for testing HP Ge detectors of the larger sizes generally suitable for the analysis of gamma radiation have been described in IEEE Std 325-1971 (Reaff 1977). In reporting the results of these tests, the count rate which was used is to be stated.

The cyclability of HP Ge detectors is to be specified in accord with the methods and procedures described in Section 2, Temperature Cyclability.

IEEE Standard Techniques for Determination of Germanium Semiconductor Gamma-Ray Efficiency Using a Standard Marinelli (Reentrant) Beaker Geometry

Sponsor

**Nuclear Instruments and Detectors Committee of the
IEEE Nuclear and Plasma Sciences Society**

680

© Copyright 1978 by

The Institute of Electrical and Electronics Enginers, Inc.

Foreword

(This Foreword is not a part of IEEE Std 680-1978, Techniques for Determination of Germanium Semiconductor Detector Gamma-Ray Efficiency Using a Standard Marinelli (Reentrant) Beaker Geometry.)

This standard defines a sample geometry and measurement techniques for determination of gamma-ray efficiencies of germanium semiconductor detectors. ANSI/IEEE Std 325-1971 (Reaff 1977), Test Procedures for Germanium Gamma-Ray Detectors, is a related document.

This standard was reviewed and balloted by the Nuclear Instruments and Detectors Committee of the IEEE Nuclear and Plasma Sciences Society.

At the time of approval of this standard, the membership of the Nuclear Instruments and Detectors Committee of the IEEE Nuclear and Plasma Sciences Society was as follows:

Contents

IEEE Standard Techniques for Determination of Germanium Semiconductor Detector Gamma-Ray Efficiency Using a Standard Marinelli (Reentrant) Beaker Geometry

1. Scope and Object

This standard for determination of gamma-ray efficiencies of germanium semiconductor detectors was developed in recognition of the increasing number of large-volume, low-activity samples being measured by gamma-ray spectroscopy. The standardized sample geometry and measurement techniques described, when used in conjunction with the relative efficiency measurement standard. ANSI/IEEE Std 325-1971 (Reaff 1977), Test Procedures for Germanium Gamma-Ray Detectors, provide a meaningful assessment of detector performance.

It is recognized that many Marinelli beaker geometries are in use. However, the object of this standard is to specify a single configuration for the sole purpose of characterizing detector performance.

2. Introduction

The Marinelli beaker[1] geometry specified herein (Figs 1 and 2) has been chosen primarily for its near-optimum design in placing the sample material as near the active detector material as feasible. Germanium semiconductor detectors are encapsulated in vacuum cryostats such that internal dimensions are not readily measurable by the user. Therefore, in this document reference is made to the cryostat external dimensions. For the purposes of this document the detector is considered to be part of a *singles spectrometer* where coincidence or Compton suppression modes are not used. The spectrometer shall be set up and calibrated as in ANSI/IEEE Std 325-1971.

3. Marinelli Beaker Standard Source (MBSS)

A Marinelli Beaker Standard Source (MBSS) consists of a Standard Marinelli Beaker (see Section 4) containing a carrier with radioactive material (see Section 5). An MBSS may be a *certified MBSS*, a *calibrated MBSS*, a *certified solution MBSS*, or a *calibrated solution MBSS*. The calibration uncertainty of the photon emission rate[2] for the filled beaker shall be not more than 3 percent unless otherwise stated.[3]

3.1 Certified MBSS. A *certified MBSS* is an MBSS that has been calibrated as to photon emission rate[2] at specified energies by a laboratory recognized as a country's National Standardizing Laboratory for radioactivity measurements[4] and has been so certified by the calibrating laboratory.

3.2 Calibrated MBSS. A *calibrated MBSS* is an MBSS that has been calibrated by comparing its photon emission rate[2] to that of a *certified MBSS*.

3.3 Certified-Solution MBSS. A *certified-solution MBSS* is a standard beaker conforming to Section 4 that contains a *certified solution* (see 3.5) as its radioactive filling material (see Section 5).

3.4 Calibrated-Solution MBSS. A *calibrated-solution MBSS* is a standard beaker conforming to Section 4 that contains as its radioactive filling material (see Section 5) a solution that has been calibrated by comparing its photon emission rate[2] at specified energies to that of a *certified solution* (see 3.5).

---

[1] Hill, R. F., Hine, G. J., and Marinelli, L. D. The Quantitative Determination of Gamma Ray Radiation in Biological Research, *American Journal of Roentgenology, Radium Therapy*, vol 63, 1950, p 160.

[2] The photon emission rate as used in this standard is the number of photons per second resulting from the decay of radionuclides in the source and is thus higher than the detected rate at the surface.

[3] International Commission on Radiation Units and Measurements. Certification of Standardized Radioactive Sources, ICRU Report No 12, Sept 15, 1968.

[4] For the United States, the US National Bureau of Standards.

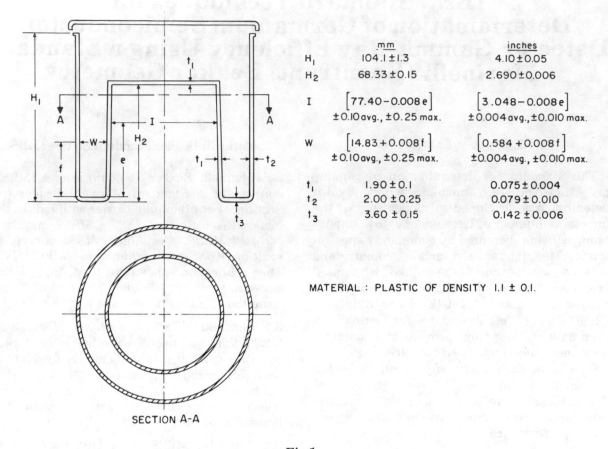

	mm	inches
H_1	104.1 ±1.3	4.10 ±0.05
H_2	68.33 ±0.15	2.690 ±0.006
I	$[77.40-0.008e]$ ±0.10 avg., ±0.25 max.	$[3.048-0.008e]$ ±0.004 avg., ±0.010 max.
W	$[14.83+0.008f]$ ±0.10 avg., ±0.25 max.	$[0.584+0.008f]$ ±0.004 avg., ±0.010 max.
t_1	1.90 ±0.1	0.075 ±0.004
t_2	2.00 ±0.25	0.079 ±0.010
t_3	3.60 ±0.15	0.142 ±0.006

MATERIAL : PLASTIC OF DENSITY 1.1 ± 0.1.

SECTION A-A

Fig 1
Standard Marinelli Beaker

3.5 Certified Solution. A *certified solution* is a liquid radioactive filling material (see Section 5) that has been calibrated by a laboratory recognized as a country's National Standardizing Laboratory for radioactivity measurements[4] and has been so certified by the calibrating laboratory.

4. The Beaker

Marinelli beakers are reentrant (inverted well) beakers. They are available in a variety of sizes for use in large volume, low level, measurements. The beaker specified herein is shown in Fig 1.[5] Fig 2 shows a schematic of a typical sample-detector geometry. The specified beaker is considered to be of 450 mL capacity. The actual volume is greater than this, but, for purposes of this standard, the beaker is to be filled to 450 mL ± 2 mL (see Section 5). The beaker specified was selected because of:

(1) High counting efficiency for the sample material used

(2) Commercial availability at low cost

(3) Common usage in many laboratories

(4) Physical convenience

[5]Potential suppliers for this beaker are Bel-Art Products, Control Molding Corporation, GA-MA Associates, and New England Nuclear.

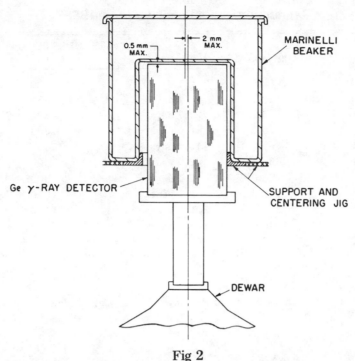

Fig 2
Marinelli Beaker with Solid State Detector

5. Radioactive Filling Material

The beaker to be used for calibration of the detector shall be filled with 450 mL ± 2 mL of a solid or liquid carrier containing uniformly dispersed radioactive material. The solid carrier is preferable for general use because it constitutes a safe and chemically inert sealed source.[6] Representative radionuclides for use in MBSSs are listed in Table 1.

The carrier shall have a mean atomic number Z of 4.0 ± 0.7 and a mean specific gravity of 1.15 ± 0.02 g·cm^{-3} for solid carriers, or 1.06 ± 0.01 g·cm^{-3} for liquid carriers. The density of the carrier used shall be stated (see Section 7(1)(d)) so that attenuation corrections may be made. The source activity should be such as not to exceed the count-rate capability of the detector system. In some systems significant spectral distortion may occur at count-rates as low as a few thousand per second.

6. The Measurement

The absolute full-energy peak efficiency E_a of a detector assembly being calibrated shall be determined with the MBSS placed over the end cap as in Fig 2, counted for a live time count interval T and measured according to 3.1 of ANSI/IEEE Std 325-1971. (See also 5.1 of ANSI/IEEE Std 325-1971). E_a is defined as:

$$E_a = A/N_s \qquad \text{(Eq 1)}$$

where

A = number of events from the MBSS registered as counts in the full-energy peak representing energy E during the live time count interval T.

N_s = number of gamma rays of energy E originating in the MBSS during the same live time interval T.

When the MBSS efficiencies are specified, the energies used in the determination shall be stated. Convenient and frequently used gamma-ray energies include those listed below. The 1332 keV gamma ray of ^{60}Co has been the most widely used for specifying efficiency and resolution. The preferred line at low energies is the 88 keV gamma ray of ^{109}Cd.

---

[6]Potential suppliers for the filled beaker are APT (Applied Physical Technology), Amersham Corporation, Isotope Products Laboratories, and New England Nuclear.

Table 1
Representative Radionuclides for MBSS*

Parent Radionuclide	Energy (keV)**	Half Life**	Typical MBSS initial emission rate in γs^{-1}
^{109}Cd	88.0	461 days	200
^{57}Co	122.1	271 days	250
^{139}Ce	165.9	137.7 days	200
^{203}Hg	279.2	46.6 days	600
^{113}Sn	391.7	115.1 days	1000
^{137}Cs	661.6	30.0 years	500
^{88}Y	898.0	106.6 days	3500
^{60}Co	1173.2	5.271 years	1500
^{60}Co	1332.5	5.271 years	1500
^{88}Y	1836.0	106.6 days	4000

*Coursey, B. M. Use of NBS Mixed-Radionuclide Gamma-Ray Standards for Calibration of Ge(Li) Detectors Used in the Assay of Environmental Radioactivity, NBS SP456, 1976, pp 173-179, US National Bureau of Standards, Washington, DC 20234.

**Private communication from Murray Martin, Oak Ridge Nuclear Data Project, May 1977.

Table 2
Total Absorption Full-Energy Peak Detection
Efficiencies Determined for Two Ge(Li) Detectors

Energy (keV)	E_a for Detector 1* (E_{rel} = 11.4%)**	E_a for Detector 2* (E_{rel} = 24.5%)**
87.7	0.025	0.0414
122.1	0.032	0.0484
165.9	0.0236	0.0368
279.2	0.0157	0.0260
391.7	0.0112	0.020
661.6	0.0068	0.0131
898	0.0050	0.0097
1173	0.0039	0.0077
1333	0.0034	0.0070
1836	0.0026	0.0054

*Detector characteristics:

	Detector 1	Detector 2
Resolution at 1.33 MeV	1.88 keV	2.04 keV
E_{rel}** at 1.33 MeV	11.4%	24.5%
Peak/Compton at 1.33 MeV	44/1	50/1
Diameter	41.2 mm	53.5 mm
Length	47.5 mm	53 mm
Diffusion depth	1000 μ	1500 μ
Core diameter	5.0 mm	9.0 mm
Active volume	56.2 cm^3	103.8 cm^3
Surface area exposed to beaker	74.8 cm^2	111.6 cm^2

**E_{rel}: total absorption detector efficiency relative to that of a 3 × 3-in (76 × 76-mm) NaI(Tl) scintillation crystal at a source to detector distance of 25.0 cm, where the peak area for the NAI(Tl) crystal is taken as $1.2 \times 10^{-3} N_s$. N_s is the total number of 1332.5 keV gamma-ray photons emitted by the source during the live counting time. [See 5.1 and 5.2 of ANSI/IEEE Std 325-1971.]

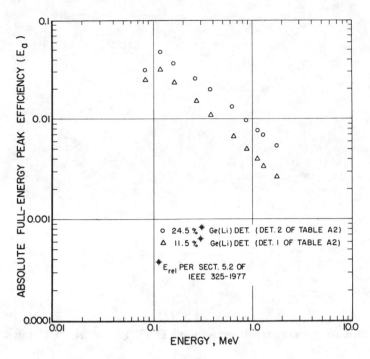

Fig 3
Variation of Absolute Full-Energy Peak Efficiency (E_a) with
Energy for Two Ge(Li) Detectors, Using an MBSS

88 keV (^{109}Cd)
122 keV (^{57}Co)
662 keV (^{137}Cs)
1332 keV (^{60}Co)

If the user wishes to extend the use of these techniques to establish a detailed calibration curve for specific applications of the detector, additional energy peaks will generally be required (see Table 1). Table 2 shows typical results for a calibrated MBSS used with two Ge(Li) detectors. The two detectors are described by their relative efficiencies as in 5.2 of ANSI/IEEE Std 325-1971. Fig 3 shows the variation of the absolute full-energy peak efficiency plotted for the two Ge(Li) detectors of Table 2 for the MBSS geometry.

Fig 4 shows a linear plot of the spectrum observed using the MBSS. Fig 5 shows the log plot of the spectrum of Fig 4. Most of the peaks are clearly observable and available for direct measurement. A possible source of error arises from the fact that the ^{88}Y single-escape peak is about 7 keV (at 1325 keV) from the 1332 keV peak of ^{60}Co. Fig 6 shows an expanded portion of the spectrum around 1332

keV for the data of Fig 4. Care should be exercised to assure that counts from the escape peak are not attributed to the peak due to the 1332 keV gamma ray.

It should also be noted that the 1173 keV and 1332 keV gamma rays of ^{60}Co are in prompt cascade and can therefore sum, resulting in errors of as much as 5 percent. Approximate correction factors may be determined from a knowledge of the peak-to-total ratios for the secondary gamma and the gamma-gamma angular correlation. A similar situation exists for all radionuclides emitting gamma rays in prompt cascade.

7. MBSS Documentation

A certificate shall be provided with each MBSS stating whether the MBSS is a *certified MBSS* (see 3.1) or a *calibrated MBSS* (see 3.2). In the case of the *certified-solution MBSS* (see 3.3), and the *calibrated-solution MBSS* (see 3.4), the certificate shall be for the solu-

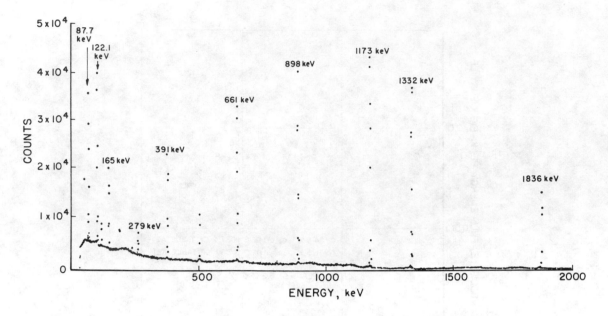

Fig 4
Linear Plot of Spectrum Observed with Ge(Li) Detector*, Using an MBSS

*Detector 2 of Table 2

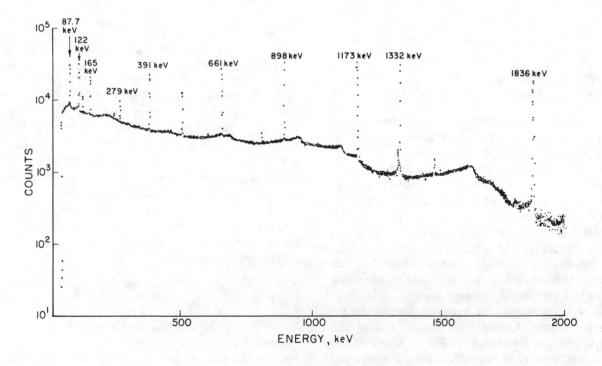

Fig 5
Log Plot of Spectrum Observed with Ge(Li) Detector*, Using an MBSS

*Detector 2 of Table 2

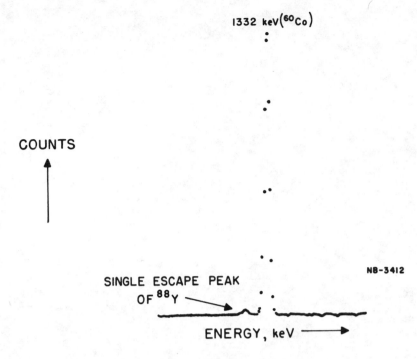

Fig 6
Expanded Portion of Fig 4 Spectrum Showing 1.33 MeV
^{60}Co Peak Together with ^{88}Y Single Escape Peak

tion alone. In any case, the certificate shall contain the following information:

(1) Radionuclides used and calibrated, together with:

(a) The photon energies and radionuclide half lives

(b) Dates corresponding to the stated photon emission rates

(c) Uncertainties in the stated photon emission rates. In the case of a *calibrated MBSS*, the errors for the *certified MBSS* against which the calibrated MBSS has been calibrated and the errors in the comparison shall be given separately. In the case of a *calibrated-solution MBSS*, the errors for the *certified solution* against which the radioactive filling material solution has been calibrated and the errors in the comparison shall be given separately

(d) Density of the carrier

(2) Calibration dates. In the case of a *calibrated MBSS*, the date of calibration against the *certified MBSS* and the date of calibration of the *certified MBSS* shall be given. In the case of a *calibrated-solution MBSS*, the date of calibration of the radioactive filling material solution against the *certified solu-*

tion and the date of calibration of the *certified solution* shall be given.

(3) Full geneology in the case of a *calibrated MBSS* or a *calibrated-solution MBSS*, including identification of the *certified MBSS* or *certified solution* against which it has been calibrated.

(4) Calibrating laboratory or organization

8. Discussion

Use of this standard in conjunction with the relative efficiency measurement (5.2 of ANSI/IEEE Std 325-1971) enables a fuller understanding of both detector performance and expected performance for measuring large volume samples.

Solid angle, source self attenuation, and gamma-ray energy are all factors in the measurement of the absolute full-energy peak efficiency. This detector calibration standard fixes the source geometry thereby allowing systematic comparison of detector effects such as window thickness, sensitive volume, and length/diameter ratio.

IEEE Std C37.98
(formerly designated IEEE Std 501-1978)

IEEE Standard
Seismic Testing of Relays

Sponsor

**Power System Relaying Committee of the
IEEE Power Engineering Society**

C37.98

© Copyright 1978 by

The Institute of Electrical and Electronics Engineers, Inc

Foreword

(This Foreword is not a part of IEEE Std 501-1978, Standard Seismic Testing of Relays.)

This is a standard for developing data related to the seismic capabilities of relays. This standard compliments ANSI/IEEE Std 344-1975, Recommended Practices for Seismic Qualification of Class 1E Equipment for Nuclear Power Generating Stations, which provides general guidelines for developing data related to seismic qualification of equipment used in nuclear generating plants.

Data must be developed by testing rather than analytical means since protective relays fall into the category of complex devices as described in Section 5 of ANSI/IEEE Std 344-1975; however, analysis may be used in data reduction, in reconciling response spectra, in justifying methods of evaluating changes, and in justifying seismic qualifications of relays of similar construction.

While aging is recognized as a potential influence on the seismic capability of relays, insufficient definition presently exists to permit realistic simulation of end-of-life for various relays. When applicable definition of the various degrading factors and methods of accelerating the effect of these factors is developed in other related standards, they should be considered in the previbration preparation.

It should be emphasized that while a primary purpose in preparing this standard was to cover the application of relays to nuclear generating plants, the standard is not restricted to this application. This standard may be applied to any area in which the seismic response of relays is a design consideration.

This standard was prepared by Working Group K1 (Seismic Relay Performance) of Subcommittee K (Relay Practices) of the Power System Relaying Committee. The members of the working group are:

D. M. Clark, *Chairman*

H. J. Calhoun	F. N. Meissner
W. F. Clark	V. S. Noonan
D. H. Colwell	M. T. Orzechowski
A. T. Giuliante	L. Scharf
R. Harkleroad	J. F. Sehring
D. H. Jackson	K. M. Skreiner
J. L. Koepfinger	J. E. Waldron
R. W. Long	F. Wolf

At the time it approved this standard, Subcommittee K (Relay Practices) had the following membership:

T. Niessink, *Chairman*   R. E. Linton, *Vice Chairman/Secretary*

A. J. Adoue, Jr	C. M. Gadsden	W. R. Roemish
E. A. Baumgartner	A. T. Giuliante	R. J. Sullivan
J. E. Benning	K. R. Gruesen	J. R. Tudor
J. Berdy	C. G. Hewitt	J. R. Turley
J. R. Boyle	J. M. Intrabartola	D. R. Volzka
R. E. Carlson	L. E. Landoll	F. Von Roeschlaub
D. M. Clark	M. D. Limerick	C. L. Wagner
D. H. Colwell	V. A. Nosko	C. W. Walker
H. Disante	G. C. Parr	J. W. Walton
M. J. Fein	R. W. Pashley	D. Zollman
	O. F. Pumphrey	

At the time it approved this standard, the Power System Relaying Committee had the following membership:

Contents

IEEE Standard
Seismic Testing of Relays

1. Scope

1.1 Introduction. This standard specifies the procedures to be used in the seismic testing of relays used in power system facilities. The standard is concerned with the determination of the seismic fragility level of relays and also gives recommendations for proof testing.

1.2 Purpose. The purpose of this standard is to establish procedures for determining the seismic capabilities of protective and auxiliary relays. These procedures employ what has been called fragility testing in ANSI/IEEE Std 344-1975, Recommended Practices for Seismic Qualification of Class 1E Equipment for Nuclear Power Generating Stations. In order to define the conditions for fragility testing of relays, parameters in three separate areas must be specified. In general they are:

(1) The electrical settings and inputs to the relay, and other information to define its conditions during the test.

(2) The change in state, deviation in operating characteristics or tolerances, or other change of performance of the relay which constitutes failure.

(3) The seismic vibration environment to be imposed during the test.

Since it is not possible to define the conditions for every conceivable application for all relays, those parameters, which in practice encompass the majority of applications, have been specified in this standard. When the application of the relay is other than as specified under any of (1), (2), and (3), or if it is not practical to apply existing results of fragility tests to that new application, then proof testing must be performed for that new case. The use of these capability data will assist in the selection of relays. One number will be used to catalog the seismic capability of a relay. This number is the zero period acceleration level; and is understood to refer to the standard response spectrum shape as defined in this standard. The capability data will help designers of generating stations, substations, and various other power system installations to incorporate the seismic capabilities of the relays into the overall design of these facilities.

1.3 Limitation of Test Results. This standard applies only to the testing of protective and auxiliary relays and does not apply to switchboards, panels, or any structure upon which the relay may be mounted. It is the responsibility of the power system facility designer to combine data on the seismic performance of the relay mounting structure and the relay to arrive at an acceptable equipment design.

2. Definitions

auxiliary relay. A relay whose function is to assist another relay or control device in performing a general function by supplying supplementary actions.

NOTES:
(1) Some of the specific functions of an auxiliary relay are:

(a) reinforcing contact current-carrying capacity of another relay or device
(b) providing circuit seal-in functions
(c) increasing available number of independent contacts
(d) providing circuit-opening instead of circuit-closing contacts or *vice versa*
(e) providing time delay in the completion of a function
(f) providing simple functions for interlocking or programming.

(2) The operating coil or the contacts of an auxiliary relay may be used in the control circuit of another

relay or other control device. For example, an auxiliary relay may be applied to the auxiliary contact circuits of a circuit breaker in order to coordinate closing and tripping control sequences.

(3) A relay which is auxiliary in its functions even though it may derive its driving energy from the power system current or voltage is a form of auxiliary relay; for example, a timing relay operating from current or potential transformers.

(4) Relays which, by direct response to power system input quantities, assist other relays to respond to such quantities with greater discrimination are not auxiliary relays; for example, fault-detector relay.

(5) Relays which are limited in function by a control circuit but are actuated primarily by system input quantities are not auxiliary relays; for example, torque-controlled relays.

biaxial test. The relay under test is subjected to acceleration in one principal horizontal axis and the vertical axis simultaneously.

dependent biaxial test. The horizontal and the vertical acceleration components are derived from a single-input signal.

fragility. Susceptibility of equipment to malfunction as the result of structural or operational limitations, or both.

fragility level. The highest level of input excitation, expressed as a function of input frequency, that an equipment can withstand and still perform the required Class 1E functions.

fragility response spectrum (FRS). A test response spectrum obtained from tests to determine the fragility level of equipment.

independent biaxial test. The horizontal and the vertical acceleration components are derived from two different input signals, which are phase incoherent.

octave. The interval between two frequencies which have a frequency ratio of two. For example, 1 to 2, 2 to 4, 4 to 8 Hz, etc.

one-third octave. The interval between two frequencies which have a frequency ratio of $2^{\frac{1}{3}}$. For example, 1 to 1.26, 1.26 to 1.59, 1.59 to 2.0 Hz, etc.

operating basis earthquake (OBE). That earthquake which could reasonably be expected to affect the plant site during the operating life of the plant; it is that earthquake which produces the vibratory ground motion for which those features of the nuclear plant necessary for continued operation without undue risk to the health and safety of the public are designed to remain functional.

operating time. The time interval from occurrence of specified input conditions to a specified operation.

protective relay. A relay whose function is to detect defective lines or apparatus or other power system conditions of an abnormal or dangerous nature and to initiate appropriate control circuit action.

NOTE: A protective relay may be classified according to its operating quantities, operating principle, or performance characteristics.

required response spectrum (RRS). The response spectrum issued by the user or his agent as part of his specifications for proof testing, or artificially created to cover future applications. The RRS constitutes a requirement to be met.

response spectrum (as applied to relays). A plot of the peak acceleration response of damped, single-degree-of-freedom bodies, at a damping value expressed as a percent of critical damping of different natural frequencies, when these bodies are rigidly mounted on the surface of interest.

safe shutdown earthquake (SSE). That earthquake which produces the maximum vibratory ground motion for which certain structures, systems, and components are designed to remain functional. These structures, systems, and components are those necessary to assure: (1) the integrity of the reactor coolant pressure boundry, (2) the capability to shut down the reactor and maintain it in a safe shutdown condition, or (3) the capability to prevent or mitigate the consequences of accidents which could result in potential offsite exposures comparable to the guideline exposures of Code of Federal Regulations, Title 10, Part 100 (December 5, 1973).

standard response spectrum (SRS). An RRS which is artificially created to cover the standard testing of relays and whose shape is defined. For example, Figs 1 and 3. The SRS may be terminated at any convenient frequency above 35 Hz.

test response spectrum (TRS, as applied to relays). The acceleration response spectrum that is constructed using analysis or derived using spectrum analysis equipment based on the actual motion of the shake table.

transitional mode. The change from the nonoperating to the operating mode, caused by switching the input to the relay from the non-operating to the operating input, or *vice versa*.

zero period acceleration (ZPA). The peak acceleration of the motion time–history which corresponds to the high-frequency asymptote on the response spectrum.

3. Test Preparation

3.1 Environmental Conditions. All relay seismic tests shall be conducted under the prevailing ambient conditions of the test laboratory.

3.2 Method of Mounting. The test specimen shall be mounted on the test fixture as it normally would be in service, using recommended mounting hardware. The effect of normal electrical connections shall be considered. The test fixture must be rigid structure to minimize amplification and spurious motion within the frequency range of the test.

3.3 Instrumentation

3.3.1 Instrumentation Required. Sufficient instrumentation shall provide:

(1) Acceleration at mounting surface of relay and table
(2) Electrical inputs to relay under test
(3) Relay output change of state
(4) Operating time of relay under test, if applicable
(5) Test response spectrum

3.3.2 Test Tolerances. All test equipment used to monitor the above tests shall be checked and calibrated periodically in accordance with an established quality control program. The accuracy of the measurements made to monitor the parameters listed in 3.3.1 shall be per the following:

(1) The measured acceleration values shall be within ± 5 percent of true value
(2) The measured electrical input values shall be within ± 5 percent of true value
(3) The measured relay output change of state detector time shall be within ± 10 percent of the true value
(4) The measured relay operating time shall be within ± 10 percent of the true value or ± 1 ms, whichever is greater
(5) The test response spectrum generation equipment shall produce a plot within ± 10 percent of the true plot.

4. Test Conditions

4.1 Selection and Preparation of Samples. A minimum of three specimens is required for the test. These specimens shall be selected from production and shall conform to manufacturer's applicable standards of quality assurance and control. They shall have all calibrated quantities recorded along with their accepted tolerances. Reference should be made to 6.3.2 through 6.3.5 of IEEE Std 323-1974, Standard for Qualifying Class 1E Equipment for Nuclear Power Generating Stations, for other previbration preparations if the relays are being considered for Class 1E functions.

4.2 State of the Relay Under Test. All relays shall be tested in the nonoperating, transition, and operating modes. For each of these tests the attached summary, Table 1, lists (by relay characteristic) the required relay settings and inputs. All relays shall be tested in the transition from the nonoperating to operating mode as specified in Table 1. This test shall be performed by applying an input equal to the nonoperating input followed in time by an acceleration equal to the fragility level established for the device in the nonoperating condition. During the seismic simulation, the relay input shall be changed to the value specified under operating input in Table 1 and the operating time measured. The percent deviation from zero acceleration test of the same operating input shall be included in the documentation.

4.3 Relay Output

4.3.1 Contact Monitoring. All multicontact relays that can have a variation of contact formations shall be tested with the following arrangement:

(1) All contacts open
(2) Half normally open and half normally closed contacts
(3) A contact arrangement that can be proven more severe

All relays that have single contacts, or groups of contacts, or both, between external terminals shall be tested such that each set of external terminals shall be monitored separately. As an alternative, all normally open contacts can be connected in parallel and all normally closed contacts can be connected in series when the relays are in the nonoperate condition. When using the alternative connection for re-

Table 1
Relay Settings and Inputs
(Nonoperating and Operating Modes)

(1) Nondirectional Current Relays

(a) Instantaneous
    Relay Setting: Minimum setting
    Relay Input: Nonoperating — 25 percent of pickup setting
         Operating — 200 percent of pickup setting

(b) Time Delay
    Relay Setting: Minimum setting, 1 time dial
    Relay Input: Nonoperating — 60 percent of pickup setting
         Operating — 200 percent of pickup setting

(c) Over/Undercurrent
    Relay Setting: Minimum overcurrent/undercurrent ratio and minimum overcurrent setting
    Relay Input: Nonoperating — Midway electrically between overcurrent and undercurrent settings
         Operating — 50 percent of undercurrent and 200 percent of overcurrent settings

(2) Directional Current Relays

(a) Phase Overcurrent Element
    Relay Setting: Same as single function current relays
    Relay Input: Same as single function current relays

(b) Phase Directional Element
    Relay Setting: Maximum sensitivity
    Relay Input: Nonoperating — 100 percent voltage with the current voltage phase relationship of unity power factor in both the tripping and nontripping direction
         Operating — 50 percent voltage with the current voltage phase relationship at maximum sensitivity angle in the tripping direction

(c) Ground Overcurrent Element
    Relay Setting: Same as single function current relays
    Relay Input: Nonoperating — 0 percent of current
         Operating — 200 percent of pickup setting

(d) Ground Directional Element
    Relay Setting: Maximum sensitivity
    Relay Input: Nonoperating — 0 percent polarizing quantity
         Operating — 200 percent of pickup current in the polarizing circuit at maximum sensitivity angle in the tripping direction; polarizing voltage to be 10 percent of continuous rating

(3) Power or Angle Directional Relays

(a) Instantaneous
    Relay Setting: Maximum sensitivity
    Relay Input: Nonoperating — 100 percent voltage, 25 percent current at maximum sensitivity angle in the tripping and nontripping direction
         Operating — 50 percent voltage, 200 percent of pickup current at maximum sensitivity angle in the tripping direction

(b) Time Delay
    Relay Setting: Maximum sensitivity, 1 time dial
    Relay Input: Nonoperating — Same as for instantaneous power or angle directional relays
         Operating — Same as for instantaneous power or angle direction relays

(4) Voltage Relays

(a) Undervoltage (Time Delay or Instantaneous)
    Relay Setting: Maximum setting or approximately 90 percent of rating whichever is lower
    Relay Input: Nonoperating — rated voltage
         Operating — 80 percent of setting

(Continued on page 11)

Table 1
(Continued)

(4) Voltage Relays

(b) Overvoltage (Time Delay or Instantaneous)

Relay Setting:	Minimum setting or approximately 110 percent of rating whichever is higher
Relay Input:	Nonoperating — 80 percent of setting
	Operating — 120 percent of setting

(c) Over/Undervoltage (Time Delay or Instantaneous)

Relay Setting:	Minimum overvoltage setting or approximately 110 percent of rating whichever is higher
	Maximum undervoltage setting or approximately 90 percent of rating whichever is lower
Relay Input:	Nonoperating — rated voltage
	Operating — 120 percent of rated for overvoltage setting followed by 80 percent of undervoltage setting

(5) Differential Relays

(a) Low Impedance

Relay Setting:	Maximum sensitivity, minimum slope, and 1 time dial if induction disk
Relay Input:	Nonoperating — 25 percent of pickup current in the differential circuit and tap current in one restraint circuit
	Operating — 200 percent of pickup setting through the differential circuit

(b) High Impedance

Relay Setting:	Maximum sensitivity and 1 time dial if induction disk
Relay Input:	Nonoperating — 0 percent of pickup voltage
	Operating — 200 percent of pickup voltage (auxiliary current relay to be tested in accordance with Section 1, Part (a))

(c) Linear Coupler Types

Relay Setting:	Acceptance test condition as defined in manufacturer's instructions
Relay Input:	Nonoperating — 25 percent of pickup current or voltage
	Operating — 200 percent of pickup current or voltage

(6) Temperature Relays

Relay Setting:	Minimum setting
Relay Input:	Thermal indication such as actual temperature or RTD resistance value to be:
	Nonoperating — 80 percent of pickup
	Operating — 120 percent of pickup
	Replica relays to be energized at:
	Nonoperating — full load current
	Operating — 150 percent of full load current

(7) Reclosing Relays

Relay Setting:	Instantaneous or minimum time delay
Relay Input:	Nonoperating — Energized in a reset condition
	Operating — Energized for a reclosure mode

(8) Synchronizing and Synchrocheck Relays

Relay Setting:	Minimum angular setting and 1 time dial delay where applicable
Relay Input:	Nonoperating — Completely deenergized
	Operating — Rated voltage on both sides at 0°

(9) Timing Relays

Relay Setting:	Minimum time delay
Relay Input:	Nonoperating — Completely deenergized
	Operating — Energized normally

(Continued on page 12)

Table 1
(Continued)

(10) Underfrequency Relays	
Relay Setting:	1 Hz below normal system frequency with a minimum delay
Relay Input:	Nonoperating — 100 percent rated voltage at normal system frequency
	Operating — 85 percent rated voltage at 2 Hz below setting

(11) Impedance Measuring Relays	
Relay Setting:	Maximum sensitivity with the reach set at 25 percent of maximum ohmic setting
Relay Input:	Nonoperating — Rated voltage and current at unity power factor in the tripping direction
	Operating — 200 percent of rated current at maximum sensitivity angle and voltage produced by an impedance of 50 percent of the relay setting

(12) Auxiliary Relays

(a) Latching Relays
  Relay Setting: Normal calibration with relay latched and unlatched
  Relay Input: Completely deenergized
(b) Contact Multipliers and Contactors
  Relay Setting: Normal calibration

Relay Input:		Case 1	Case 2
	Nonoperating —	0 percent of rating	100 percent of rating
	Operating —	80 percent of rating (ac)	0 percent of rating
		100 percent of rating (dc)	

(13) Current or Voltage Balance Relays	
Relay Setting:	Maximum sensitivity and minimum time delay
Relay Input:	Nonoperating — Three-phase, balanced current or voltage at rated magnitude
	Operating — An unbalanced condition representing 120 percent of the sensitivity setting.

lays in the operating mode, the reverse configuration shall be used, that is, all normally open in series and all normally closed in parallel.

4.3.2 Output Loading. The source voltage for all contact loadings shall be 125 V dc or rated, whichever is less.

If a contact circuit between a pair of external terminals has a seal-in or holding coil function, the contact load circuit shall be such that it draws twice rated seal-in or holding current from the source voltage. The load shall have a series L/R ratio of 0.04 s. This load shall be applied for both nonoperating and operating mode tests. For all contacts not having a seal-in or holding coil function, the contact load shall be as shown in Table 2.

The output load of all relays having a solid-state output shall be 1 A resistive from a 125 V dc source or as rated, whichever is less.

4.3.3 Control Power. If the relay uses external control power, it shall be applied during test at rated voltage.

4.3.4 Relay Operate Time. Relay operate time shall be the time measured from initiation of electrical input to completion of 2 ms sustained output minus 2 ms.

4.4 Maintenance and Adjustment of the Relay During Test. At the completion of any given fragility test, it is permissible to inspect the relay and make any adjustments to insure the relay's integrity for the next series of tests. Any adjustments made shall be recorded as per Section 6.

Table 2
Contact Loading During Tests

	Contact Status During Test	Load
Nontransition Mode	Open	2 A or more $L/R = 0.04$ s
	Closed	25 mA or less
Transition Mode	Open to Closed	25 mA or less
	Closed to Open	25 mA or less

5. Test Methods

5.1 Introduction. The principal aim of this standard, as stated in 1.2, is to establish standard methods to generate data which define the seismic capability of protective and auxiliary relays. These methods employ what has been called Fragility Testing in IEEE Std 344-1975, and they have also been referred to as generic testing in the industry. These data will aid the designer of a nuclear power plant in the selection of relays for his application since their seismic capability will be catalogued and can be compared based on performance under standardized tests.

The use of a relay for a given function in any one application requires that it be qualified for that specific nuclear power plant. The qualification of relays for such specific applications is only a secondary aim of this standard since it is not practical to perform fragility testing for every conceivable set of operating conditions and seismic environments. It may be possible to use the fragility data described above in the qualification of a relay if the electrical settings, the failure criteria, and the seismic conditions employed in fragility testing satisfy the requirements of the specific application. If the fragility data cannot be extended or otherwise justified to be applicable, it shall be necessary to qualify the relay using proof testing for that specific application. For these reasons this standard: (a) gives specific direction in establishing standards for fragility testing, and (b) it also provides guidance for proof testing.

5.2 Fragility Testing. In order to define the seismic environment for standardized fragility tests a number of practical considerations must be taken into account.

(1) The seismic environments, both in terms of the levels and frequencies of vibration, vary from one site to another and from one relay mounting location to another.

(2) It is desirable that the capability data of relays which are generated in fragility testing be applicable to as many different locations as is practical.

(3) One approach is to construct an SRS shape which is sufficiently broad-band in its frequency content that it envelopes, within practical limitations, the individual requirements for all locations. Using this SRS it is possible with a minimum of testing to meet most seismic requirements. In this standard this is the preferred approach. This method of fragility testing is described in 5.2.1.1. In some cases, this approach has practical disadvantages which require that an alternative method of fragility testing be made available to the relay manufacturer or user. The reasons for this are briefly described in the following.

(4) The width in terms of frequency of the region of maximum amplification of acceleration of the broad-band SRS of (3) is much greater than the typically narrow-band region of the RRS for each specific location. In practice it has been found that the peak input acceleration (ZPA) during such a broad-band test is much higher relative to the level of the amplified region of the response spectrum, compared to that of the actual narrow-band requirement in any specific application. In other words, the broad-band SRS shape of (3) produces a severe overtest. Therefore, in the interests of producing a relatively simple test, a severe seismic requirement is imposed which differs significantly from that which applies for any one location. It is practical, therefore, to consider a second method of fragility testing as

13

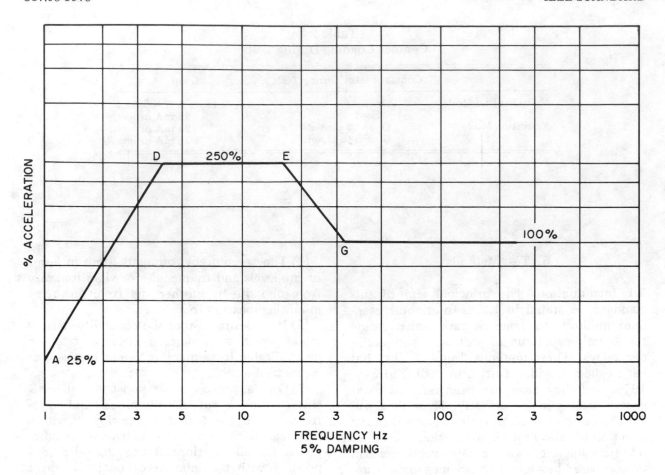

Fig 1
Multifrequency Broad-Band Standard Response Spectrum Shape

an alternative which imposes a more realistic seismic environment.

(5) The second approach is to divide the broad-band SRS shape into a series of narrower spectrum shapes which more closely represent the requirements of specific locations. This alternative is described in 5.2.1.2. By testing separately for each of the component peaks, capability data will be developed which meet the intent of the broad-band SRS shape while imposing a seismic environment which more realistically fits the actual requirements. The shaker-table input motion to generate each component peak may be put on a tape and thus by testing for each peak in sequence essentially only one test is required.

5.2.1 Fragility Test Options

5.2.1.1 *Broad-Band Multifrequency Fragility Testing.* Repeatable multifrequency input motions shall be used in the fragility testing. It is the test's objective to produce an FRS which envelopes the SRS shape using a bi-axial input motion. The method of achieving an acceptable biaxial test is described in 5.2.2.

Fig 1, the SRS shape (at 5 percent damping), is defined by four points:

point A = 1.0 Hz and an acceleration equal to 25 percent of the ZPA

point D = 4.0 Hz and 250 percent of the ZPA

point E = 16.0 Hz and 250 percent of the ZPA

point G = 33.0 Hz and a level equal to the ZPA.

14

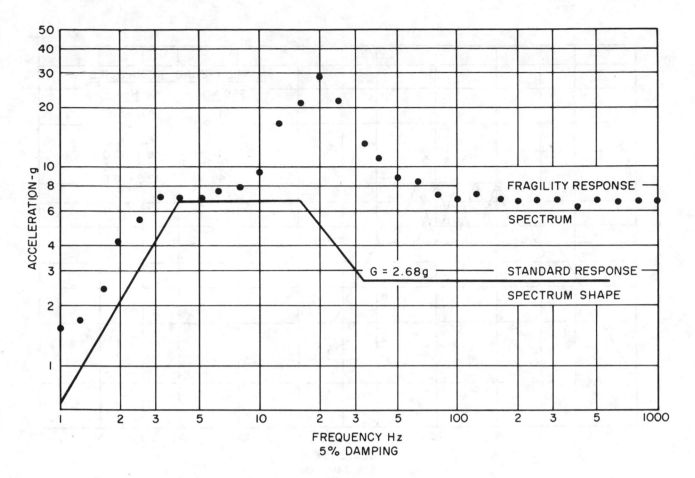

**Fig 2
Fitting the Standard Response Spectrum Shape to a Fragility Response Spectrum**

The range of maximum amplification of acceleration, 4.0 to 16.0 Hz has been designed to most realistically match the range of peak acceleration input to the relays by the equipments and panels on which they are mounted. Below 4.0 Hz, it is possible to encounter building frequencies to as low as 1.5 Hz. The resulting panel motions would probably be enveloped by the line *AD* since the amplification of panels at these low frequencies is small. Above 16 Hz, there are equipment and panel resonances; however, the seismic energy input in this range is generally reduced and, therefore, the motions would probably be enveloped by the line *EG*.

The horizontal and vertical shaker-table input components for the test should be equal within the capability of the test equipment. For the equalization of the input motion, the multifrequency input signals should be spaced at $^1/_3$ octave frequency intervals or less. The duration of the test should at least meet the requirements of 5.2.3.

The FRS will in all probability not have a shape which exactly matches that of the SRS of Fig 1. A relay will given an acceleration "*g*" rating based on the lower value of the horizontal or vertical ZPA, or point *G* of Fig 1, when the FRS has completely enveloped the shape of Fig 1. Fig 2 illustrates the use of this method of seismically rating a relay. In this example, note that the point *G* on the SRS, defines the "*g*" rating of the relay at 2.68; see 5.2.5.

5.2.1.2 Narrow-Band Multifrequency Fragility Testing. Repeatable multifrequency input

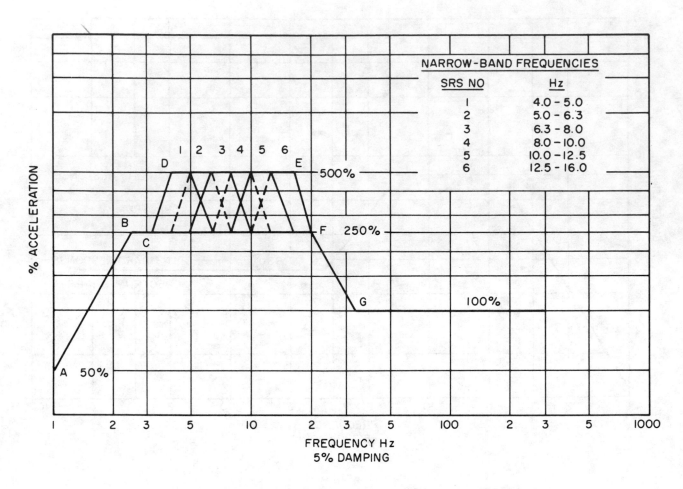

Fig 3
Multifrequency Narrow-Band Standard Response Spectrum Shape

motions shall be used in the fragility testing. It is the test's objective to produce a series of narrow-band FRS which envelope the series of narrow-band standard response spectrum shapes using biaxial input motions. The method of achieving an acceptable biaxial test is described in 5.2.2.

In dividing the broad-band curve of Fig 1 to produce a number of component narrow-band curves, a practical compromise must be made. On the one hand, as the width of the peak is reduced the test approaches the realism of an actual real-world requirement, and on the other, as the number of curves increases the sophistication of the test method escalates. It is recommended that as a maximum the amplified region 4.0 to 16.0 Hz is divided into six component peaks each $1/3$ octave in width, as shown by the standard response spectrum

shapes in Fig 3 (at 5 percent damping). If a more conservative simulation is selected, the test may be performed by lumping pairs of narrow peaks to produce three separate input motions, or lumping peaks in two groups of three to produce two separate input motions. The break-point frequencies which define the component peaks are as follows: 4.0, 5.0, 6.3, 8.0, 10.0, 12.5, and 16.0 Hz. The SRS shape is defined by the following points:

point A = 1.0 Hz and an acceleration equal to 50 percent of the ZPA

point B = 2.5 Hz and 250 percent times the ZPA

point C = 3.2 Hz and 250 percent times the ZPA

point D = 4.0 Hz and 500 percent times the ZPA

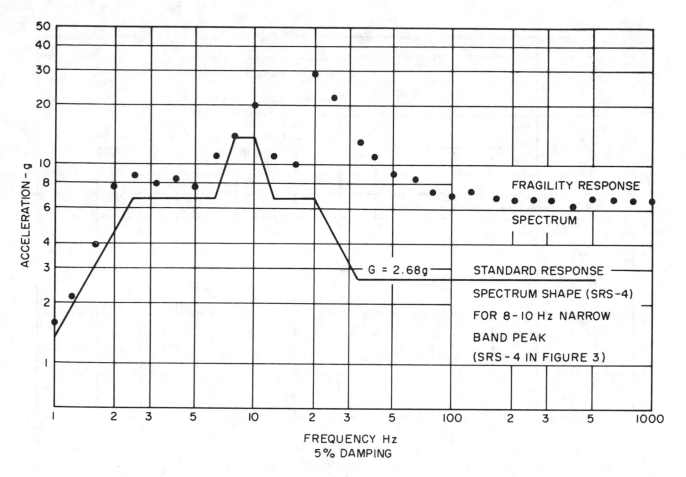

Fig 4
Fitting the Narrow-Band Standard Response Spectrum Shape (SRS-4)
to a Fragility Response Spectrum

point E = 16.0 Hz and 500 percent times the ZPA

point F = 20.0 Hz and 250 percent times the ZPA

point G = 33.0 Hz and a level equal to the ZPA.

The narrow spectrum shape of Fig 3 has a broad-band shoulder under the narrow peaks which is half the level of the peaks and ranges from 2.5 to 20 Hz. This shoulder is maintained for each peak. Its purpose is to envelope secondary peaks due to other lesser modes of vibration which occur in many applications. The side slopes of the narrow peaks are such that they meet the broadened shoulder at the next $^1/_3$ octave interval frequency point.

The horizontal and vertical shaker-table input

components for the test should be equal within the capability of the test equipment. For the equalization of the input motion the multi-frequency input signals should be spaced at $^1/_3$ octave frequency intervals or less. The duration of each narrow-band test should at least meet the requirements of 5.2.3.

The FRS for each component peak will in all probability not have a shape which exactly matches that in Fig 3. A relay will be given an acceleration "g" rating based on the lower value of the horizontal or vertical ZPA, or point G of Fig 3, when the FRS has completely enveloped that component peak SRS shape of Fig 3. The actual "g" rating for the relay will be the smallest of the values obtained from the test samples for each given narrow-band SRS. Fig 4 illustrates the use of this method of

17

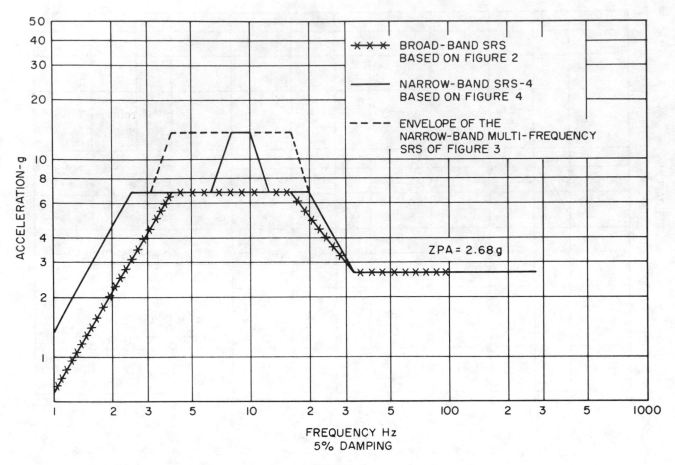

Fig 5
Comparison of the Two Multifrequency Testing Methods

seismically rating a relay, by establishing the "*g*" level of point *G* for a test designed to envelope the 8 to 10 Hz component peak. In this example note that the point *G* on the SRS defines the "*g*" rating of the relay at 2.68; see 5.2.5.

5.2.2 Multi-Axis Testing. Seismic ground motion occurs simultaneously in all directions in a random fashion. The direction of test input motion should, therefore, be in all three principal axes simultaneously. At the present time, however, two-axis test facilities are limited and three-axis facilities are nonexistent; therefore, two satisfactory biaxial alternatives are allowed.

It is the test's objective to develop a FRS which reproduces the standard response spectrum shapes of Fig 1 or Fig 3 using a biaxial multifrequency input motion. The horizontal and vertical levels of the components of the

motion shall be equal within the capability of the test equipment.

When phase independent biaxial inputs are used, relays shall be tested in two orientations. Relays shall be rotated 90° about the vertical axis for the second orientation.

When identical biaxial inputs are used in the two axes, which is always the case with a single thruster inclined at 45° to the horizontal plane, four tests shall be made. First with the inputs in phase; second, with one input 180° out of phase; third, with relays rotated 90° about the vertical axis and the inputs in phase; and, finally with the relay orientation as in the third step but with one input 180° out of phase. When a single inclined thruster is used, the 180° out of phase input is achieved by rotating relays 180° about the vertical axis for tests two and four.

5.2.3 SSE and OBE, and the Duration of Testing. The duration of each test designed to reproduce the SRS shape shall be at least 15 s in duration.

Prior to taking credit for tests at the rated "*g*" level (the SSE level), relays shall be tested five or more times to at least one-half of the rated level (the OBE level). The duration of each OBE test shall at least equal the 15 s duration of the SSE test. Credit may be taken for any test, preceding the SSE test if shown to be greater than or equal in severity to the required OBE's. Fragility test duration, by their nature, will normally far exceed the preceding duration requirement.

5.2.4 Determination of Failure

5.2.4.1 *Output Change of State.* Failure is an unauthorized electrical output change of state in the output circuit lasting 2 ms or more as determined by a device having a reset time of less than 200 μs and an operating point of 50 percent of applied voltage. In addition, the unauthorized operation of a defined external auxiliary device operated by the relay under test will also be considered as a failure.

5.2.4.2 *Operability.* The inability of the relay to perform during transitional testing shall be classified as a failure.

5.2.4.3 *Structural Damage.* Structural members of the relay exhibiting any fracture damage or plastic yielding that might cause misalignment or failure of the relay shall be classified as test failure.

5.2.4.4 *Post Test.* Failure of the relay to perform within twice accepted tolerances as stated in 4.1 after the fragility testing is completed without readjustment shall be classified as a test failure.

5.2.5 Application of "*g*" Rating Data. Seismic data produced by testing a series of relays can be presented in catalog data by referring to the "*g*" level at the ZPA and the method of test, that is broad-band or narrow-band.

NOTES:

(1) The "*g*" rating is based on the lower value of the horizontal or vertical components and not on the resultant of the two.

(2) The minimum fragility level of the test samples described in 4.1 shall be used.

Thus for industry use, the SRS of Fig 1 or Fig 3 becomes the TRS for the application engineer. The application engineer can select seismically qualified relays by obtaining from the catalog data sheets the "*g*" values at the ZPA which can be used to construct a TRS similar to Fig 1 or Fig 3. Based on his unique relay requirement, he can compare the needed seismic environment as described in a specific RRS to the prequalified seismic environment (derived from Fig 1 or Fig 3) and select qualified relays accordingly. The additional requirement in the qualification of relays is that the standardized failure criteria employed in the fragility testing matches those, or can be extended or otherwise justified to apply, in his specific application, and that the electrical inputs to the relay during fragility testing are compatible with his application.

An additional note of explanation is necessary in comparing the SRS shapes of Figs 1 and 3. In the examples which illustrate the use of the SRS shapes, Figs 2 and 4, both relays have a "*g*" rating of 2.68. Because the narrow-band SRS shape has a much higher ratio of the peak response (points *D* to *E* in Fig 3) to the ZPA (point *G*) compared to that in Fig 1, the two 2.68 "*g*" relays are not equivalent. It can be seen that the relay which was tested using the narrow-band approach has been subjected to a much higher seismic environment and, consequently, it would be the superior relay; see Fig 5. In summary, therefore, when two relays have been tested to the two methods their capability cannot be compared based on the ZPA alone, rather the entire SRS shape must be considered.

5.3 Proof Testing

5.3.1 Introduction. A proof test shall be used to qualify relays for a particular application or to a particular requirement using the acceptable methods described in IEEE Std 344-1975. As such, a detailed testing specification must be provided. Items to be considered in this specification are as follows:

5.3.1.1 *Seismic Input.* The selection of the seismic input should achieve the following.

(1) A TRS which envelopes the RRS using single- or multiple-frequency input as required to provide a conservative test table motion.

(2) A peak input acceleration equal to or greater than the ZPA of the RRS. The TRS shall be plotted to that frequency which gives a clear indication of the ZPA.

(3) No frequencies shall be input above the ZPA frequency of the RRS.

(4) A duration equal to a minimum of 15 s

in each test.

(5) At least 5 OBE level tests should precede the SSE level test.

5.3.1.2 *Rigid Mounting.* When relay equipment to be tested is rigidly mounted to the test table, the seismic input to the shaker shall be such that the TRS envelopes the RRS applicable at the mounting position of the relay in its normal service location. The mounting location RRS shall be obtained either from test data or calculation of accelerations imposed on the relay in its normal service location.

5.3.1.3 *Determination of Failure.* For proof testing, the detailed specification shall define what constitutes failure. This will not necessarily be the same as the standardized failure criteria described in 5.2.4 used in fragility testing.

6. Documentation

6.1 General. The documentation for each device shall show its performance when subjected to seismic accelerations.

6.2 Test Report. The test report shall contain the following information:

(1) Dates of test

(2) Reference to QA procedures followed

(3) Test specimen description (identification and specification)

(4) Test facility name and location

(5) Test facility and equipment hookup description

(6) Test equipment certification

(7) The general test description shall be given to clearly demonstrate that all requirements of Section 5 have been met

(8) Test data in the form of acceleration versus frequency curves, tables, and photographs and records of any adjustments made to the relay during the test

(9) Test facility temperature, humidity, and pressure ranges during tests

(10) Results, conclusions, and analysis of failures and specific device limitations

(11) Approved signatures and date.

7. Generalization of Test Results

It is desirable to minimize the testing requirements of qualifying relays for seismic conditions, however, generalization of test results must be made with good engineering judgement and periodically reviewed for significant design changes that could affect the qualification of an entire line of relays, and shall be documented by analysis. There are obvious situations where the test results on a particular model of relay may be valid information for many relays of the same type, provided the mechanism directly associated with the device's output, its electrical rating, and its packaging are justified to be equivalent. A particular relay may consist of several subassemblies. A test on the complete package would be directly applicable to the relay using only one or more of the identical subassemblies; however, the tests on a relay incorporating only one of these subassemblies would not be valid for the complete relay.

These considerations should be made before the specimens are selected for testing.

ANSI
N13.4-1971

American National Standard
for the Specification of Portable
X- or Gamma-Radiation Survey Instruments

Secretariat for N13
Atomic Industrial Forum, Inc.

Secretariat for N42
Institute of Electrical and Electronics Engineers, Inc.

Approved December 9, 1971
American National Standards Institute

N13.4

Foreword

(This Foreword is not a part of American National Standard for the Specification of Portable X- or Gamma-Radiation Survey Instruments, N13.4-1971.)

The Joint Subcommittee N13/42 which was responsible for the development of this standard was established by authority of the Chairmen of American National Standards Committees N13 and N42 to represent the interests of the respective parent committees. Since the Joint Subcommittee comprises 50 percent manufacturer and 50 percent user membership, the standard includes the additional feature of incorporating the experience of both the manufacturer and the user.

The American National Standards Committee on Radiation Protection, N13 and American National Standards Committee on Instrumentation, N42, which reviewed and approved this standard, had the following representatives at the time of approval.

American National Standards Committee N13

J.W. Healy, *Chairman*   J. Sohngen, *Secretary*

Organization Represented	*Name of Representative*
American Chemical Society	Ira B. Whitney
American Conference of Governmental Industrial Hygienists	D. E. VanFarowe
American Health Physics Society	L. J. Cherubin
American Industrial Hygiene Association	Wilbur Speicher
American Insurance Association	Carl Karlson
American Mutual Insurance Alliance	William J. Uber
American Nuclear Society	James E. McLaughlin
American Public Health Association	Dr. Simon Kinsman
American Society for Testing and Materials	Dr. L. B. Gardner
American Society of Mechanical Engineers	H. J. Larson
Association of State and Territorial Health Officers	*Representation Vacant*
Electric Light & Power Group	Marvin Sullivan
Industrial Medical Association	Dr. W. T. Doran
Institute of Nuclear Materials Management	Ken Okolowitz
International Association of Government Labor Officials	Dr. Morris Kleinfeld
International Brotherhood of Electrical Workers	Edward J. Legan
Manufacturing Chemists Association, Inc.	P. W. McDaniel
National Bureau of Standards	Thomas P. Loftus
National Safety Council	Dr. Hugh F. Henry
Underwriters' Laboratories, Inc.	Leonard H. Horn
Uranium Operators Association	L. W. Swent
U. S. Atomic Energy Commission	Edward J. Vallario
U. S. Bureau of Labor	John P. O'Neill
U. S. Public Health Service	John Villforth
Members-at-Large	William O. Chatfield
	Donald Fleckenstein
	Duncan A. Holaday
	Remus G. McAllister

American National Standards Committee N42

Louis Costrell, *Chairman*   David C. Cook, *Secretary*

Organization Represented	*Name of Representative*
American Chemical Society	Louis P. Remsberg, Jr
American Conference of Governmental Industrial Hygienists	Jesse Lieberman
American Industrial Hygiene Association	W. H. Ray
American Nuclear Society	W. C. Lipinski
	Thomas Mulcahey (*Alt*)
American Society of Mechanical Engineers	R. C. Austin
American Society of Safety Engineers	*Representation Vacant*
American Society for Testing and Materials	John L. Kuranz
	Jack Bystrom (*Alt*)
Atomic Industrial Forum	*Representation Vacant*

The Joint Subcommittee N13/42 which was responsible for developing the standard on performance specifications for Portable Rate Meters consisted of the following personnel.

Contents

American National Standard for the Specification of Portable X- or Gamma-Radiation Survey Instruments

1. Scope

This standard does not provide a series of performance requirements but rather the means of stating or describing performance. The recommendations contained herein apply to specifications for portable X- or gamma-radiation survey instruments, as defined in Section 2 below. Each of these instruments includes a detector and visual, analog and/or digital type readout. In recognition of the practical difficulties inherent in exposure measurements in the energy region above a few MeV, two broad energy ranges have been delineated:

(1) ≤ 1.5 MeV
(2) > 1.5 MeV

The photon energy from ^{60}Co, which is widely used for test and calibration, is conveniently near the upper limit of the low range.

2. Definitions

Use of technical terminology within this standard is consistent with the definitions in the American National Standard Glossary of Terms in Nuclear Science and Technology, N1.1-1967. The following additional terms have been defined for use within this document.

2.1 portable X- or gamma-radiation survey instrument. An instrument with a self-contained energy source (for example, batteries) designed to measure exposure rate while being carried. Such instruments may also have the capability to measure integral exposure, but instruments with the capability of measuring integral exposure only are specifically excluded from this definition.

2.2 switching transients. Sudden excursions of the meter which occur when the range switch is changed from one position to the next.

2.3 effective energy. The energy of mon-ochromatic photons which undergoes the same percentage attenuation in a specified filter as the heterogeneous beam under consideration. Aluminum is the filter specified for photon energies less than, or equal to, 100 keV, copper for photon energies between 100 keV and 1.5 MeV, and lead for photons with energies greater than 1.5 MeV.

3. Specifications

Specifications shall be stated in writing for each instrument and shall be available upon request to bona fide customers or prospective customers for examination. Stated specifications shall be based on the results of tests of representative production models to which the specifications apply. Tests shall be those as given in Section 5 of this standard with any deviations clearly stated. Where tests are not specified (for example, as for rf effects) but have been performed, the testing method should be described along with the result. If no tests are made, the specification should be listed and followed by the phrase "Not Tested."

3.1 Mechanical.

3.1.1 *Weight.* The weight of the complete instrument package including batteries shall be stated in pounds and kilograms.

3.1.2 *Controls.* All externally accessible controls shall be specified as to type (for example, toggle, rotary, etc) and function.

3.1.3 *Dimensions.* The maximum overall height, width, and length of the fully assembled package shall be stated in inches and centimeters. An illustration should be supplied to aid in interpretation of the dimensions.

3.1.4 *Readout.*

3.1.4.1 *Meter.* The scale length shall be stated in inches and centimeters along with the meter type (for example, taut band, dual movement, etc).

3.1.4.2 *Digital.* The type of presentation (for example, neon tube, electromechanical, etc), number of digits displayed, and height of individual digits in inches and centimeters shall be stated.

3.1.4.3 *External.* If an external signal output is provided, the type of connector, signal characteristics and impedance shall be stated.

3.1.4.4 *Other.* No specifications.

3.1.5 *Markings.*

3.1.5.1 *Case Markings.* The following should appear on the exterior surface:

(1) Function designation (for example, exposure ratemeter)

(2) Manufacturer and model number

(3) Serial number

(4) Geometric center of detector

(5) Function designation of all externally accessible controls

3.1.5.2 *Meter Markings.* Meter markings shall be in units of exposure per unit time (for example, mR/h) or exposure (for example, mR), or absorbed dose, as applicable. This does not preclude the use of supplementary markings.

3.1.5.3 *Battery Check.* If a battery check is provided, it shall be stated whether the check accurately reflects the end point voltages. (see 5.10.)

3.2 Detector.

The type (for example, scintillation, ion chamber, etc), dimensions of sensitive volume, sensitive volume in cm^3, wall material and density in mg/cm^2 should be specified. If the instrument has an external probe, its size and connecting-cable length and type should be specified.

3.3 System Operational Specifications.

3.3.1 *Operating Range.* The dynamic range for each scale and the type of readout (for example, linear, logarithmic) shall be stated.

3.3.2 *System Accuracy.* The ability of the instrument to correctly measure exposure rates over its entire range for the standard set of exposure conditions (see 5.1) shall be given as ± percent of reading for logarithmic readouts or ± percent of full scale of the scales or digits being read for instruments with linear scales.

3.3.3 *Spectral and Angular Dependence.* The change in response as a function of both photon energy (or spectrum) and angle of incidence shall be specified over the stated energy range of the instrument. A graphical or tabular presentation should be used and the energy to which the response is normalized and the method by which this energy is obtained should be stated.

3.3.4 *Temperature Influences.* The operating temperature limits shall be specified for all ranges tested along with the change in response as a function of temperature. These data shall be provided for the instrument exclusive of energy source as well as for the instrument equipped with energy sources, such as batteries, recommended by the manufacturer. The temperature influences for linear readouts should be expressed as percent of full scale change per °C or as total percent change of full scale over a specified temperature range (for example, ± percent of full scale from X to Y°C). For logarithmic or other non-linear readouts, the temperature influences should be expressed as percent of initial reading over a specified temperature range. Storage extremes shall also be stated.

3.3.5 *Other Environmental Effects.* Humidity dependence, pressure dependence, and maximum geotropism shall be stated. In particular, any inherent effects, such as pressure dependence of unsealed ionization chambers, shall be explicitly noted. Storage extremes shall also be stated.

3.3.6 *Switching Transients.* The magnitude and duration of switching transients shall be given.

3.3.7 *Warmup Time.* The warmup time (that is, the interval after application of power during which the instrument will not meet the accuracy specifications as given in 3.3.2) shall be stated.

3.3.8 *Response Time.* The time required for the reading to reach 90 percent of the final value for a step exposure increase from zero to the midpoint shall be specified for each range or decade.

3.3.9 *Response to Other Radiation.* The effects of coexistent fields of beta and photon radiation shall be given. The minimum alpha and beta particle energy required to make the penetration shall be expressed in MeV. A statement of the effects of neutron and non-ionizing (for example, rf, uv) radiations should be given along with any extracameral volume effects.

3.3.10 *Exposure Rate Limitations.*

(1) *Ratemeters.* The exposure rate above full scale at which the instrument fails to give a full scale or greater reading shall be stated. Where a specific value has not been determined, the maximum exposure rate to which the instrumentation has been tested shall be so stated.

(2) *Integrating Mode.* The steady state exposure rate at which the response drops by 10 percent shall be stated. Where a specific value has not been determined, the maximum to which the instrument has been tested shall be stated.

3.3.11 *Battery Lifetime.* The battery lifetime in terms of hours of continuous operation in a field of <0.1 mR/h and end-point voltage shall be stated along with the size, number, and type of battery (for example, mercury, carbon-zinc, etc.)

3.3.12 *Drift.* The up or downscale drift from zero after warmup shall be stated in units of equivalent exposure rate-drift per unit time.

3.3.13 *Zero Adjust.* If the instrument has a zero adjust, the ability to zero the instrument in a radiation field should be stated.

4. Operating and Maintenance Manual

An operating and maintenance manual shall be provided for each instrument and shall contain as a minimum the following:

() A complete circuit and block diagram with all parts identified and appropriately labeled

(2) Parts list with all values specified

(3) Operating instructions

(4) Detailed specifications in accordance with this standard

(5) Graphical and/or tabular presentations of energy dependence, pressure dependence, etc should be included

(6) Maintenance procedures, including waveforms, voltages, and calibration techniques

5. Testing Procedures

Except as otherwise stated, all testing procedures shall be carried out within a normal laboratory environment in which extremes of environment will not prevail. Where necessary, results will be normalized to NTP (normal temperature and pressure = 20°C and 760 torr (1.01 bar)) and 50 percent relative humidity.

5.1 System Accuracy. An instrument shall be exposed to photon fluxes with known spectral distribution and exposure rates of approximately one-fifth, one-half, and four-fifths the indicated full-scale or decade range. This shall be done for all ranges or decades and the error determined by the following equations:

For logarithmic readouts

Percent error =
$$\left(\frac{\text{Meter Reading} - \text{True Exposure Rate}}{\text{True Exposure Rate}} \right) \times 100 \quad \text{(eq 1)}$$

For linear readouts

Percent error =
$$\left(\frac{\text{Meter Reading} - \text{True Exposure Rate}}{\text{Full-scale Reading}} \right) \times 100 \quad \text{(eq 2)}$$

The maximum error for each decade or range shall be stated.

5.2 Spectral and Angular Dependence. For energies below about 1.5 MeV, the ratio of indicated to true exposure or exposure-rate as a function of energy shall be obtained at several effective energies over the operating energy range specified for the instrument. For the energy region below 200 keV, the ratio of indicated to true exposure or exposure-rate should be measured at appropriate increments. Increments of ≤ 25 keV should be used. For the energy region above 200 keV but less than about 1.5 MeV, photon energies of 662 keV ([137] Cs) or 1.25 * MeV ([60] Co) should be used. For energies above about 1.5 MeV, the source and energy of the photons shall be given (for example, Van de Graaff, $^{19}F(p,\alpha)^{16}N$, Eγ - 6.3,7.1 MeV). At least one energy in the

---

* Effective energy for the 1.17 MeV and 1.33 MeV photon emission from [60] Co.

range 1.5 – 10 MeV shall be used. Extrapolations shall not be made above the highest energy nor below the lowest energy for which tests are performed.

To determine angular dependence, the instrument or detector or source, as applicable, shall be rotated through at least two perpendicular planes using the center of the sensitive volume of the detector as axis of rotation, and the ratio of indicated to true exposure rate obtained for at least thirty degree increments and three energies, one in each third of the specified operating energy range of the instrument. For symmetrical instruments or detectors, this rotation need not be accomplished over the entire 360 degree ranges, but should be representative of the response as a whole.

5.3 Temperature Influences. Temperature influences, as specified in 3.3.4, shall be obtained over the stated temperature range of the instrument by comparing at 10°C intervals the response of the instrument to a flux providing approximately midscale response at 20°C. The instrument should be permitted to come to thermal equilibrium at each temperature before data are taken.

5.4 Other Environmental Influences.

5.4.1 Humidity dependence shall be determined at a constant temperature of 20±5°C by placing the instrument in a constant relative humidity of at least 90 percent for at least four hours and comparing the response to a photon flux providing approximately half-scale reading at 50 percent relative humidity.

5.4.2 Geotropism shall be determined by rotating the instrument through a full 360 degrees in each of two planes normal to each other and perpendicular to the ground, and the maximum effect noted.

5.4.3 Pressure dependence shall be determined over the range 0.6 – 1.06 bar (520 – 800 torr). The instrument shall be allowed to equilibrate at each pressure and the response to a flux providing aproximately half-scale

reading at 1.01 bar (760 torr) noted and normalized to the reading at 1.01 bar.

5.5 Switching Transients. The range switch shall be changed stepwise in both directions, and the magnitude and the recovery time as described in 5.7 of switching transients noted; the meter reset shall not be used for this test.

5.6 Warmup Time. Warmup time as described in 3.3.7 shall be determined at 20±5°C.

5.7 Recovery or Decay Time. The instrument shall be removed from a radiation field providing a full-scale or decade reading, and the time required for the scale or decade reading to return to 10 percent of full-scale or decade shall be noted.

5.8 Response Time. Response time specified in 3.3.8 shall be determined by measuring the time for the entire system to reach 90 percent of the midscale reading when the instrument is exposed to a step change in flux sufficient to provide a midscale or mid-decade reading. This test shall be performed on all ranges or decades, as applicable.

5.9 Exposure Rate Limitations. The exposure or exposure rate above which the instrument fails to give a full-scale response should be checked by placing the instrument in an appropriate radiation flux and increasing the intensity until the effect is noted or a level at least 100 times the maximum range of the instrument has been reached. This should be done on all ranges.

5.10 Battery Lifetime. Battery lifetime shall be determined with fresh batteries installed in the instrument. The instrument shall be turned on in a field < 0.1 mR/h and left on continuously. At intervals, the response of the instrument should be tested with a flux providing approximately a half-scale reading; the point in time at which the instrument response no longer meets all performance specifications should be used to determine battery lifetime and corresponding end-point voltage.

ANSI
N13.10-1974

American National Standard
Specification and Performance of
On-Site Instrumentation for
Continuously Monitoring Radioactivity in Effluents

Atomic Industrial Forum, Inc.
Institute of Electrical and Electronics Enginers, Inc.

Approved September 19, 1974
American National Standards Institute

N13.10

© Copyright 1974 by

The Institute of Electrical and Electronics Engineers, Inc.

Foreword

The Joint Subcommittee of American National Standards Committees N13 and N42, which was responsible for the development of this standard, was established by authority of the Chairmen of American National Standards Committees N13 and N42 to represent the interests of the respective parent committees. The Joint Subcommittee comprises 50 percent manufacturer and 50 percent user membership; thus this standard includes the additional feature of incorporating the experience of both the manufacturer and the user.

The American National Standards Committee on Radiation Protection, N13, which, together with the American National Standards Committee on Instrumentation, N42, reviewed and approved this standard, had the following membership at the time of approval:

M. E. Wrenn, *Chairman*　　　　　　　　　　　**James E. Sohngen,** *Secretary*

Organization Represented	*Name of Representative*
American Chemical Society	Ira B. Whitney
American Conference of Governmental Industrial Hygienists	D. E. Van Farrowe
American Industrial Hygiene Association	H. Wilbur Speicher
American Insurance Association	Karl H. Carlson
American Mutual Insurance Alliance	William J. Uber
American Nuclear Society	James E. McLaughlin
American Public Health Association	Simon Kinsman
	Gerald S. Parker (*Alt*)
American Society for Testing and Materials	L. B. Gardner
	J. H. Bystrom (*Alt*)
	A. N. Tschaeche (*Alt*)
American Society of Mechanical Engineers	H. J. Larson
Association of State and Territorial Health Officers	Sherwood Davies
Atomic Industrial Forum	James E. Sohngen
Electric Light and Power Group	Marvin Sullivan
	Gordon A. Olson (*Alt*)
Environmental Protection Agency	David S. Smith
Health Physics Society	L. J. Cherubin
	John J. Ferry (*Alt*)
Industrial Medical Association	William Doran, Jr
Institute of Electrical and Electronics Engineers	D. G. Pitcher
Institute of Nuclear Materials Management	Ken Okolowitz
	Robert Budd (*Alt*)
International Association of Governmental Labor Officials	Jacqueline Messite
	Phillip M. Bourland (*Alt*)
International Brotherhood of Electrical Workers	Paul R. Shoop
Manufacturing Chemists Association, Inc	P. W. McDaniel
	Paul Estey (*Alt*)
National Bureau of Standards	Thomas P. Loftus
	Thomas Hobbs (*Alt*)
National Safety Council	Hugh F. Henry
	Julian B. Olishifski (*Alt*)
	Harry Rapp (*Alt*)
Uranium Operators Association	L. W. Swent
	R. T. Zitting (*Alt*)
US Atomic Energy Commission	Edward J. Vallario
	Walter Cool (*Alt*)
US Bureau of Labor	John P. O'Neill
	G. Walker Daubenspeck (*Alt*)
US Public Health Service	Richard F. Boggs
	Roger G. Bostrom (*Alt*)
Individual Members	William O. Chatfield
	Donald C. Fleckenstein
	Duncan A. Holaday
	Remus G. McAllister

The American National Standards Committee on Instrumentation, N42, had the following membership at the time of approval:

L. Costrell, *Chairman*　　　　　　　　　　　**D. C. Cook,** *Secretary*

Organization Represented	*Name of Representative*
American Chemical Society	L. P. Remsberg, Jr
American Conference of Governmental Industrial Hygienists	J. Lieberman

Contents

American National Standard
Specification and Performance of
On-Site Instrumentation for
Continuously Monitoring Radioactivity in Effluents

1. Introduction

The release of radioactivity from nuclear facilities to the environment generally is monitored by installed instrumentation. The objective of such instrumentation is to measure the quantity or rate, or both, of release of radionuclides in the effluent streams and to provide documentation useful for scientific and legal purposes. This standard applies to continuous monitors that measure normal releases, detect inadvertent releases, show general trends, and annunciate radiation levels that have exceeded predetermined values.

2. Scope

This standard provides recommendations for the selection of instrumentation specific to the continuous monitoring and quantification of radioactivity in effluents released to the environment. The effluent streams considered may contain radioactive gases, liquids, particulates, or dissolved solids singly or in combination. This standard specifies detection capabilities, physical and operating limits, reliability, and calibration requirements and sets forth minimum performance requirements for effluent monitoring instrumentation. Unless otherwise specified, the criteria herein refer to the total system. This standard applies only to monitoring during routine operation that includes abnormal releases. Emergency situations, where additional performance capability will be required, are a matter of separate consideration.

Also outside the scope of this standard are sample extraction and laboratory analyses, normally used for intercomparison between monitor calibrations and laboratory analyses, and other applications for continuous monitoring instrumentation such as environmental monitoring or process control. Sampling techniques are covered in ANSI N13.1-1969, Guide to Sampling Airborne Radioactive Materials in Nuclear Facilities.

3. Definitions

These definitions are restricted to the purpose of this standard.

accuracy. The degree of agreement with the true value of the quantity being measured.

NOTE: Accuracy is subject to the influence of unknown systematic errors.

authorities. Any governmental agencies or recognized scientific bodies which by their charter define regulations or standards dealing with radiation protection.

calibrate. Adjustment of the system and the determination of system accuracy using one or more sources traceable to the NBS (National Bureau of Standards).

check. The use of a source to determine if the detector and all electronic components of the system are operating correctly.

detector. Any device for converting radiation flux to a signal suitable for observation and measurement.

effluent. The liquid or gaseous waste streams released to the environment.

extracameral effect. Apparent response of an

instrument caused by radiation on any other portion of the system than the detector.

in line. A system where the detector assembly is adjacent to or immersed in the total effluent stream.

off line. A system where an aliquot is withdrawn from the effluent stream and conveyed to the detector assembly.

on site. Location within a facility that is controlled with respect to access by the general public.

plate out. A thermal, electrical, chemical, or mechanical action that results in a loss of material by deposition on surfaces between sampling point and detector.

precision. The degree of agreement of repeated measurements of the same property, expressed in terms of dispersion of test results about the mean result obtained by repetitive testing of a homogenous sample under specified conditions. The precision of a method is expressed quantitatively as the standard deviation computed from the results of a series of controlled determinations.

primary calibration. The determination of the electronic system accuracy when the detector is exposed in a known geometry to radiation from sources of known energies and activity levels traceable to the NBS.

quality assurance. All those planned and systematic actions necessary to provide adequate confidence that a system or component will perform satisfactorily in service.

response time. The time interval from a step change in the input concentration at the instrument inlet to a reading of 90 percent (nominally equivalent to 2.2 time constants) of the ultimate recorded output.

secondary calibration. The determination of the response of a system with an applicable source whose effect on the system was established at the time of a primary calibration.

sensitivity. The minimum amount of contaminant that can repeatedly be detected by an instrument.

system. The entire assembled equipment excluding only the sample collecting pipe.

4. Factors Influencing Selection of Instrumentation

4.1 Effluent Stream Factors.

4.1.1 *Radiological Characteristics of the Effluent Stream.* The radiological characteristics of an effluent stream influence system capability requirements. The concentration of each radionuclide present with its particular half-life and type (alpha, beta, and photon) and the energy of radiation emitted must be considered in the selection of detectors. Alpha and beta radiation may be particularly subject to energy absorption in the detector housing of the effluent materials, thus influencing the sensitivity of the system. Alpha and low-energy beta radiation will not be measured at all if there is any appreciable thickness of material between the effluent and the detector. If a detector housing or the wall of the effluent line separate the effluent from the detector, the absorption of photons and charged particle equilibrium must be considered. In addition, a delay between sampling and measurement may significantly affect the detection capability of the system for radionuclides with short half-lives.

4.1.2 *Physical Characteristics of the Effluent.* Physical characteristics of a gaseous or liquid effluent that may influence system capability include temperature, pressure, humidity, size and number of suspended particles, and flow rate. Since the sensitivity of detection may be related to one or more of these physical characteristics, they must be considered so that the sample characteristics are accurately related to those of the effluent stream.

The size and distribution of suspended particles in the effluent will affect the choice and location of the sampling device and detector. It should be recognized that the density or accumulation of collected particles might cause self-absorption losses and that the detector could become contaminated.

Effluent flow rate may also govern the type of sampling device, particularly for an off-line monitoring system. Since most monitoring systems directly measure the concentration of radioactivity in an effluent stream, the stream flow rates must be determined accurately to derive the total released activity. And since most monitoring systems measure only a portion of the radioactivity in the effluent stream, a representative sample must be assured.

4.1.3 *Chemical Characteristics of the Effluent.* The major chemical characteristics that may influence system capability include sample plate out, corrosiveness, and combustibility. Plate out can result in differences in the concentrations of radionuclides in the sample and in the effluent. The corrosiveness of the sample can damage the system components such as the detector, sampler, filter medium, piping and pump. Damage of any of these could result in a nonrepresentative sample and eventual system failure. If an effluent contains combustible or highly reactive materials, care must be taken to prevent conditions that could permit explosion or combustion.

4.2 Environmental Factors.

4.2.1 *Temperature.* Both the detector and the electronic portions of an effluent monitoring system may be influenced by temperature variations. The effects may vary from minor calibration shifts to severe degradation of performance and, in some instances, permanent damage. When selecting monitoring instrumentation, therefore, it is particularly important to consider the manufacturer's specifications in light of the anticipated ambient thermal operating extremes. Additionally, other heat loads, such as those from nearby instruments or from direct sunlight, must be taken into account.

4.2.2 *Mechanical.* Mechanical effects such as shock, vibration, pressure, and noise may adversely affect the system operation. Instrument locations should be chosen to minimize these effects. When other technical or practical considerations dictate location in an area where the instrument may be subjected to such forces, appropriate measures such as shock mounting or sound baffling should be considered.

4.2.3 *Chemical.* While effluent monitors are rarely located in atmospheres containing corrosive or highly reactive chemicals, it should be recognized that small amounts of certain chemicals (for example, H_2SO_4, HNO_3, HF) can lead to deterioration and failure of electronic components. Where such chemicals are expected to be present, precautionary measures such as potting, hermetically sealing components, or providing appropriately rated enclosures should be employed.

4.2.4 *Ambient Ionizing Radiation.* The ambient level of external penetrating radiation, even if below that of personnel protection concern, may cause an unacceptable response in an effluent monitoring system. While the detector is usually the most sensitive element, certain circuits or components may also be affected. Such effects are minimized by shielding, anticoincidence, or other compensating techniques, or some combination of these. High-intensity short-duration radiation fields (for example, criticality spikes) may cause an adverse effect on system response and should be considered.

4.2.5 *Humidity.* Instruments may have to be located in areas with high relative humidities. This can result in measurement error, and in some cases hygroscopic detectors or components can be damaged.

4.2.6 *Other.* Such effects as power variations, high current contact closures, power transients, and, in some cases, magnetic and radio frequency fields may damage equipment or cause spurious readings and alarms and should be considered in system installation, particularly in system cable routing. Also, the effects of atmospheric dust should be considered.

4.3 Standards, Regulation, and Public Responsibility. The careful selection of an effluent monitoring system will assist in the quantitative evaluation of compliance with applicable federal, state, and local regulations. If applicable regulations require the monitoring of radioactivity in effluent streams below the measurement capability of monitoring instruments within the scope of this standard, sensitivity can be enhanced by recourse to sample extraction and laboratory analysis.

4.4 Calibration. Ease and relevance of calibration, as well as instrument stability, are important factors affecting the selection of the system. The primary initial calibration should encompass the entire system, including the detector and sample collector and should be performable after the system has been installed. Secondary calibration and periodic maintenance of the partial system should be possible without using primary calibration techniques.

5. Effluent Monitoring Systems

5.1 General Considerations. Effluent monitors are conveniently classified according to the location of the detector with respect to the effluent stream. When the detector is immersed in or otherwise directly monitors the effluent stream, the system is categorized as *in line.* A system in which a portion of the primary effluent stream is diverted through a bypass loop, or similar mechanism for monitoring by the detector, is categorized as *off line.*

In-line systems are those where the detector looks directly at the effluent stream and requires no external fluid mover or sampling line. Plate out, detector contamination, and nonuniform mixing can create errors and therefore should be considered. In-line systems are primarily useful for determining the gross activity of the effluent stream. The lack of concentrating mechanisms, geometry, and background shielding considerations of in-line systems are such that high sensitivity is difficult to achieve. Direct calibration of in-line systems is also difficult.

Off-line systems, although more complex and subject to problems associated with representativeness of sample, provide a higher degree of sensitivity by concentrating or separating various components (for example, particulates, radioiodines). Additionally, off-line detectors may be used with less shielding and optimum geometry. Those off-line systems that collect and concentrate the sample may also provide a sample convenient for subsequent laboratory analysis. Calibration of the detector-readout system is relatively simple, but the sample collector and mover also need calibration.

Fixed and moving filter monitors are two types of off-line systems. The fixed filter system, which is mechanically simple, removes particulate activity (or occasionally other constituents) from the effluent by drawing an aliquot through a filter medium. The detector(s) continuously and directly monitor(s) the buildup of radioactivity on this filter as collection takes place. In the moving filter system, a tape or strip filter is commonly used. The filter may be advanced either stepwise or in continuous mode through the sample chamber. The detector may directly monitor the filter as collection takes place, or it may be located away from the collection point to monitor the filter after it has advanced to the detector location. Moving filter systems require a rather complex filter advancement mechanism, but they are less subject to filter loading. They may be subject to leakage, mechanical failure, and tape breakage. Electrostatic or thermal precipitation may also be used to deposit concentrates of airborne radioactive particulates in defined geometries before a detector.

Another important consideration is the buildup of background level due to plate out or settling of radioactive material in sight of the detector. This commonly occurs in flow-through systems, in line and off line, as most of the liquid monitoring systems are. Since the background level determines the maximum sensitivity of the system, the ease of cleaning and other maintenance of the detector assembly are important.

5.2 Selection Criteria. When choosing an effluent monitoring system, consideration should be given to the sensitivity, energy response, response time, dynamic range, and accuracy of the system for the radionuclide(s) of interest. The environmental factors discussed in Section 4.2 should also be considered.

The presentation and storage of data are an important aspect of the total system capability and deserve special consideration. Other factors such as remote or local readout, digital versus analog readout, compatibility with other systems, and alarm modes should also be considered.

5.3 Specification of Performance. The specification of performance and the method and units of characterizing performance include a specific method, procedure, form, or other means of determining or observing a given operational characteristic. The intent of this section is to provide guidance to the manufacturer to achieve uniformity in the reporting of instrument capabilities.

5.3.1 *Detection Capabilities.*

5.3.1.1 *Type of Radiation.* The type(s) of ionizing radiation(s), (for example, alpha, beta, and photon) that the instrument is capable of measuring shall be clearly stated along with an appropriate energy range for each specific type (see Section 5.3.1.2).

5.3.1.2 *Energy.* The range of energies for each specific type of ionizing radiation measured shall be given in units of megaelectronvolts. The change in response as a function of energy shall be stated or presented graphically. For alpha and beta detectors, the minimum particle energy detected as well as the thickness of the detector wall window in milligrams per square centimeter shall be given.

5.3.1.3 *Range.* The dynamic detection range of the system shall be expressed as a concentration in units of microcuries per milliliter or curies per cubic meter, as appropriate, referenced to a specific nuclide. Saturation or other irregular responses at concentrations above the stated upper detection limits shall be indicated.

5.3.1.4 *Sensitivity.* Maximum sensitivity or minimum detectable level shall be stated in terms of the smallest concentration of a specific nuclide measurable at a given confidence level in a stated time period at a given flow rate, where applicable, under specified background radiation conditions.

5.3.1.5 *Accuracy.* The accuracy of the reading over its stated range shall be given as percent of reading for logarithmic or digital readouts or percent of full scale of the scale being read for linear readouts.

5.3.1.6 *Precision.* The precision of the instrument shall be given as percent deviation up and down scale from the mean reading at the 95 percent (2 sigma) confidence level for a mid-scale or mid-decade reading.

5.3.1.7 *Response Time.* The electronic time constant of the system shall be stated. For off-line systems, the time required for the sample to travel from the instrument inlet port to the detector at a specified flow rate shall be stated. For moving filter particulate monitors, a curve of activity buildup versus time should be stated. The buildup coordinate should be expressed in percentages from 0 to 100 percent for the system response to nuclides in equilibrium deposited on the filter paper.

5.3.1.8 *Radiation Alarm.* The abnormal and high level radiation alarm capabilities shall be described, including time to alarm, latching capability, the form output mode (that is relays or driver transistor), the range of alarm set points, and reproducability of alarm points. When two alarms are provided, the interaction between alarms, if any, shall be specified.

5.3.2 *Physical and Electrical Operating Limits.*

5.3.2.1 *Temperature.* The operating temperature range in degrees Celsius shall be stated along with temperature dependence. Temperature dependence shall be expressed as a percent change in readout per degree Celsius over a range of 0 to 60°C or as a total percent change over that range. If the range exceeds the design operating temperature range it shall be stated. The time to reach thermal equilibrium shall also be stated.

5.3.2.2 *Pressure.* If the system operation is affected by ambient pressure variations, the magnitude of that effect shall be stated over a range of 500 to 800 torrs. If the system or certain components of the system are specifically designed for pressurized operation (for example, a detector for insertion in a pressurized pipe), the design pressure and safety factor shall be stated.

5.3.2.3 *Relative Humidity.* Effects from humidity shall be expressed in percent relative humidity. The capability of a system to resist moisture and corrosive atmospheres should be stated. A statement of storage and handling requirements should be included.

5.3.2.4 *Electrical Effects.*

5.3.2.4.1 *Switching Transients.* The duration of switching transients shall be stated in terms of the length of time in seconds required for the instrument to indicate the final value within its stated accuracy on the most sensitive scale or decade.

5.3.2.4.2 *Power Variations.* The effect on instrument accuracy of line voltage and frequency variations shall be stated over a range of ± 15 percent from the design values.

5.3.2.4.3 *Power Transients.* The length of time in seconds for the instrument to indicate the final value within its stated accuracy on the most sensitive scale after a total power interruption of 1 second shall be stated.

Additionally, any effects on alarms shall be noted.

5.3.2.4.4 *Magnetic Fields.* The effect on instrument accuracy due to an external magnetic field shall be stated.

5.3.2.5 *Mechanical Effects.* The minimum vibration level (of loading) or acceleration that perturbs readings or damages components shall be specified over a frequency range of 1 to 30 Hz.

5.3.2.6 *Power Requirements.* The operating power requirements shall be stated. For ac-powered instruments this statement shall include the operating voltage range, frequency, waveform, number of phases, and power and current requirements.

5.3.2.7 *Background Radiation.* The response of the instrument to ambient beta, photon, and neutron fluxes should be stated. Instrument response should be referenced to or normalized to 0.8 MeV (^{90}Sr ^{90}Y average energy) for beta fluxes, 1.2 MeV (^{60}Co) for photon fluxes, and 5 MeV (AmBe) for neutron fluxes. The response to natural airborne radioactivity (that is, radon in equilibrium with daughters) up to a concentration of 5×10^{-9} μCi/ml should be stated. Alternatively, the response of the instrument to background may be referenced to isotopes comparable to its intended use. The minimum level of detectability for the instrument should be specified with consideration given to interferences from ambient sources; for example, the detection limit might be stated as X μCi/ml in an ambient photon field of P mR/h with a photon energy of 1.2MeV, and Y μCi/ml in an ambient photon field of O mR/h with a photon energy of 1.2 MeV.

5.3.3 *System Reliability.*

5.3.3.1 *Attainment of Reliability.* The methods of attaining system reliability should be stated. These may include the choice of components, power duration factors, and parallel or redundant circuitry.

5.3.3.2 *Expendable Components.* The rated lifetime of expendable components (for example, brushes, batteries, and recorder paper) shall be stated, along with a statement of the method for determining the need for replacement, where applicable.

5.3.3.3 *Radiation Damage.* Radiation degradation or failure of the system due to external radiation fields should be stated.

5.3.3.4 *Extracameral Effects.* Extra-

cameral effects and the intensity at which these occur should be stated.

5.3.3.5 *Malfunction Alarm.* The method(s) of communicating mechanical or electrical malfunctions, together with an evaluation of possible failure modes that would not be annunciated, should be described.

5.3.3.6 *System Warranty.* The warranty term shall be clearly defined.

5.3.3.7 *Quality Assurance.* The quality control procedures used in the manufacture of effluent monitoring systems shall be available.

5.3.4 *Calibration.* The exact conditions of calibration shall be specified, including the calibration point(s), range, and traceability to the NBS. The calibration shall encompass the entire system.

5.4 Standards of Performance. The intent of this section is to provide minimum standards of performance for user instrumentation. This does not preclude different or more stringent standards to meet specific requirements.

5.4.1 *Detection Capability.* The quantity or concentration of a radionuclide or mixture of radionuclides in an effluent stream is most commonly determined by quantification of the corpuscular radiation associated with radioactive decay. In some cases, specific photon radiations, bremsstrahlung, or daughter product radiations may be used for this purpose. Occasionally, nonradiological procedures may be used. While this standard is primarily concerned with quantification by radiological methods, the use of other methods such as wet chemistry, chromatography, and mass spectrometry is not precluded, providing these methods meet the standards of performance cited below.

5.4.1.1 *Detection in Gaseous Streams.* Instruments designed to continuously monitor radioactivity in gaseous effluent streams shall have a minimum level of detectability for the nuclide(s) in question as given in Table 1, under the column labeled "Gaseous." If radionuclides are monitored collectively with no differentiation of individual species, the smallest applicable level of detectability should be chosen from the column labeled "Gaseous," Table 1, based on nuclides known to be present. Where greater sensitivity is required, sample collection together with laboratory analysis shall be considered.

Nuclide (in order of increasing A)	Effluent Stream	
	Gaseous (μCi/ml)	Liquid (μCi/ml)
^{3}H	5×10^{-6}	—
^{14}C	5×10^{-7}	—
^{24}Na	5×10^{-11}	5×10^{-7}
^{32}P	2×10^{-11}	—
^{41}A	2×10^{-7}	4×10^{-7}
^{51}Cr	8×10^{-10}	3×10^{-6}
^{54}Mn	4×10^{-7}	1×10^{-6}
^{56}Mn	2×10^{-10}	1×10^{-6}
^{58}Co	2×10^{-11}	9×10^{-7}
^{59}Fe	2×10^{-11}	5×10^{-7}
^{60}Co	8×10^{-11}	3×10^{-7}
^{64}Cu	4×10^{-10}	2×10^{-6}
^{65}Zn	2×10^{-11}	1×10^{-6}
^{84}Br	2×10^{-11}	3×10^{-7}
^{85}Kr	3×10^{-7}	8×10^{-5}
^{85m}Kr	2×10^{-7}	3×10^{-7}
^{87}Kr	2×10^{-7}	2×10^{-7}
^{88}Kr	2×10^{-7}	2×10^{-7}
^{88}Rb	2×10^{-11}	1×10^{-6}
^{89}Rb	2×10^{-11}	2×10^{-7}
^{89}Sr	3×10^{-12}	—
^{90}Sr	4×10^{-12}	—
^{90}Y	3×10^{-11}	2×10^{-3}
^{91}Sr	9×10^{-11}	3×10^{-7}
^{91}Y	2×10^{-11}	1×10^{-4}
^{92}Sr	4×10^{-11}	4×10^{-7}
^{92}Y	2×10^{-11}	2×10^{-6}
^{95}Zr	6×10^{-11}	4×10^{-7}
^{95}Nb	8×10^{-10}	4×10^{-7}
^{99}Mo	7×10^{-11}	1×10^{-6}
^{103}Ru	8×10^{-10}	8×10^{-7}
^{105}Ru	2×10^{-10}	1×10^{-6}
^{129}Te	1×10^{-10}	8×10^{-6}
^{129m}Te	9×10^{-11}	7×10^{-6}
^{129}I	8×10^{-10}	2×10^{-4}
^{131}I	4×10^{-12}	3×10^{-7}
^{132}Te	2×10^{-11}	3×10^{-7}
^{132}I	3×10^{-12}	1×10^{-7}
^{133}I	3×10^{-12}	3×10^{-7}
^{133}Xe	5×10^{-7}	6×10^{-6}
^{133m}Xe	4×10^{-7}	2×10^{-6}

Nuclide (in order of increasing A)	Effluent Stream	
	Gaseous (μCi/ml)	Liquid (μCi/ml)
^{134}Te	4×10^{-7}	4×10^{-7}
^{134}I	2×10^{-11}	1×10^{-7}
^{134}Cs	4×10^{-12}	2×10^{-7}
^{135}I	3×10^{-11}	3×10^{-7}
^{135}Xe	5×10^{-6}	3×10^{-7}
^{135m}Xe	3×10^{-7}	4×10^{-7}
^{136}Cs	6×10^{-12}	1×10^{-7}
^{137}Cs	5×10^{-12}	4×10^{-7}
^{138}Xe	2×10^{-7}	1×10^{-7}
^{138}Cs	2×10^{-11}	2×10^{-7}
^{140}Ba	3×10^{-12}	6×10^{-7}
^{140}La	3×10^{-11}	2×10^{-7}
^{144}Ce	8×10^{-12}	3×10^{-6}
^{144}Pm	2×10^{-12}	2×10^{-5}
^{147}Pm	2×10^{-11}	4×10^{-7}
^{210}Po	2×10^{-12}	—
^{220}Rn	1×10^{-8}	—
^{222}Rn	1×10^{-8}	—
^{226}Ra	2×10^{-12}	—
^{232}Th	2×10^{-12}	—
^{233}U	2×10^{-12}	—
^{234}U	2×10^{-12}	—
^{235}U	2×10^{-12}	—
^{238}U	2×10^{-12}	—
^{238}Pu	2×10^{-12}	—
^{239}Pu	2×10^{-12}	—

NOTES: (1) The numerical values in Table 1 refer to the sensitivities for detection of an effluent stream consisting solely of the nuclide as stated and under conditions specified in accordance with Section 5.3.2.7. The numerical values apply at the detector locations. In the event that mixtures of nuclides are present, the sensitivity will vary greatly from nuclide to nuclide. The detection limits for the mixture would be dependent on the relative sensitivities and the detection limits for a given pure nuclide may not be obtainable.

(2) This table assumes continuous monitoring with the levels cited measurable at the 95 percent confidence level within a 4 hour period. It represents what is reasonably obtainable consistent with state-of-the-art measurement.

(3) These values represent current minimum standards. Improved sensitivities are always encouraged and should be used when improved state-of-the-art and commercial availability are realized.

Table 1
Minimum Levels of Detectability for Radionuclides in Effluent Streams

5.4.1.2 *Detection in Liquid Streams.* Instruments designed to continuously monitor radioactivity in liquid effluent streams shall have a minimum level of detectability for the nuclide(s) in question as given in Table 1, under the column labeled "Liquids." If radionuclides are monitored collectively with no differentiation of individual species, the smallest applicable level of detectability shall be chosen from the column labeled "Liquid," Table 1, based on the nuclides known to be present.

5.4.2 *Range.* The dynamic range of the instrument shall be at least 10^4 minimum detectable levels and should be stated in units of microcuries per milliliter relative to a given isotope. The range of the instrument should overlap with the range of emergency instrumentation.

5.4.3 *Sensitivity.* The sensitivity of an effluent monitoring system should also be stated in terms of a signal count rate associated with a specific nuclide detectable at the 95 percent confidence level in the presence of a specified background count rate for the convenience of the user. The following formula can be used to calculate the signal count rate at a 95 percent confidence level for a given count rate using the system efficiency and a typical background:

$$n_s = 2\sqrt{n_b/2\,RC}$$

where

n_s = signal count rate in counts per minute

n_b = background count rate in counts per minute

$2RC$ = two times the instrument time constant in the range or decade associated with the background count rate in units of minutes

and the number 2 is the constant associated with the 95 percent confidence level.

5.4.4 *Accuracy.* The instrument error shall not exceed ± 20 percent of reading over the upper 80 percent of its dynamic range where the error is defined as:

$$\text{percent error} = \frac{R_t - R_r}{R_t} \times 100 \text{ percent}$$

where

R_r = indicated quantity

R_t = true quantity

For this determination, the indicated quantity R_r shall be taken from the most accurate readout format, for example, recorder output or analog-to-digital converter. It is recognized that the usual instrument panel meters may significantly contribute to the error (particularly in logarithmic devices), but these are rarely used as the primary quantitative readout. Typical time constants for given count rates are illustrated in Fig 1.

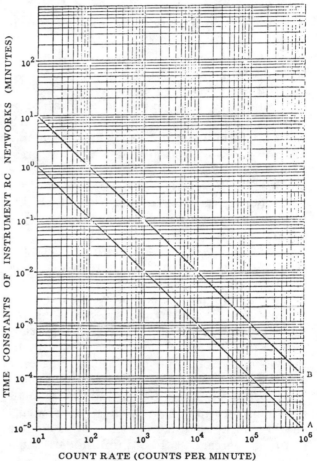

(Curve *A* depicts a 1 min time constant at a count rate of 10 counts per minute; curve *B* depicts a 10 min time constant at 10 counts per minute.)

Fig 1
Instrument Time Constants at Specific Count Rates for a Five-Decade Logarithmic Count Rate Meter

5.4.5 *Precision.* The reproducibility of a system for any given measurement over its stated range should be within ± 10 percent at the 95 percent confidence level for a mid-scale or mid-decade reading.

5.4.6 Response Time. The response time of the electronic system shall not be less than that required to maintain background readings within the required accuracy.

5.4.7 Physical, Mechanical, and Electrical Requirements.

5.4.7.1 Temperature. Regardless of the temperature of the effluent stream, the instrument system should be capable of operating with less than 5 percent change in calibration or response over a temperature range of 0 to 60°C. Where greater temperature extremes are expected or greater accuracy is required, protection from the environment shall be provided.

5.4.7.2 Pressure. The instrument system should be capable of continuous operation under ambient pressures from 500 to 800 torrs as a minimum with no change in calibration or response. When the system is to be placed in a reactor containment vessel, it shall be capable of continuous operation under ambient pressures from 500 to 2500 torrs.

5.4.7.3 Humidity. The instrument shall be capable of continuous operation in relative humidities of 10 to 95 percent within the accuracy required.

5.4.7.4 Other Environmental Effects. Instrument systems and their component parts shall be resistant to corrosive atmospheres when they are used in such atmospheres.

5.4.7.5 Power Requirements. Voltage and frequency variations of ± 15 percent within the design values shall result in reading variations of no greater than ± 5 percent at the minimum detectable level.

5.4.7.6 Electrical Effects. Radio frequency and microwave signals of $\leq 10 \ \mu W/cm^2$ shall result in reading variations of no greater than ± 5 percent at the minimum detectable level.

5.4.7.7 Mechanical Effects. Accelerations of $\leq 1 \ g$ in each of three mutually orthogonal axes over the frequency range 1 to 33 Hz shall result in reading variations of no greater than ± 5 percent at the minimum detectable level.

5.4.8 Radiation Alarm. The instrument system shall be equipped with an alarm capable of being externally set to annunciate at any point over the stated range. The alarm should be both audible and visible and should be capable of adjustment and reset without removing the instrument from service.

5.4.9 Failure Alarm. The instrument shall be equipped with an alarm device that provides appropriate notification at any time the system is incapable of monitoring radiation. A latching contact shall be provided to indicate loss of signal detector operating voltage or circuit power.

5.4.10 Calibration. A thorough primary calibration of the entire system shall be performed at least once using a radionuclide (liquid, solid, or gaseous) of known concentration. The radionuclide shall permit calibrating the range of energy and rate capabilities intended for the system. Traceability to the NBS shall be maintained. The calibration shall be related to a secondary source or method which will be used for periodic in-plant recalibration. The source-to-detector geometry shall be maintained identical to that established during the primary calibration. The surface dimensions of the secondary source shall be the same as the detector window.

Systems where the concentration of radionuclides changes significantly between the sample point and the detector, as in off-line particulate monitoring, shall be tested by using radioactive material or a known tracer such as dioctyl phthalate (DOP) to determine the loss in the sample lines.

Flow-rate-measuring devices associated with the system shall be calibrated to determine actual flow rate at the conditions of temperature and pressure under which the system will be operated. These flow-rate devices shall be periodically recalibrated.

A check source that is actuated remotely should be installed for integrity checks of the detector and the associated electrical system. A calibrated electrical signal should be provided to verify the circuit alignment.

6. Bibliography

[1] ALEXANDER, J. M. A continuous monitor for prompt detection of airborne plutonium. *Health Physics*, vol 12, 1966, p 553.

[2] BRAMSON, P. E. A system for continuous monitoring of potential dose rate to the gastrointestinal tract from drinking water. *Health Physics*, vol 18, 1970, p 523.

[3] BUTLER, M. J., and VEALE, F. L. The use of Geiger-Mueller tubes for the measurement of X- and gamma-radiation up to 1 MeV. *Radiological Monitoring of the Environment,* B. C. Godhold and J. K. Jones, Eds. New York: Pergamon Press, 1965, pp 187-199.

[4] COWPER, G., and OSBORNE, R. V. Measurement of tritium in air in the presence of gamma radiation. *Radiation Protection,* pt I, W. S. Snyder, Ed. New York: Pergamon Press, 1968, pp 285-293.

[5] CLARKE, N. T., and PEARCE, N. Instruments for plutonium monitoring. *Nuclear Electronics,* vol III, 1962, pp 403-427.

[6] EGOROV, I. M., *et al. Handbook of Recording Instruments for Ionizing Radiation,* Israel Program for Scientific Translations, Jerusalem, Israel, 1967. (Translated from the Russian. Originally published by Atomizdat, Moscow, USSR, 1965.)

[7] FULLER, A. B. Design considerations for exhaust systems involving radioactive particulates. *Proceedings of the 8th AEC Air Cleaning Conference,* Oak Ridge, TN, 1963, pp 73-82.

[8] GUPTON, E. D. *Alpha Air Monitor for* $^{239}Pu,$ Oak Ridge National Laboratories, Oak Ridge, TN, Tech Memo ORNL-TM-2011, 1967.

[9] INMAN, J. E., and KNOWLES, D. J. *Description of Facility Radiation and Contamination Alarm Systems Installed in the High Radiation Level Examination Laboratory Building 3525,* Oak Ridge National Laboratories, Oak Ridge, TN, Tech Memo ORNL-TM-1251, 1965.

[10] ISHIHARA, T. Environmental radiological monitoring system at nuclear installations. *Health Physics,* vol 13, 1967, p 549.

[11] JOSHI, C. W., INGLE, N. A., and JATHAR, N. B. Design concepts of a centralized automatic air monitoring system for a large radioisotope complex. *Proceedings of the 1968 Radiation Protection Monitoring Symposium.* Vienna: IAEA, 1969, pp 471-476.

[12] KNOWLES, D. J., DANFORTH, H. P., and CHASE, L. H. *Description of the Cell Ventilation Monitor for the Fission Products Development Laboratory Building 3517,* Oak Ridge National Laboratories, Oak Ridge, TN, Tech Memo ORNL-TM-1384, 1965.

[13] MANNESCHMIDT, J. F., *Recent Advances in Monitoring of Radioactive Gaseous Wastes at ORNL,* Oak Ridge National Laboratories, Oak Ridge, TN, Tech Memo ORNL-TM-1569, 1969.

[14] RAGUPATHY, S., and VOHRA, K. G. A multichannel analysis system for low levels of fission products in reactor coolants. *Proceedings of the 1965 Nuclear Electronics Conference.* IAEA Proceeding Series, 1966, pp 333-339.

[15] TATTERSALL, D. Gaseous effluent monitoring at CEGB nuclear power stations. *Proceedings of the 1963 Radiological Monitoring of the Environment Symposoim.* New York: Pergamon Press, 1965, pp 93-102. Compare with BODDY, K. Environmental surveys around research reactors. *Proceedings of the Radiological Monitoring of the Environment Symposium.* New York: Pergamon Press, pp 249-258.

[16] THUKRAL, S. P., SINGH, A. N., and NAIR, N. B. General review of the air monitoring instruments used at Bhabha Atomic Research Center. *Proceedings of the 1968 Radiation Protection Monitoring Symposium.* Vienna: IAEA, 1969, pp 451-458.

[17] THUKRAL, S. P., and VOHRA, K. G. Continuous monitor for radioactive fallout in air. *Nuclear Instruments and Methods,* vol 44, 1966, p 73.

[18] VANNE, J. P., and DE RUS, E. M. M. Experience with the use of a continuous alpha air monitor, using large area proportional counters and an improved psuedo coincidence circuitry. *Radiation Protection,* pt 2, *Proceedings of the First IRPA Symposium.* New York: Pergamon Press, 1968, pp 997-1001.

[19] ANSI N13.1-1969, Guide to Sampling Airborne Radioactive Materials in Nuclear Facilities.

[20] WOODWARD, W. J. *An Improved Air Monitor in a Standard Nuclear Instrument Module,* E. I. du Pont de Nemours and Company, DP-1260, 1971.

[21] BISBY, H. Radiometric techniques

and instrumentation for in-line process monitoring. *Progress in Nuclear Engineering,* ser IV, *Technology, Engineering, and Safety,* vol 5, 1963, pp 251-282.

[22] RIEL, G. K., and DUFFEY, D. Underwater gamma ray spectrometers for monitoring environmental water at nuclear power stations. *IEEE Transactions Nuclear Science,* vol NS-14, 1968, p 275.

[23] Environmental Instrumentation Group. *Instrumentation for Environmental Monitoring — Radiation,* vol 3, May 1, 1972, Lawrence Berkeley Laboratory, University of California, Berkeley, CA, LBL 1.

[24] COSTRELL, L. *Standard Nuclear Instrument Modules,* TID-20893; revision 2, Jan 1968; also revision 3, 1969.

[25] *CAMAC: A Modular Instrumentation System for Data Handling,* US Atomic Energy Commission, TID-25875, July 1972.

[26] PLINER, G. A. *Review of Effluent Monitors at U.S. Nuclear Power Stations,* Environmental Studies Section, Nuclear Facilities Branch, Division of Environmental Radiation, Bureau of Radiological Health, US Public Health Service, Department of Health, Education, and Welfare, May 1969.

[27] KNOWLES, D. J. Liquid and gaseous waste-effluent sampling and monitoring. *Nuclear Safety,* vol 7, no. 1, Fall 1965.

[28] HOUSER, B. L. *Indexed Bibliography on Environmental Monitoring for Radioactivity,* Oak Ridge National Laboratories, Oak Ridge, TN, ORNL-NSIC-101.

[29] *Air Sampling Instruments for Evaluation of Atmospheric Contaminants,* 4th ed. American Conference of Government Industrial Hygienists (ACGIH)[1]

[30] BLEHER, G. L., and HOLLOWAY, D. J. Selection of monitors for airborne particulate radioactivity. Presented at the 19th Annual Meeting of the American Nuclear Society, Chicago, Ill, June 10-14, 1973.

[31] SHIPP, R. L., Jr. *Development of Two Systems for Monitoring Alpha-Emitting Particulates in Radioactive Gas Disposal Stacks,* Oak Ridge National Laboratory, operated by the Union Carbide Corporation, Nuclear Division for the US Atomic Energy Commission, Tech Memo ORNL-TM-3165, DEC 29, 1970.

[32] GALLMAN, R. A. *Evaluation of Several Methods of Detecting Uranium Particulates in a Disposal Stack at the Oak Ridge Y-12 Plant,* Oak Ridge Y-12 Plant, Oak Ridge, TN, prepared for the US Atomic Energy Commission under US Government Contract W-7405 eng. 26, Y-1866.

[33] FILSS, P. (β-γ) chamber for nuclide determination in off-gas samples, *Nuclear Instruments and Methods,* vol 108, 1973, pp 471-475.

[34] *Health Physics Operational Monitoring,* C. A. Willis and J. S. Handloser, Eds, vols 1, 2, and 3. New York: Gordon and Breach, 1972.

[35] *Proceedings of the 1972 Radiation Safety and Protection in Industrial Applications Symposium,* US Department of Health, Education, and Welfare, Food and Drug Administration, Bureau of Radiological Health, Rockville, MD, DHEW Publication (FDA) 73-8012, BRH/DEP 73-3, Oct 1972.

[36] SCHINDLER, R. E., HAMMER, R. R., BLACK, D. E., and LAKEY, L. T. Development of a continuous sampler-monitor for the loft containment atmosphere. *Proceedings of the 11th AEC Air Cleaning Conference,* 1970, pp 653-667.

[37] SCHULTZ, R. J. Application of the continuous sampler monitor (CSM) as a hazard evaluation instrument. *Proceedings of the 11th AEC Air Cleaning Conference.* 1970, pp 668-679.

[38] *Current Practices in the Release and Monitoring of* [131]*I and NRTS, Hanford, Savannah River, and ORNL,* Oak Ridge National Laboratories, Oak Ridge, TN, ORNL-NSIC-3, 1964.

_____
[1]PO Box 1937, Cincinnati, OH, 45200.

ANSI
N42.4-1971

American National Standard for High Voltage Connectors for Nuclear Instruments

Secretariat
Institute of Electrical and Electronics Engineers, Inc.

Approved April 12, 1971
American National Standards Institute, Inc.

N42.4

© Copyright 1971 by

The Institute of Electrical and Electronics Engineers, Inc.

Foreword

(This Foreword is not a part of American National Standard for High Voltage Connectors for Nuclear Instruments, N42.4-1971.)

This standard was proposed by American National Standards Committee N42 on Nuclear Instrumentation and adopted by ANSI in order to provide for interchangeability of safe high voltage connectors in nuclear instrument applications. The connectors are of the "safe" type in that the pin and socket contacts are well and securely recessed in the connector housing so that hand or body contact of the unmated connector with rated voltage applied will not result in electrical shock. It will be noted that the connectors covered by this standard are essentially identical to the NIM (Nuclear Instrumental Module) standard high voltage connectors specified in AEC Report TID-20893.

Suggestions for improvement gained in the use of this standard will be welcomed. They should be sent to the Secretariat for N42, Institute of Electrical and Electronics Engineers, 345 East 47th Street, New York, N. Y. 10017.

The American National Standards Committee N42 on Radiation Instrumentation had the following personnel the time it approved this standard:

Louis Costrell, *Chairman*   David C. Cook, *Secretary*

Organization Represented	*Name of Representative*
American Chemical Society	Louis P. Remsberg, Jr
American Conference of Governmental Industrial Hygienists	Jesse Lieberman
American Industrial Hygiene Association	W. H. Ray
American Nuclear Society	W. C. Lipinski
	Thomas Mulcahey (*Alt*)
American Society of Mechanical Engineers	R. C. Austin
American Society of Safety Engineers	*Representation Vacant*
American Society for Testing and Materials	John L. Kuranz
	Jack Bystrom (*Alt*)
Atomic Industrial Forum	*Representation Vacant*
Electric Light and Power Group	G. S. Keeley
	G. A. Olson (*Alt*)
Health Physics Society	J. B. Horner Kuper
	Robert L. Butenhoff (*Alt*)
Institute of Electrical and Electronics Engineers	Louis Costrell
	Lester Kornblith, Jr
	J. J. Loving
	J. Forster (*Alt*)
Instrument Society of America	M. T. Slind
	J. E. Kaveckis (*Alt*)
Manufacturing Chemists Association	Mont G. Mason
National Electrical Manufacturers Association	Theodore Hamburger
Oak Ridge National Laboratory	Frank W. Manning
Scientific Apparatus Makers Association	Robert Breen
Underwriters' Laboratories	Leonard Horn
U. S. Atomic Energy Commission, Division of Biology and Medicine	Hodge R. Wasson
U. S. Atomic Energy Commission, Division of Reactor Development and Technology	Paul L. Havenstein
	W. E. Womac (*Alt*)
U. S. Department of the Army, Materiel Command	Abraham E. Cohen
U. S. Department of the Army, Office of Civil Defense Mobilization	Carl R. Siebentritt, Jr
	Ronald H. Sandwina (*Alt*)
U. S. Department of Commerce, National Bureau of Standards	Louis Costrell
U. S. Department of Health, Education, and Welfare, Public Health Service	Henry J. L. Rechen, Jr
	Roger Schneider (*Alt*)
U. S. Naval Research Laboratory	D. C. Cook
Members-at-Large	O. W. Bilharz
	S. H. Hanauer
	John M. Gallagher, Jr
	Voss A. Moore
	R. F. Shea

American National Standard
for High Voltage Connectors
for Nuclear Instruments

1. Scope

This standard is applicable to coaxial high voltage connectors on nuclear instruments for dc applications up to 5000 volts and ac applications up to 3500 volts rms at 60 Hz. The connectors may also be used at higher frequencies provided the operating voltage is appropriately reduced.

2. Standard Connector

2.1 General. The connectors shall be in accordance with Figs. 1 and 2. These connectors are of the "safe" type in that the pin and socket contacts are well and securely recessed in the connector housings so that hand or body contact of the unmated connector with rated voltage applied will not result in electrical shock.

2.2 Mechanical Integrity. The mechanical construction shall be such that, even where the center wire loosens or develops slack, the 0.238/0.262 inch (6.05/6.65 mm) dimension of the plug and the 0.188/0.208 inch (4.78/5.28 mm) dimension of the receptacle shall be maintained when a force of up to 6 pounds (27 newtons) is applied to the contact in either direction.

2.3 Connector Mating. In the mated condition the longitudinal force of the spring of the coupling mechanism shall exceed the pressure exerted by the sealing gasket by an amount necessary to ensure butting of the outer contacts at the reference plane. In the mated condition the resistance between the center contact pin and the center contact socket shall not exceed 2.1 milliohms and the outer contact resistance shall not exceed 1.5 milliohms. Gage tests shall be as follows:

Center Contact Socket Gage Test

Oversize Test Pin. . . . . 0.057 inch (1.45 mm) min dia
Number of Insertions. 1
Insertion Depth — (For All Tests). . . . . 0.125 inch (3.18 mm) min at gaging dia
Insertion Force Test: Steel Test Pin Dia. . . . . 0.054 inch (1.37 mm) min
Test Pin Finish. . . . . 16 microinch (0.00041 mm)
Insertion Force. . . . . 2 pounds (9 newtons) max
Withdrawal Force Test: Steel Test Pin Dia. . . . . 0.052 inch (1.32 mm) max
Test Pin Finish. . . . . 16 microinch (0.00041 mm)
Withdrawal Force. . . . . 2 ounces (0.56 newtons) min

Outer Spring Contact Gage Test

Test Ring ID. 0.319 inch (8.10 mm) max, 16 microinch (0.00041 mm) finish
Insertion Force. . . . . 5 pounds (22 newtons) max when inserted min of 0.093 inch (2.36 mm)
All spring members shall contact 0.324 inch (8.23 mm) min ring within 0.050 inch (1.27 mm) of their tip ends.

2.4 Insulation. The insulation shall be polytetrafluorethylene.

2.5 Finish. Center contacts shall be gold plated with a minimum thickness of 0.0001 inch (0.0025 mm) *with no silver underplate.* All other parts shall be finished so as to provide a connector that meets the requirements of this standard.

2.6 Insulation Resistance. The insulation resistance shall be 10^{12} ohms minimum at 50 percent relative humidity at 25°C.

2.7 Dielectric Withstanding Voltage. Must withstand 10,000 volts dc and 5000 volts rms, 60 Hz, for 1 minute.

2.8 Altitude/Corona. Corona level shall be 350 volts minimum at 70,000 feet (21,300 m) altitude with cable length of 60 inches (1.52 m).

2.9 Retention Engagement and Disengagement Forces.

(1) Minimum cable retention force shall be as follows:

For Crimp Types { 50 pounds (224 newtons) for cables 0.175 - 0.199 inch (4.44 - 5.05 mm) od 60 pounds (269 newtons) for cables 0.200 - 0.249 inch (5.08 - 6.32 mm) od 75 pounds (336 newtons) for cables 0.250 inch (6.35 mm) od and larger

For Non-crimp Types - 40 pounds (179 newtons)

(2) Minimum bayonet sleeve retention force shall be as follows:
100 pounds (448 newtons)

(3) Maximum force to engage and disengage shall be as follows:

Longitudinal Force... 3 pounds (13 newtons)
Torque...60 inch ounces (0.42 newton meter)

2.10 Durability. A connector assembly shall be capable of being mated and unmated at least 500 times with no evidence of physical damage which could affect the mechanical or electrical performance of the connector.

2.11 Pressure/Vacuum Seal. The pressurized feed thru shall have a leakage rate of not more than 1×10^{-9} milliliters per second of helium tracer gas when one side is at 1 atmosphere pressure, 15 pounds per square inch ($1.03 \times 10^5 \text{N/m}^2$), at 25° C. and the other side is at a "vacuum" of not greater than 0.03937 inches (1.000 mm) of mercury absolute (not greater than 1 torr).

2.12 Corrosion. Connectors shall be exposed to a 5 percent salt solution at 35°C for 48 hours. After exposure, the connectors shall be washed, shaken and lightly brushed and then permitted to dry for 24 hours at 40°C. Connectors shall then show no sign of corrosion or pitting and shall meet the specifications herein regarding maximum force to engage and disengage.

2.13 Moisture Resistance. Connectors shall meet the following moisture resistance test: Connectors shall be exposed to 95 percent relative humidity at 40°C for 96 hours. Connectors shall then be removed to an environment of 50 percent relative humidity at 25°C. Within 5 minutes after removal from the 95 percent relative humidity, 40°C environment, the insulation resistance shall be not less than 10^{12} ohms.

2.14 Temperature Range. The connector shall be capable of operating within the specifications herein over the temperature range of -65°C to $+200$°C.

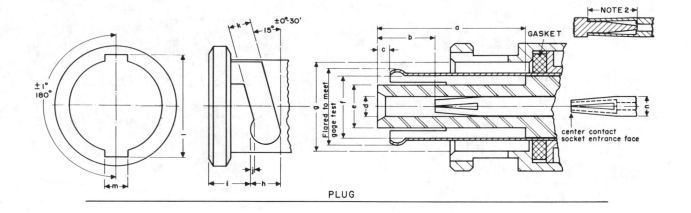

PLUG

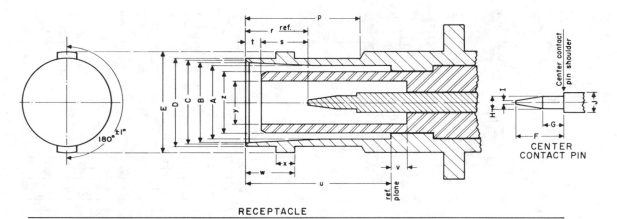

RECEPTACLE

REF	INCHES		MILLIMETERS ✱	
	MIN.	MAX.	MIN.	MAX.
a	0.628	0.632	15.96	16.05
b	0.238	0.262	6.05	6.65
c	0.046	0.064	1.17	1.63
d	0.082	—	2.08	—
e	0.180	0.186	4.57	4.72
f	0.264	—	6.71	—
g	0.385	0.390	9.78	9.91
h	0.124	—	3.15	—
i	0.180	0.184	4.57	4.67
j	0.018	0.022	0.46	0.56
k	0.091	0.097	2.31	2.46
l	0.463	0.473	11.76	12.01
m	0.091	0.097	2.31	2.46
n	0.081	0.083	2.06	2.11
p	0.427	—	10.85	—
r	0.249	0.286	6.32	7.26
s	0.188	0.208	4.78	5.28
t	0.061	0.078	1.55	1.98
u	0.626	0.630	15.90	16.00
v	0.064	0.086	1.63	2.18
w	0.204	0.208	5.18	5.28
x	0.075	0.081	1.90	2.06
y	0.190	0.196	4.83	4.98
z	—	0.260	—	6.60

REF.	INCHES		MILLIMETERS ✱	
	MIN.	MAX.	MIN.	MAX.
A	0.319	0.321	8.10	8.15
B	0.328	0.333	8.33	8.46
C	0.347	0.357	8.81	9.07
D	0.378	0.382	9.60	9.70
E	0.432	0.436	10.97	11.07
F	0.207	0.214	5.26	5.44
G	0.130	—	3.30	—
H	0.052	0.054	1.32	1.37
I	0.015	0.025	0.38	0.64
J	0.081	0.083	2.06	2.11

✱ SEE NOTE 1.

NOTES:
1. THE MILLIMETER DIMENSIONS ARE DERIVED FROM THE ORIGINAL INCH DIMENSIONS.
2. CENTER CONTACT SOCKET MUST ACCEPT MAXIMUM LENGTH CENTER CONTACT PIN OF 0.214 INCH (5.44 mm) AS SHOWN. INTERNAL DESIGN OPTIONAL.

Fig. 1 Mating Dimensions

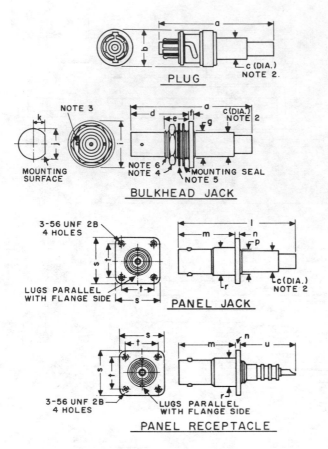

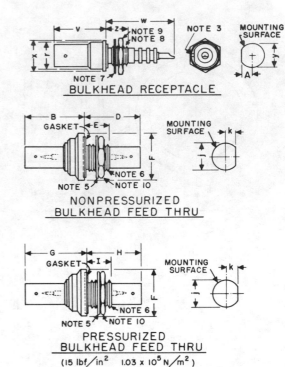

NOTES:

1. THE MILLIMETER DIMENSIONS ARE DERIVED FROM THE ORIGINAL INCH DIMENSIONS.
2. THE MAXIMUM DIAMETER SPECIFIED INCLUDES CLAMPING HARDWARE.
3. BAYONET LUGS TO BE IN LINE WITH MOUNTING FLAT AS INDICATED.
4. JAM NUT 0.625 ±0.005 INCH (15.88 ±0.13mm) ACROSS FLATS X 0.109 ±0.005 INCH (2.77 ±0.13mm) THICK.
5. LOCKWASHER, INTERNAL, 0.630 ±0.005 INCH (1.60 ±0.13mm) MAX. O.D. X 0.022 ±0.005 INCH (0.56 ±0.13mm) THICK.
6. THREAD – 1/2-28 UNEF-2A.
7. LOCKWASHER, INTERNAL, 0.510 ±0.005 INCH (12.95 ±0.13mm), MAX. O.D. X 0.022 ±0.005 INCH (0.56 ±0.13mm) THICK.
8. THREAD – 3/8-32 UNEF-2A.
9. JAM NUT 0.500 ±0.005 INCH (12.70 ±0.13mm) ACROSS FLATS X 0.125 ±0.005 INCH (3.18 ±0.13mm) THICK.
10. HEXAGONAL NUT 1/2-28 UNEF-2B, 0.109 ±0.005 INCH (2.77 ±0.13mm) THICK.

REF.	INCHES		MILLIMETERS✱	
	MIN.	MAX.	MIN.	MAX.
a	—	2.500	—	63.50
b	—	0.593	—	15.06
c	—	0.500	—	12.70
d	0.796	0.816	20.22	20.73
e	0.349	—	8.87	—
f	0.120	0.130	3.05	3.30
g	—	0.600	—	15.24
j	0.677	0.697	17.20	17.70
J	0.495	0.500	12.57	12.70
k	0.215	0.218	5.46	5.54
l	—	2.500	—	63.50
m	0.840	0.870	21.34	22.10
n	0.075	0.095	1.90	2.41
p	—	0.600	—	15.24
r	—	0.440	—	11.18
s	0.682	0.692	17.32	17.58
t	0.496	0.504	12.60	12.80
u	0.812	0.874	20.62	22.20
v	0.745	0.775	18.92	19.68
w	0.985	1.047	25.02	26.59
x	0.495	0.505	12.57	12.83
y	0.360	0.380	9.14	9.65
z	0.320	—	8.13	—
A	0.153	0.156	3.89	3.96
B	—	1.072	—	27.23
D	—	1.300	—	33.02
E	0.349	—	8.86	—
F	0.682	0.692	17.32	17.58
G	—	1.500	—	38.10
H	—	1.500	—	38.10
I	0.625	—	15.88	—

✱ SEE NOTE 1.

Fig. 2 Outline Dimensions

ANSI
N42.5-1965
(Reaffirmed 1971)

American National Standard Bases for GM Counter Tubes

Approved September 20, 1965 as N3.1-1965,
reaffirmed and redesignated June 16, 1971 as N42.5-1965
American National Standards Institute, Inc.

ANSI
N42.6-1965
(Reaffirmed 1971)

American National Standard Interrelationship of Quartz-Fiber Electrometer Type Dosimeters and Companion Dosimeter Chargers

Approved September 20, 1965 as N3.2-1965,
reaffirmed and redesignated June 16, 1971 as N42.6-1965
American National Standards Institute

Secretariat

Institute of Electrical and Electronics Engineers, Inc.

N42.5
N42.6

© Copyright 1971 by

The Institute of Electrical and Electronics Engineers, Inc.

Foreword

(This Foreword is not a part of ANSI N42.5-1965 or ANSI N42.6-1965.)

American National Standards N42.5-1965 and N42.6-1965 were originally developed within the N3 Sectional Committee of the American Standards Association, and approved by ASA as American Standards N3.1-1965 and N3.2-1965. In 1966 ASA underwent a reorganization that not only resulted in a change of name to the United States of America Standards Institute but also produced a restudy of the scopes and responsibilities of many Standards Committees. In the field of nuclear standards, it was decided to broaden the scope of Standards Committee N42 on Nuclear Instruments so that it would include the responsibilities previously assigned to Standards Committee N3 on Nuclear Instrumentation. Both committees, sponsored by the Institute of Electrical and Electronics Engineers, had a substantial overlapping of representation, and the work of the two groups was closely related. During 1967 membership in the two committees was merged. In 1970 Standards Committee N42, under the new Procedures of what had become the American National Standards Institute, was required to review its five-year-old standards to determine whether they should be revised, withdrawn, or reaffirmed. The Committee found that both documents remained valid and up-to-date, and its recommendation to reaffirm was forwarded to ANSI's Board of Standards Review. Following a period for public review and comment, the Board of Standards Review reaffirmed both standards and redesignated them as N42 documents.

Suggestions for improvement gained in the use of these standards will be welcomed. They should be sent to the Secretariat for N42, Institute of Electrical and Electronics Engineers, 345 East 47th Street, New York, N.Y. 10017.

At the time it reaffirmed these two standards, American National Standards Committee on Radiation Instrumentation had the following personnel:

Louis Costrell, *Chairman*   David C. Cook, *Secretary*

Organization Represented	Name of Representative
American Chemical Society	Louis P. Remsberg, Jr
American Conference of Governmental Industrial Hygienists	Jesse Lieberman
American Industrial Hygiene Association	W. H. Ray
American Nuclear Society	W. C. Lipinski
	Thomas Mulcahey (*Alt*)
American Society of Mechanical Engineers	R. C. Austin
American Society of Safety Engineers	*Representation Vacant*
American Society for Testing and Materials	John L. Kuranz
	Jack Bystrom (*Alt*)
Atomic Industrial Forum	*Representation Vacant*
Electric Light and Power Group	G. S. Keeley
	G. A. Olson (*Alt*)
Health Physics Society	J. B. Horner Kuper
	Robert L. Butenhoff (*Alt*)
Institute of Electrical and Electronics Engineers	Louis Costrell
	Lester Kornblith, Jr
	J. J. Loving
	J. Forster (*Alt*)
Instrument Society of America	M. T. Slind
	J. E. Kaveckis (*Alt*)
Manufacturing Chemists Association	Mont G. Mason
National Electrical Manufacturers Association	Theodore Hamburger
Oak Ridge National Laboratory	Frank W. Manning
Scientific Apparatus Makers Association	Robert Breen
Underwriters' Laboratories	Leonard Horn
U. S. Atomic Energy Commission, Division of Biology and Medicine	Hodge R. Wasson
U. S. Atomic Energy Commission, Division of Reactor Development and Technology	Paul L. Havenstein
	W. E. Womac (*Alt*)
U. S. Department of the Army, Materiel Command	Abraham E. Cohen

American National Standard Bases for GM Counter Tubes

Standard bases for GM counter tubes shall be as follows:

(1) 3-Pin "Pee Wee" base (JETEC-A3-1) with the following pin connections:

Pin No.	Connection
1	Cathode
2	Anode
3	No Connection

(2) 4-Pin "Medium" base (JETEC-A4-9) with the following pin connections:

Pin No.	Connection
1	No Connection
2	Anode
3	No Connection
4	Cathode

American National Standard
Interrelationship of Quartz-Fiber Electrometer Type Dosimeters and Companion Dosimeter Chargers

1. Purpose and Scope

The purpose of this standard is to specify interrelating mechanical and electrical properties so that quartz-fiber dosimeters may be used with any charger. Other characteristics peculiar to these devices but not affecting the interrelationship between chargers and dosimeters are purposely omitted.

2. Quartz-Fiber Dosimeters

2.1 Zero Set Voltages. The voltage required for zero set shall be in the range of 140 to 195 volts.

2.2 Distance to Charging Electrode. The distance from the end of the dosimeter to the charging electrode shall be 0.220 ± 0.030 inch in the charging position, and 0.150-inch minimum in the normal position (see Fig. 1).

2.3 Force to Close Switch in Dosimeter. The charging electrode shall make electrical contact with the internal electrometer under an applied force of from $2\frac{1}{2}$ to $5\frac{1}{2}$ pounds over a temperature range of $-40°F$ to $+150°F$.

2.4 Charging Recess Diameter. The diameter of the charging recess shall be 0.365 ± 0.005 inch to a depth of at least 0.190 inch.

The diameter of the charging recess shall be a minimum of 0.360 inch to a depth of at least 0.260 inch.

2.5 Outside Diameter. The dosimeter shall be a uniform right cylinder and have an outside diameter, exclusive of the clip, of 0.625-inch maximum and 0.495-inch minimum.

2.6 Length. The dosimeter shall not exceed 4.5 inches in length.

2.7 Center Electrode Diameter. The charging electrode shall have an outside diameter of from 0.040 to 0.080 inch. If a tube, the electrode shall have a wall thickness of at least 0.010 inch and the tube shall have a closed end.

2.8 Electrode Concentricity. The charging electrode shall be coaxial with the charging recess to within 0.005 inch.

3. Dosimeter Charger

3.1 Charger Voltage. The charging potential shall be continuously variable between 100 and 220 volts and shall be of positive polarity.

3.2 Pedestal Height. The overall active height of the pedestal from the top of the charging electrode to the top of the bottoming feature shall be 0.375 ± 0.075 inch.

3.3 Bottoming Feature. As a dosimeter is moved downward over the charging contact assembly, forces in excess of 8 pounds shall cause the dosimeter barrel to bottom with the nonmoving mounting of the charging contact assembly.

3.4 Charging Electrode Force. The charging electrode shall be capable of applying a force of between $6\frac{1}{2}$ and 8 pounds to the charging pin of the dosimeter. In no case shall it be possible to apply more than 8 pounds to the charging pin of the dosimeter.

3.5 Light Switch Force. The force required to activate the light switch (if included) shall be between 0.250 and 1.5 pounds.

3.6 Pedestal Diameter. The outside diameter of the charging pedestal shall be 0.345 ± 0.005 inch.

3.7 Charging Electrode Diameter. The diameter of the contact end of the charging electrode shall be 0.075 ± 0.025 inch.

3.8 Charging Electrode Height. The charging electrode or cap shall extend above the top of the charging pedestal no less than 0.005 inch and no more than 0.040 inch.

3.9 Electrode Concentricity. The charging electrode shall be coaxial with the pedestal within 0.005 inch.

3.10 Light Switch Travel. The light shall be turned on when a dosimeter is placed in position on the charging contact assembly and is moved downward a distance no greater than 0.125 inches.

3.11 Charging Recess. If the pedestal is recessed in a well, the minimum inside diameter of the charging socket well shall be 0.630 inch.

3.12 Contact. The contact resistance between any two surfaces designed to be at the same electrical potential shall never exceed one megohm.

4. Dosimeter and Charger

4.1 Contact. During the normal charging operation, the dosimeter and charger shall provide for electrical contacts between both the center electrodes and the shells. or cases of each. At no time during the operation shall the contact resistance between the dosimeter and charger exceed one megohm.

4.2 Light. The charging unit shall permit the position of the indicating element on the scale of the dosimeter to be clearly resolved. If the charger has self-contained illumination, this criterion shall apply over the life of the battery utilized.

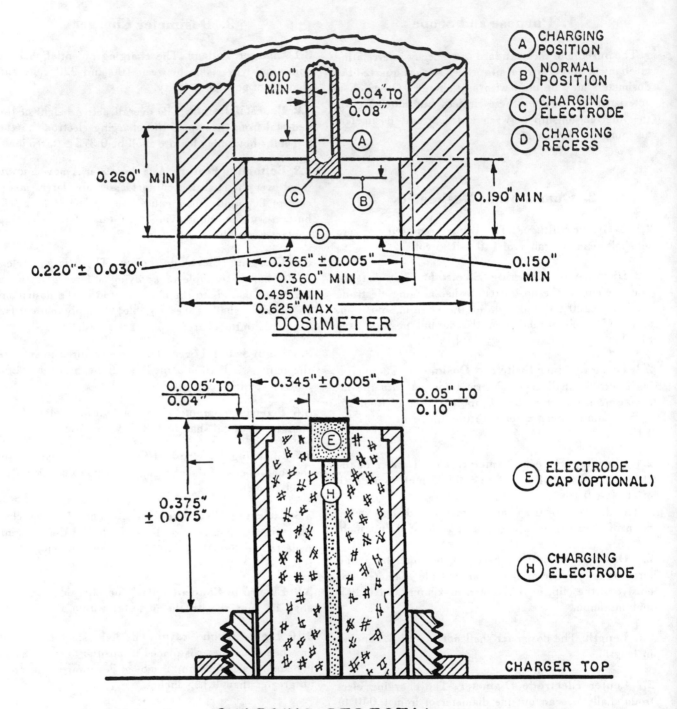

Fig. 1
Dimensions for Radiological
Dosimeter and Dosimeter Charger

ANSI
N42.13-1978

American National Standard
Calibration and Usage of
"Dose Calibrator" Ionization Chambers for
the Assay of Radionuclides

Secretariat for N42

Institute of Electrical and Electronics Engineers, Inc

Approved April 10, 1978

American National Standards Institute

N42.13

American National Standard

An American National Standard implies a consensus of those substantially concerned with its scope and provisions. An American National Standard is intended as a guide to aid the manufacturer, the consumer, and the general public. The existence of an American National Standard does not in any respect preclude anyone, whether he has approved the standard or not, from manufacturing, marketing, purchasing, or using products, processes, or procedures not conforming to the standard. American National Standards are subject to periodic review and users are cautioned to obtain the latest editions.

CAUTION NOTICE: This American National Standard may be revised or withdrawn at any time. The procedures of the American National Standards Institute require that action be taken to reaffirm, revise, or withdraw this standard no later than five years from the date of publication. Purchasers of American National Standards may receive current information on all standards by calling or writing the American National Standards Institute.

Foreword

(This Foreword is not a part of American National Standard Calibration and Usage of "Dose Calibrator" Ionization Chambers for the Assay of Radionuclides, ANSI N42.13-1978.)

This standard is the responsibility of American National Standards Committee N42 on Radiation Instrumentation. Committee N42 delegated the development of the standard to its Subcommittee N42.2 which in turn assigned the task to its Working Group N42.2.2. Drafts were reviewed by the members of the Committee N42, Subcommittee N42.2, and Working Group N42.2.2 as well as by other interested parties, and the comments received were utilized in producing the standard as finally approved and issued. The standard was approved by N42 letter ballot of January 26, 1977, with modifications as submitted to the N42 membership on June 24, 1977.

At the time it approved this standard, the American National Standards Committee on Radiation Instrumentation had the following personnel:

Louis Costrell, *Chairman*   **David C. Cook,** *Secretary*

Organization Represented	*Name of Representative*
American Chemical Society	*Vacant*
American Conference of Governmental Industrial Hygienists	Jesse Lieberman
American Industrial Hygiene Association	*Vacant*
American Nuclear Society	Frank W. Manning
American Society of Mechanical Engineers	P. E. Greenwood
American Society of Safety Engineers	*Vacant*
Atomic Industrial Forum	*Vacant*
Health Physics Society	J. B. Horner Kuper
	Robert L. Butenhoff (*Alt*)
Institute of Electrical and Electronics Engineers	Louis Costrell
	D. C. Cook (*Alt*)
	A. J. Spurgin (*Alt*)
	J. Forster (*Alt*)
Instrument Society of America	M. T. Slind
	J. E. Kaveckis (*Alt*)
Lawrence Berkeley Laboratory	D. A. Mack
Manufacturing Chemists Association	*Vacant*
National Electrical Manufacturers Association	Theodore Hamburger
Oak Ridge National Laboratory	Frank W. Manning
	D. J. Knowles (*Alt*)
Scientific Apparatus Makers Association	*Vacant*
U.S. Department of the Army, Materiel Command	Abraham E. Cohen
U.S. Department of Commerce, National Bureau of Standards	Louis Costrell
U.S. Civil Defense Preparedness Agency	Carl R. Siebentritt, Jr
U.S. Energy Research and Development Administration	Hodge R. Wasson
U.S. Naval Research Laboratory	David C. Cook
Members-at-Large	J. G. Bellian
	O. W. Bilharz
	S. H. Hanauer
	John M. Gallagher, Jr
	Voss A. Moore
	R. F. Shea
	E. J. Vallario

Membership of Subcommittee N42.2:

Carl W. Seidel, *Chairman*
J. M. Robin Hutchinson, *Secretary*

Ron Althardt	Allen Goldstein	F. X. Masse
Karl Amlauer	P. C. Heidt	Gerald Martin, Jr
E. U. Buddemeyer	D. D. Hoppes	Paul Murphy
Lucy Cavallo	D. R. Horrocks	Ralph Nuelle
R. F. Coley	A. Jarvis	M. J. Oestmann
B. M. Coursey	D. S. Kearns	Edward Rapkin
R. Dayton	S. M. Kim	J. Ring
Roger Ferris	Y. Kobayshi	Patricia Vacca
James Gibbs	William MacIntyre	S. S. Yaniv
	W. B. Mann	

Membership of Working Group N42.2.2:

Frank X. Masse, *Chairman*

R. L. Ayres	W. B. Mann	C. W. Seidel
J. M. R. Hutchinson	Gerald Martin, Jr	S. S. Yaniv
	Paul Murphy	

Contents

American National Standard Calibration and Usage of "Dose Calibrator" Ionization Chambers for the Assay of Radionuclides

1. Introduction

The wide range of "calibrator"-type instruments currently being used primarily for radionuclide assay in nuclear medicine indicates the need for a standard for uniformity in measurement and test techniques. Such devices are composite systems consisting of an ionization chamber integrally coupled to appropriate electronic circuitry which converts the ionization current to a readout in units of activity. The principles of operation of the ionization chamber are well summarized in the NCRP Report No 58 [1][1] and will not be repeated here. Wide activity range and stability are useful characteristics of ionization chambers in this application. The advantages of this type of system for radionuclide assay include ease of use and interpretation.

2. Scope

2.1 This standard covers the technique for the quantification of the activity of identified radionuclides using any of a variety of ionization chambers currently available for this purpose. Application of the standard is limited to instruments that incorporate well-type ionization chambers as detectors.

2.2 This standard provides a method for obtaining measurements that are accurate to within ± 10 percent and reproducible to within ± 5 percent [usually for sources of more than 100 μCi (3.7×10^6 Bq)]. The standard is also intended to assure continuing performance of the apparatus within these specifications. For purposes of this standard, accuracy and reproducibility are described in 4.6.

[1] Numbers in brackets correspond to those of the references listed in Section 7 of this standard.

3. Definitions

accuracy. The accuracy, usually described in terms of overall uncertainty, is the estimate of the overall possible deviation from the "true" value. As used in this standard, the overall uncertainty is a total of the estimated errors itemized in Section 5 plus the random error of the measurement.

calibration. The process of determining the numerical relationship, within an overall stated uncertainty, between the observed output of a measurement system and the value, based on standard sources, of the physical quantity being measured.

shall. Shall indicates a recommendation that is necessary or essential to meet requirements of this standard.

should. Should indicates an advisory recommendation that is to be applied when practicable.

simulated sources. Simulated sources usually contain long-lived radionuclides, alone or in combination, that are chosen to simulate, in terms of photon or particle emission, a short-lived radionuclide of interest.

standard sources. The term standard sources is a general term used to refer to the standard sources of (1) and (2) below.

(1) **national radioactivity standard source.** A calibrated radioactive source prepared and distributed as a standard reference material by the U.S. National Bureau of Standards.

(2) **certified radioactivity standard source.** A calibrated radioactive source, with stated accuracy, whose calibration is certified by the source supplier as traceable to the National Radioactivity Measurements System [2].

4. Procedure

4.1 General. Instruments shall be installed and operated in accordance with the manufacturer's instructions.

4.2 Initial Calibrations. Instruments shall be calibrated with identified radionuclide sources of known activity and established purity. As described in 4.4, calibrations should be performed with standard sources of each radionuclide of interest, if at all feasible.

4.2.1 Geometry. The dependence of the assay on the geometrical configuration and composition of the source container must be taken into consideration in the calibration procedure. Most manufacturers have adopted a calibration geometry using a nominal 30 mL multidose vial with 20 mL of contents, and standard sources of this description are generally available (usually in plastic containers). Positioning of such vials in the detector well is usually reproducible for such systems. Correction factors or new calibrations must be obtained for assaying radionuclides in containers of different sizes or shapes. Such correction factors may be determined by measurement of a fixed amount of a given radionuclide in containers of different geometry, with any necessary adjustment to the volume using the appropriate carrier solution. Correction factors supplied by the manufacturer should also be checked as described above.

4.2.2 Activity Ranges. Calibration of the equipment should cover as completely as practicable the activity ranges for which it will be used, particularly those ranges of activity of radionuclides to be administered to patients. Whenever measurements in the low microcurie range are attempted, background corrections are imperative.

4.2.3 Energy Range. Calibration shall be performed over the energy range of proposed application.

4.2.4 Accuracy and Reproducibility. The calibration procedures should be such that the accuracy and reproducibility of measurements made with the calibrated instrument will be within the limits stated in 4.6.

4.3 Standard Sources. Suitable standard sources characterized as to radionuclide purity and activity shall be used for routine calibration of the equipment. Correction for decay of a standard source since the time of standardization should be applied if more than 2 percent of a half-life has expired.

4.3.1 Geometry. Ideally, to avoid the necessity for corrections, the geometry of the standard source should be identical to the geometry of the source to be assayed. Source manufacturers now offer standard sources that conform to the calibration geometry described in 4.2.1.

4.3.2 Activity Range. A suitable range of activities should be available for use. The selection of standard sources should taken into consideration the accuracy required over the ranges of activity of radionuclides to be administered to patients.

4.3.3 Energy Range. A suitable range of photon emission energies should be covered in the selection of standard sources. ^{125}I (0.03 MeV) (0.05 × 10^{-13}J), ^{57}Co (0.12 MeV) (0.19 × 10^{-13}J), ^{133}Ba (0.36 MeV) (0.58 × 10^{-13}J), and ^{137}Cs (0.66 MeV) (1.06 × 10^{-13}J), are representative of radionuclides emitting photons in the energy range typically used in nuclear medicine.

4.4 Assay. Radionuclides shall be assayed in a properly calibrated instrument using an appropriate setting or module. The activity of a radionuclide for which no setting or module is available may also be accurately measured relative to a standard source of the same radionuclide using any setting or module which yields a high enough reading to give reproducible results.

4.5 Performance Testing. Regular testing of the instrument performance is required to assure the accuracy of assays.

4.5.1 Reference Source Checks. Calibration checks using at least one long-lived reference source (for example, ^{137}Cs) shall be performed and logged on each work shift during which the instrument is used. This check shall be repeated whenever sample readings are not within 10 percent of their anticipated assay. It is suggested that at least two such reference sources be used [for example, 100 to 200 μCi (3.70 to 7.40 × 10^6 Bq) of ^{137}Cs, and 1 to 5 mCi (40 to 180 × 10^6 Bq) of ^{57}Co with appropriate correction for decay]. These sources could be alternated each day of use to test the instrument's performance over a range of photon energies and source activities.

4.5.2 Linearity Check. A convenient high-activity-range linearity check is outlined in

6.2. This check should be performed and logged at intervals not to exceed 3 months.

4.5.3 Background Checks. Background checks should be performed and logged daily, at least at the radionuclide settings to be used that day. These checks will serve to detect either contamination or faulty operation.

4.5.4 Response Check at Various Settings. Measurement of the long-lived reference source at settings for several radionuclides of interest will yield readings that should be reproducible over a period of time. Such readings serve as suitable checks on the stability of the instrument for measurements of radionuclides for which calibrations have been established, but for which standard sources are not always available for use. This check should be performed and logged daily utilizing the sources described in 4.5.1 to check the response of the instrument for the calibration of radionuclides the user anticipates assaying on that day.

4.5.5 Frequency of Calibration. Annually, following repair, and following extended periods of nonutilization, calibrations using standard sources of additional radionuclides of covering the energy and activity ranges of interest shall be performed and logged.

4.5.6 Supplementary Calibrations. As standard sources of additional radionuclides of interest become available, the instrument should be calibrated against such standard sources, particularly if such radionuclides are intended for human administration.

4.6 Accuracy and Reproducibility. Following are minimum requirements in terms of accuracy and reproducibility for such instruments.

4.6.1 Accuracy. The accuracy of the instruments, when used with the source geometry recommended by the manufacturer, at activity levels above 100 μCi (3.7×10^6 Bq) shall be such that the measured activity of a standard source as defined in 3(1) or (2) shall be within $\pm$ 10 percent of the stated activity of that source. Accuracy of measurements of activity levels below 100 μCi (3.7×10^6 Bq) may not fall within the $\pm$ 10 percent limits and should be determined for each instrument on which such assays will be performed.

4.6.2 Reproducibility. The reproducibility (or random error of the measurement) shall be such that all of the results in a series of ten consecutive measurements on a source of great-

er than 100 μCi (3.7×10^6 Bq) in the same geometry shall be within $\pm$ 5 percent of the average measured activity for that source, assuming no decay corrections over the measurement period are required.

4.6.3 Corrective Action. If the accuracy or reproducibility requirements are not met, it shall be recalibrated or repaired and recalibrated. If the instrument exhibits erratic performance, it shall be repaired and recalibrated.

5. Sources of Error

5.1 Following are common sources of error in the assay of radionuclides with ion chambers:

5.1.1 Errors in calibration of the standard source.

5.1.2 Variations in geometries of the sample to be assayed (see 4.2.1).

5.1.3 Variations in radiation background (particularly for low activity measurements).

5.1.4 The presence of radionuclidic impurities (see 6.3).

5.1.5 Changes in attenuation due to variations in container wall thickness or material (see 6.4, 6.5).

5.1.6 Nonuniformity of radioactivity distribution (see 6.6, 6.7).

6. Precautions

Following are some of the major areas in which discrepancies have been experienced with equipment of this type.

6.1 Assay of a Radionuclide for Which No Standard or Calibration Setting Is Available. The user must consider all gamma-ray and other photon emissions (including bremsstrahlung) and all beta-particle contributions to radiation emitted from the container when assaying a radionuclde for which no setting or module is provided, or for which no standard source is available. An understanding of the energy response function of the ion chamber is also necessary, particularly where energies of less than 150 keV (0.240×10^{-13} J) are present in the decay scheme. In general, the manufacturer should be consulted for advice on such measurements. (See [1] for further assistance in this regard).

6.2 Nonlinearity Effect. A nonlinearity effect at high activity levels is characteristic of all equipment of this type. To guard against errors in this regard, instruments should be checked against activities at the upper range of proposed use for a given radionuclide. If it is not practical to obtain a standard source of such high activity, measure the highest activity likely to be used (for example, total elution from a fresh generator); divide, and measure both parts (compensating for volume changes by adding carrier or a suitable diluent as necessary). Compare total of parts with the original reading. Repeat division as necessary until extent of nonlinearity is established. (Appropriate radiation safety precautions to minimize exposure and contamination must be observed in handling such quantities of radioactive material). Such nonlinearity problems have usually been noted above 100 μCi (3.7×10^6 Bq). If such nonlinearity is detected above a given level of activity, measurements at or above this level should not be depended upon. If accurate measurements above such levels are necessary, correction factors for each radionuclide and geometry should be developed.

6.3 Radionuclidic Impurities. The presence of radionuclidic impurities may result in large assay errors, particularly during measurement of short-lived radionuclides several half-lives after initial preparation. Determination of the photon energy spectrum with a photon spectrometry system may be necessary if accurate assays are to be performed whenever the presence of radionuclidic impurities may be a problem.

6.4 Beta-Particle Emitters. When measuring beta-particle-emitting radionuclides in an instrument of this type, the container becomes extremely important. Measurements on sources of the same radionuclide and activity will vary greatly with container composition (for example, glass versus plastic) and wall thickness. Such measurements depend on a measure of bremsstrahlung produced by deceleration of the beta particles in the container material. Reproducible measurements require consistent container selection and consistency in the manner in which the instrument is used.

6.5 Low-Energy Photon Emitters. Low-energy photon emitters (for example, ^{125}I) may be assayed incorrectly unless care is taken in the selection of the source container. The wall thickness of the container plus the thickness of the interior wall of the chamber may represent a significant attenuation factor for low-energy photons. Wide variations in solution

volume or container composition may also lead to erroneous results due to variations in absorption of such low-energy photons in the solution or the container.

6.6 Dissolved Gaseous Radionuclides. The user must be alert to a possible source of error in the measurement of radioactive solutions that tend to be unstable to the extent that part of the radionuclide may be present in a gaseous phase (for example, ^{133}Xe in saline). Readout results will be strongly dependent on the partitioning of the radionuclide between the gaseous and solution phases that occur in such a situation. Questionable readings in this regard may be checked by removing the liquid from the vial through the rubber septum with a syringe. Measurement of the vial after liquid removal will provide an estimate of the quantity of gaseous activity present.

6.7 Plate Out of Radionuclides. Care must be taken in the measurement of radionuclides that show a tendency to plate out of solution onto the walls or the cap of the container. This phenomenon will greatly affect the measurement due to both change in geometry and change in internal absorption factors. Repeating the measurement on the vial after the liquid has been removed will yield data that may be used to determine the net activity removed or to estimate the fraction plated out.

6.8 Simulated Sources. Although they may be useful as a check source, simulated sources in general are not recommended for activity-calibration purposes. Such sources, which are usually a mixture of long-lived radionuclides chosen to yield an approximation of the photon spectrum of the radionuclide they simulate, may not yield accurate calibration data in terms of ionization current. Also, their component parts may decay at different rates.

7. References

[1] National Council on Radiation Protection and Measurements. A Handbook of Radioactivity Measurements Procedures. NCRP Report No 58, Washington DC, 1977.

[2] CAVALLO, L.M., COURSEY, B.M., GARFINKEL, S.B., HUTCHINSON, J.M.R., and MANN, W.B. Needs for Radioactivity Standards and Measurements in Different Fields. *Nuclear Instruments and Methods*, vol 112, pp 5-18, 1973.

ANSI
N42.14-1978

American National Standard
Calibration and Usage of Germanium
Detectors for Measurement of
Gamma-Ray Emission of Radionuclides

Secretariat for N42

Institute of Electrical and Electronics Engineers, Inc

Approved April 10, 1978

American National Standards Institute

© Copyright 1978 by

American National Standards Institute

N42.14

American National Standard

An American National Standard implies a consensus of those substantially concerned with its scope and provisions. An American National Standard is intended as a guide to aid the manufacturer, the consumer, and the general public. The existence of an American National Standard does not in any respect preclude anyone, whether he has approved the standard or not, from manufacturing, marketing, purchasing, or using products, processes, or procedures not conforming to the standard. American National Standards are subject to periodic review and users are cautioned to obtain the latest editions.

CAUTION NOTICE: This American National Standard may be revised or withdrawn at any time. The procedures of the American National Standards Institute require that action be taken to reaffirm, revise, or withdraw this standard no later than five years from the date of publication. Purchasers of American National Standards may receive current information on all standards by calling or writing the American National Standards Institute.

Foreword

(This Foreword is not a part of American National Standard Calibration and Usage of Germanium Detectors for Measurement of Gamma-Ray Emission of Radionuclides, ANSI N42.14-1978.)

This standard is the responsibility of American National Standards Committee N42 on Radiation Instrumentation. Committee N42 delegated the development of the standard to its Subcommittee N42.2 which in turn assigned the task to its Working Group N42.2.1. Drafts were reviewed by the members of Committee N42, Subcommittee N42.2, and Working Group N42.2.1 as well as by other interested parties and the comments received were utilized in producing the standard as finally approved and issued. The standard was approved by N42 letter ballot of June 24, 1977.

Other relevant standards are:

ANSI/IEEE Std 300-1969, Test Procedure for Semiconductor Radiation Detectors.

ANSI/IEEE Std 301-1976, Test Procedures for Amplifiers and Preamplifiers for Semiconductor Radiation Detectors for Ionizing Radiation.

ANSI/IEEE Std 325-1971 (Reaff 1977), Test Procedures for Germanium Gamma-Ray Detectors.

ANSI/IEEE Std 645-1977 Test Procedures for High Purity Germanium Detectors for Ionizing Radiation (Supplement to ANSI/IEEE Std 325-1971).

IEEE Std 680-1978, Techniques for Determination of Germanium Semiconductor Detector Gamma-Ray Efficiency Using a Standard Marinelli (Reentrant) Beaker Geometry.

At the time it approved this standard, the American National Standards Committee on Radiation Instrumentation had the following personnel:

Louis Costrell, *Chairman* **David C. Cook,** *Secretary*

Membership of Subcommittee N42.2:

Carl W. Seidel, *Chairman*

J. M. Robin Hutchinson, *Secretary*

Ron Althardt	Allen Goldstein	F. X. Masse
Karl Amlauer	P. C. Heidt	Gerald Martin, Jr
E. U. Buddemeyer	D. D. Hoppes	Paul Murphy
Lucy Cavallo	D. R. Horrocks	Ralph Nuelle
R. F. Coley	A. Jarvis	M. J. Oestmann
B. M. Coursey	D. S. Kearns	Edward Rapkin
R. Dayton	S. M. Kim	J. Ring
Roger Ferris	Y. Kobayshi	Patricia Vacca
James Gibbs	William MacIntyre	S. S. Yaniv
	W. B. Mann	

Membership of Working Group N42.2.1:

Gerald Martin, *Chairman*

R. F. Coley
D. D. Hoppes
F. X. Masse

Contents

American National Standard
Calibration and Usage of Germanium Detectors for Measurement of Gamma-Ray Emission of Radionuclides

1. Introduction

1.1 The purpose of this standard is to provide a standardized basis for the calibration and usage of germanium detectors for measurement of gamma-ray emission rates of radionuclides. This standard is intended for use by knowledgeable persons who are responsible for the development of correct procedures for the calibration and usage of germanium detectors.

1.2 A typical gamma-ray spectrometry system consists of a germanium detector (with its liquid nitrogen cryostat, preamplifier, and possibly a high-voltage filter) in conjunction with a detector bias supply, linear amplifier, multichannel analyzer, and data readout device, for example, a printer, plotter, oscilloscope, or typewriter. Gamma rays interact with the detector to produce pulses which are analyzed and counted by the supportive electronics system.

1.3 A source emission rate for a gamma ray of a selected energy is determined from the counting rate in a full-energy peak of a spectrum, together with the measured efficiency of the spectrometry system for that energy and source location. It is usually not possible to measure the efficiency directly with emission-rate standards at all desired energies. Therefore a curve or function is constructed to permit interpolation between available calibration points.

2. Scope

This standard establishes methods for calibration and usage of germanium detectors for the measurement of gamma-ray emission rates of radionuclides. It covers the energy and full-energy peak efficiency calibration as well as the determination of gamma-ray energies in the 0.06 to 2 MeV energy region and is designed to yield gamma-ray emission rates with an uncertainty of ± 3 percent.[1] This standard applies primarily to measurements which do not involve overlapping peaks, and in which peak to continuum considerations are not important.

3. Definitions

certified radioactivity standard source. A calibrated radioactive source, with stated accuracy, whose calibration is certified by the source supplier as traceable to the National Radioactivity Measurements System [4].

---

[1] Uncertainty U is given at the 68 percent confidence level; that is, $U = \sqrt{\Sigma \sigma_i^2 + 1/3 \Sigma \delta_i^2}$ where δ_i are the estimated maximum systematic uncertainties, and σ_i are the random uncertainties at the 68 percent confidence level [1][2]. Other methods of error analysis are in use [2], [3].

[2] Numbers in brackets correspond to those of the references listed in Section 7 of this standard.

check source. A radioactivity source, not necessarily calibrated, which is used to confirm the continuing satisfactory operation of an instrument.

correlated photon summing. The simultaneous detection of two or more photons originating from a single nuclear disintegration.

FWHM (full width at half maximum). The full width of a gamma-ray peak distribution measured at half the maximum ordinate above the continuum.

national radioactivity standard source. A calibrated radioactive source prepared and distributed as a standard reference material by the U.S. National Bureau of Standards.

random summing. The simultaneous detection of two or more photons originating from the disintegration of more than one atom.

resolution, gamma ray. The measured FWHM, after background subtraction, of a gamma-ray peak distribution, expressed in units of energy.

total detection efficiency. The ratio of the total (peak plus Compton) counting rate to the gamma-ray emission rate.

NOTE: The terms *standard source* and *radioactivity standard* are general terms used to refer to the sources and standards of National Radioactivity Standard Source and Certified Radioactivity Standard Source.

4. Text of Standard

4.1 Preparation of Apparatus. Follow the manufacturer's instructions for setting up and preliminary testing of the equipment. Observe all of the manufacturer's limitations and cautions. All tests described in Section 6 should be performed before starting the calibrations, and all corrections shall be made when required. A check source should be used to check the stability of the system at least before and after the calibration.

4.2 Calibration Procedure

 4.2.1 Energy Calibration. Determine the energy calibration (channel number versus gamma-ray energy) of the detector system at a fixed gain by determining the full-energy peak channel numbers from gamma rays emitted from appropriate radioactivity sources. Determine nonlinearity correction factors as necessary [5], [6].

4.2.2 Efficiency Calibration

 4.2.2.1 Accumulate an energy spectrum using calibrated radioactivity standards at a desired and reproducible source-to-detector distance. At least 20 000 net counts should be accumulated in each full-energy gamma-ray peak of interest using National or Certified Radioactivity Standard Sources, or both (see 6.1).

 4.2.2.2 For each standard source, obtain the net count rate (total count rate of region of interest minus the Compton continuum count rate and, if applicable, the ambient background count rate within the same region) in the full-energy gamma-ray peak, or peaks, using a tested method that provides consistent results (see 6.2, 6.3, and 6.4).

 4.2.2.3 Correct the standard source emission rate for decay to the count time of 4.2.2.2.

 4.2.2.4 Calculate the full-energy peak efficiency E_f as follows:

$$E_f = \frac{N_p}{N_\gamma}$$

where
 E_f = full-energy peak efficiency (counts per gamma ray emitted)
 N_p = net gamma-ray count in the full-energy peak (counts per second)[3] (see 4.2.2.2)
 N_γ = gamma-ray emission rate (gamma rays per second)

If the standard source is calibrated as to activity, the gamma-ray emission rate is given by

$$N_\gamma = AP_\gamma$$

where
 A = number of nuclear decays per second
 P_γ = probability per nuclear decay for the gamma ray

 4.2.2.5 Plot, or fit to an appropriate mathematical function, the values for full-energy peak efficiency (determined in 4.2.2.4) versus gamma-ray energy (see 6.5).

4.3 Measurement of Gamma-Ray Emission Rate of the Sample

 4.3.1 Place the sample to be measured at the source-to-detector distance used for efficiency calibration (see 6.6).

 4.3.2 Accumulate the gamma-ray spectrum,

---

[3] Any other unit of time is acceptable provided it is used consistently throughout.

recording the count duration.

4.3.3 Determine the energy of the gamma rays present by use of the energy calibration obtained under, and at the same gain as, 4.2.1.

4.3.4 Obtain the net count rate in each full-energy gamma-ray peak of interest as described in 4.2.2.2.

4.3.5 Determine the full-energy peak efficiency for each energy of interest from the curve or function obtained in 4.2.2.5.

4.3.6 Calculate the number of gamma rays emitted per unit time for each full-energy peak as follows:

$$N_\gamma = \frac{N_p}{E_f}$$

When calculating a nuclear transportation rate from a gamma-ray emission rate determined for a specific radionuclide, a knowledge of the gamma-ray probability per decay is required [7], [8], that is,

$$A = \frac{N_\gamma}{P_\gamma}$$

4.4 Performance Testing

4.4.1 The following system tests should be performed on a regularly scheduled basis. The frequency for performing each test will depend on the stability of the particular system as well as on the accuracy and reliability of the required results. Where health or safety are involved, much more frequent checking may be appropriate. A range of typical frequencies for noncritical applications is given below for each test.

4.4.1.1 Check the system energy calibration (typically daily to semiweekly) using two or more gamma rays whose energies span at least 50 percent of the calibration range of interest. Correct the energy calibration, if necessary.

4.4.1.2 Check the system count rate reproducibility (typically daily to weekly) using at least one long-lived radionuclide. Correct for radioactive decay if significant decay (> 1 percent) has occurred between checks.

4.4.1.3 Check the system resolution (typically weekly to monthly) using at least one gamma-ray emitting radionuclide [9].

4.4.1.4 Check the efficiency calibration (typically monthly to yearly) using a National or Certified Radioactivity Standard (or Standards) emitting gamma rays of widely differing energies.

4.4.2 Record the results of the performance checks or plot them on a control chart, or both. Appropriate action shall be taken when the measured value falls outside the predetermined limits.

4.4.3 In addition, the above performance checks (see 4.4.1) should be made after an event (such as power failures or repairs) which might lead to potential changes in the system.

5. Sources of Error

Other than Poisson-distribution errors, the principal sources of error (and typical magnitudes) in this method are:

5.1 The calibration of the standard source, including errors introduced in using a standard radioactivity solution, or aliquot thereof, to prepare another (working) standard for counting (typically ± 3 percent).

5.2 The reproducibility in the determination of net full-energy peak counts (typically ± 2 percent).

5.3 The reproducibility of the positioning of the source relative to the detector and the source geometry (typically ± 3 percent).

5.4 The accuracy with which the full-energy peak efficiency at a given energy can be determined from the calibration curve or function (typically ± 3 percent).

5.5 The accuracy of the live-time determinations and pile-up corrections (typically ± 2 percent).

6. Precautions and Tests

6.1 Random Summing and Dead Time
6.1.1 Precaution

6.1.1.1 The shape and length of pulses used can cause a reduction in peak areas due to random summing of pulses at rates of over a few hundred per second [10]. Sample count rates should be low enough to reduce the effect of random summing of gamma rays to a level where it may be neglected, or one should use pile-up rejectors and live-time circuits, or reference pulser techniques, of verified accuracy at the required rates.

6.1.2 Test

6.1.2.1 If the maximum total count rate (above the amplifier noise level) ever used is less than 1000 s^{-1} and the amplifier time constant is less than 5 μs, this test need not be performed. Otherwise, accumulate a ^{60}Co spectrum at the highest total count rate used for gamma-ray-emission rate determinations until at least 100 000 counts are collected in the 1.332 MeV full-energy peak. Record the counting time, or live time (if the use of live time constitutes a part of the correction method). The source may be placed at any convenient distance from the detector.

6.1.2.2 Evaluate the net counting rate (subtract any background) for the 1.332 MeV peak including any methods employed to correct for pile-up and dead-time losses.

6.1.2.3 Without moving the ^{60}Co source, introduce a ^{57}Co source, or any other source with no gamma rays emitted with an energy greater than 0.4 MeV. Position the added source so that the total counting rate is increased 20 percent or more.

6.1.2.4 Erase the first spectrum, and accumulate another spectrum for the same length of time as in 6.1.2.1. The same live time may be used, if the use of live time constitutes at least a part of the correction method.

6.1.2.5 Evaluate and correct the net counting rate for the 1.332 MeV peak as in 6.1.2.2. For the correction method to be acceptable, the corrected count rate shall differ from that in 6.1.2.2 by no more than 1 percent.

6.2 Peak Evaluation

6.2.1 Precaution. Many methods [9], [11], [12], [13] exist for specifying the full-energy peak area and removing the contribution of any continuum under the peak. Within the scope of this standard, various methods give equivalent results if they are applied consistently to the calibration standards and the sources to be measured, and if they are not sensitive to moderate amounts of underlying continuum. A test of the latter point is a required part of the standard.

6.2.2 Test

6.2.2.1 Accumulate a spectrum of ^{137}Cs until at least 100 000 counts are recorded in the 0.662 MeV peak. The source may be placed at any convenient distance from the detector.

6.2.2.2 Determine the peak area with the analysis method to be tested. Subtract any background contribution.

6.2.2.3 Remove the ^{137}Cs source.

6.2.2.4 Without erasing the original spectrum, add a continuum in the 0.662 MeV peak region by accumulating a ^{60}Co spectrum at any rate that does not alter the system gain by more than 5 percent.

6.2.2.5 Continue the counting until the total counts within the channels defining the 0.662 MeV peak are at least three times the original peak counts. (The continuum contribution is at least two times the original peak counts.)

6.2.2.6 Analyze the 0.662 MeV peak with the method under test.

6.2.2.7 Subtract from the evaluated peak area any contribution introduced by room background at 0.662 MeV during the counting periods. The net peak area shall not deviate more than 2 percent from that in 6.2.2.2 for the evaluation method to be acceptable.

6.3 Correlated Photon Summing Correction

6.3.1 When another gamma ray or x ray is emitted in cascade with the gamma ray being measured, in many cases a multiplicative correlated summing correction C must be applied to the net full-energy-peak count rate if the sample-to-detector distance is 10 cm or less. The correction factor is expressed as

$$C = \frac{1}{\Pi_i^n (1 - q_i \epsilon_i)}$$

where

C = correlated summing correction to be applied to the measured count rate

n = number of gamma or x rays in correlation with gamma ray of interest

i = identification of correlated photon

q_i = fraction of ith correlated photon in correlation with the gamma ray of interest

ϵ_i = *total* detection efficiency of ith correlated photon

Correlated summing correction factors for the primary gamma rays of radionuclides ^{60}Co, ^{88}Y, ^{46}Sc are approximately 1.09 and 1.03 for a 65 cm^3 detector at 1 cm and at 4 cm sample-to-detector distances, respectively, and approximately 1.01 for a 100 cm^3 detector at a 10 cm sample-to-detector distance. The q_i must be obtained from the nuclear decay scheme, while the ϵ_i, which are slowly-varying functions of the energy, can be measured or calculated [14], [15], [16].

6.3.2 A similar correction must be applied

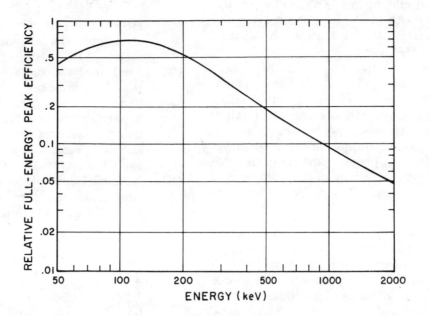

Fig 1
Typical Efficiency Versus Energy

when a weak gamma ray occurs in a decay scheme as an alternate decay mode to two strong cascade gamma rays with energies that total to that of the weak gamma ray [16]. The correction is over 5 percent for the 0.40 MeV gamma ray of ^{75}Se when a source is counted 10 cm from a 65 cm³ detector. Other common radionuclides with similar-type decay schemes, however, do not require a correction of this magnitude. For example, ^{47}Ca (1.297 MeV), ^{59}Fe (1.292 MeV), ^{144}Pr (2.186 MeV), ^{187}W (0.686 MeV), and ^{175}Yb (0.396 MeV) require corrections between 0.990 and 0.998 when counted at 4 cm from a 65 cm³ detector.

6.4 Correction for Decay During the Counting Period

6.4.1 If the value of a full-energy peak counting rate is determined by a measurement that spans a significant fraction of a half-life, and the value is assigned to the beginning of the counting period, a multiplicative correction F_b must be applied,

$$F_b = \frac{\lambda t}{1 - e^{-\lambda t}}$$

where

F_b = decay during count correction (count rate referenced to beginning of counting period)

t = elapsed counting time

λ = radionuclide decay constant $\left(\dfrac{\ln 2}{T_{1/2}}\right)$

$T_{1/2}$ = radionuclide half-life

t and $T_{1/2}$ must be in the same units of time (F_b = 1.01 for $t/T_{1/2}$ = 0.03).

6.4.2 If under the same conditions the counting rate is assigned to the midpoint of the counting period, the multiplicative correction F_m will be essentially 1 for $t/T_{1/2}$ = 0.03 and 0.995 for $t/T_{1/2}$ = 0.5. If it need be applied, the correction to be used is

$$F_m = \frac{\lambda t}{1 - e^{-\lambda t}}\ e^{\frac{-\lambda t}{2}}$$

6.5 Efficiency Versus Energy Function or Curve. The expression or curve showing the variation of efficiency with energy (see Fig 1 for an example) must be determined for a particular detector [17], [18], [19], and must be checked for changes with time as specified in the standard. If the full energy

11

range covered by this standard is to be used, calibrations should be made at least every 0.1 MeV from 0.06 to 0.30 MeV, about every 0.2 MeV from 0.3 MeV to 1.4 MeV, and at least at one energy between 1.4 MeV and 2 MeV. Radionuclides emitting two or more gamma rays with well-established relative gamma-ray probabilities may be used to better define the form of the calibration curve or function. A calibration with the same radionuclides that are to be measured should be made whenever possible and may provide the only reliable calibration when a radionuclide with cascade gamma rays is measured very close to the detector.

6.6 Source Geometry. A gamma ray undergoing even small-angle scattering is lost from the narrow full-energy peak, making the full-energy peak efficiency sensitive to the source or container thickness and composition. For most accurate results, the source to be measured must duplicate, as closely as possible, the calibration standards in all aspects (for example, shape, physical, and chemical characteristics, etc). If this is not practicable, appropriate corrections must be determined and applied.

7. References

[1] WAGNER, S. How to Treat Systematic Errors in Order to State the Uncertainty of a Measurement. Bericht FMRB 31/69, Physikalisch-Technische Bundesanstalt Forschungs-und MeBreaktor Braunschweig, Nov 1969.

[2] EISENHART, C. Expression of the Uncertainties of Final Results. *Science*, vol 160, p 1201, Jun 1968.

[3] International Commission on Radiation Units and Measurements. Certification of Standardized Radioactive Sources, ICRU Report No 12, Sept 15, 1968.

[4] CAVALLO, L. M., COURSEY, B. M., GARFINKEL, S. B., HUTCHINSON, J.M.R., and MANN, W. B. Needs for Radioactivity Standards and Measurements in Different Fields. *Nuclear Instruments and Methods*, vol 112, p 5, 1973.

[5] HELMER, R. G., et al. Precise Comparison and Measurement of Gamma-Ray Energies with a Ge(Li) Detector. *Nuclear Instruments and Methods*, vol 96, p 173, 1972.

[6] HEATH, R. L. Gamma-Ray Spectrum Catalogue Ge(Li) and Si(Li) Spectrometry. ANCR 1000-2, vol 2, Mar 1974.

[7] Nuclear Data Project, Oak Ridge National Laboratory, Ed. *Nuclear Data Sheets*. New York: Academic Press; and other compilations using the Evaluated Nuclear Structure Data File (ENSDF) maintained at ORNL for the US Nuclear Data Network.

[8] National Council on Radiation Protection and Measurements. A Manual of Radioactivity Measurement Procedures, NCRP Report No 58, Washington, DC, 1978.

[9] ANSI/IEEE Std 325-1971 (Reaff 1977), Test Procedures for Germanium Gamma-Ray Detectors.

[10] COHEN, E. J. Live Time and Pile-Up Corrections for Multichannel Analyzer Spectra. *Nuclear Instruments and Methods*, vol 121, p 25, 1974.

[11] KOKTA, L. Determination of Peak Area. *Nuclear Instruments and Methods*, vol 112, p 245, 1973.

[12] YULE, H. P. Computation of Lithium-Drifted Germanium Detector Peak Areas for Activation Analysis and Gamma-Ray Spectrometry. *Analytical Chemistry*, vol 40, p 1480, 1968.

[13] HEYDORN, K., and LADA, W. Peak Boundary Selection in Photopeak Integration by the Method of Covell. *Analytical Chemistry*, vol 44, p 2313, 1972.

[14] HEATH, R. L. Scintillation Spectrometry Gamma-Ray Spectrum Catalogue. Idaho Operations Office, US Atomic Energy Commission, IDO-16408, 1957.

[15] HEATH, R. L. Scintillation Spectrometry Gamma-Ray Spectrum Catalogue. 2nd Ed., Atomic Energy Commission Research and Development Report, IDO-16880, 1964.

[16] McCALLUM, G. J., and COOTE, G. E. Influence of Source-Detector Distance on Relative Intensity and Angular Correlation Measurements with Ge(Li) Spectrometers. *Nuclear Instruments and Methods*, vol 130, p 189, 1975.

[17] DEBERTIN, K., et al. Efficiency Calibration of Semiconductor Spectrometers — Techniques and Accuracies. *Proceed-*

ings of ERDA Symposium on X- and Gammay-Ray Sources and Applications, CONF-760539, May 1976, p 59, available from the National Technical Information Service, US Department of Commerce, Springfield, VA 22161.

[18] HIRSCHFELD, A. T., et al. Germanium Detector Efficiency Calibration with NBS Standards, *Proceedings of ERDA Symposium on X- and Gamma-Ray Sources and Applications,* CONF-760539, May 1976, p 90, available from the National Technical Information Service, US Department of Commerce, Springfield, VA 22161.

[19] McNELLES, L. A., and CAMPBELL, J. L. Absolute Efficiency Calibration of Coaxial Ge(Li) Detectors for the Energy Range 160-1330 keV. *Nuclear Instruments and Methods,* vol 109, p 241, 1973.

8. General References

[1] HELMER, R. G., et al. Gamma-Ray Energy and Intensity Measurements with Ge(Li) Spectrometers. The Electromagnetic Interaction in Nuclear Spectroscopy, Edited by W.D. Hamilton, North Holland American Elsevier Publishing Companies, 1975, Chapter 17, p 775.

[2] MEYER, R. A., et al. Current Research Relevant to the Improvement of γ-Ray Spectroscopy as an Analytical Tool. *Proceedings of ERDA Symposium on X- and Gamma-Ray Sources and Applications,* CONF-760539, May 1976, p 40, available from the National Technical Information Service, US Department of Commerce, Springfield, VA 22161.

[3] GUNNINK, R., et al. A Working Model for Ge(Li) Detector Counting Efficiencies. *Proceedings of ERDA Symposium on X- and Gamma-Ray Sources and Applications,* CONF-760539, May 1976, p 55, available from the National Technical Information Service, US Department of Commerce, Springfield, VA 22161.

[4] ADAMS, F., and DAMS, R. Applied Gamma-Ray Spectrometry, 2nd Ed., London: Pergamon Press, Ltd., 1970.

ANSI
N322-1977

American National Standard
Inspection and Test Specifications for
Direct and Indirect Reading Quartz
Fiber Pocket Dosimeters

Secretariat for N13

Health Physics Society

Secretariat for N42

Institute of Electrical and Electronics Engineers, Inc

Approved November 19, 1975

American National Standards Institute

Published by

The Institute of Electrical and Electronics Engineers, Inc
345 East 47th Street, New York, NY 10017

N322

American National Standard

An American National Standard implies a consensus of those substantially concerned with its scope and provisions. An American National Standard is intended as a guide to aid the manufacturer, the consumer, and the general public. The existence of an American National Standard does not in any respect preclude anyone, whether he has approved the standard or not, from manufacturing, marketing, purchasing, or using products, processes, or procedures not conforming to the standard. American National Standards are subject to periodic review and users are cautioned to obtain the latest editions.

CAUTION NOTICE: This American National Standard may be revised or withdrawn at any time. The procedures of the American National Standards Institute require that action be taken to reaffirm, revise, or withdraw this standard no later than five years from the date of publication. Purchasers of American National Standards may receive current information on all standards by calling or writing the American National Standards Institute.

Foreword

(This Foreword is not a part of American National Standard Inspection and Test Specifications for Direct and Indirect Reading Quartz Fiber Pocket Dosimeters, N322-1977.)

This standard was processed as an American National Standard by American National Standards Institute Committees N13 and N42. It defines the procedures to be used by manufacturers in testing direct and indirect reading quartz fiber pocket dosimeters relative to the performance specifications described in ANSI N13.5-1972, Performance Specifications for Direct Reading and Indirect Reading Pocket Dosimeters for X- and Gamma Radiation. ANSI N42.6-1971, Interrelationship of Quartz-Fiber Electrometer Type Dosimeters and Companion Dosimeters Chargers, is also relevant.

The memberships of the Committees, Subcommittee and Working Group at the time of approval of this standard were as listed below.

American National Standards Committee N13

American National Standards Committee N42

This standard was prepared under the direction of the Health Physics Society Standards Committee.

Edward J. Vallario, *Chairman*

The Working Group responsible for the preparation of this standard was:

Contents

American National Standard
Inspection and Test Specifications for
Direct and Indirect Reading Quartz
Fiber Pocket Dosimeters

1. Introduction

Pocket dosimeters are used to provide estimates of personnel exposures to X and gamma rays. Since operational decisions may be based on information provided by pocket dosimeters, it is essential that reliability by the key factor in their design, fabrication, and testing. The test and inspection specifications presented here apply to the manufacture of both direct and indirect reading quartz fiber pocket dosimeters. The manufacturer should use sufficient care in the manufacturing process to assure quality and maximum reliability consistent with the state of the art so that the dosimeters meet the requirements of continuous use for long periods under varied and severe environmental conditions.

The manufacturer is responsible for performing the inspections and tests described in this standard to assure that each dosimeter of that model (1) meets the requisite standards of performance and workmanship as specified in ANSI N13.5-1972, Performance Specifications for Direct Reading and Indirect Reading Pocket Dosimeters for X and Gamma Radiation, and (2) conforms to the certification for that model provided to the user.

The user is encouraged to perform certain of the inspections and tests at intervals considered appropriate for his specific applications. As a minimum, the electrical leakage and calibration tests specified in 7.2.1 and 8.1.1 should be performed.

2. Scope

This standard defines the procedures to be used by manufacturers in testing direct and indirect reading quartz fiber pocket dosimeters relative to the performance specifications described in ANSI N13.5-1972. The testing procedures apply to those dosimeters designed to respond to X and gamma radiation with energies extending from 20 keV up to 3 MeV.

This specification does not include test requirements for direct reading dosimeter chargers and is only partially applicable to charger readers for indirect reading dosimeters.

3. Equipment Requirements

3.1 Radiation Sources. The testing of dosimeters over the energy range of 20 keV to 3 MeV requires that a variety of radiation energies be available. Fluorescence X-rays are useful for the 20 keV to 100 keV region, and heavily filtered X-rays can also be used in the 20 keV to 250 keV region. Isotopic sources which can be used include:

Cobalt-60	— 1175 keV, 1333 keV
Cesium-137	— 662 keV
Gadolinium-153	— 99 keV
Americium-241	— 60 keV.

A radiation field is required of sufficient intensity to provide the maximum exposure rate for which the manufacturer certifies accurate operation of the dosimeter. The energy of the radiation, as well as the method for determining the radiation, shall be stated by the manufacturer. A range of exposure rates extending from 100 R/h to 10 000 R/h is required for some of the tests. This exposure range can be at any energy in the energy range specified for the dosimeter. All tests of the dosimeter, except for radiation rate effects (See 8.2) and those to determine proper operation at the maximum exposure rate certified by the manufacturer, shall be carried out using exposure rates that are at least a factor of 1000 below the maximum exposure rate specified for the dosimeter.

Exposure rates at all energies shall be known to within ± 5 percent. The calibration of the exposure fields shall be relatable to standards established and maintained by the National Bureau of Standards.

3.2 Environmental Chamber. Environmental chambers are required in which the dosimeter can be exposed under unusual environmental conditions of temperature, pressure, and humidity covering a temperature range of −10°C (14°F) to 50°C (122°F), a pressure range of 6.08 × 10⁴ Pascals (Pa) (456 mmHg) to 12.16 × 10⁴ Pa (912 mmHg), and a relative humidity up to 90 percent at 50°C (122°F).

3.3 Light Source. A diffused light source with an intensity of at least 55 lux (5 ft candles) is required.

4. Test Requirements

4.1 Each dosimeter shall be given the inspections and tests decribed in 7.2.1 and 8.1.1. Other tests shall be performed in accordance with statistical sampling procedures. These procedures shall be followed to insure that production batches are within error limits as stated in this specification. The procedures to be used shall be the result of specific agreements between the user and the manufacturer. ANSI Z1.4-1971, (MIL-STD-105D) Sampling Procedures and Tables for Inspection by Attributes is recommended as a guide for the development of acceptance criteria based on statistical sampling.

4.2 The term "normal" conditions or room conditions, as applied herein, describes:

Temperature Range:	15 to 25°C (59 to 77°F)
Relative Humidity Range:	40 to 80 percent
Pressure Range:	8.93 × 10⁴ to 10.67 × 10⁴ Pa (670 to 800 mmHg).

4.3 Unless otherwise specified, tolerances on environmental conditions are ± 2°C (± 3.6°F) for temperature and ± 10 percent for other parameters.

5. Inspection for Workmanship

The external appearance of the dosimeter is an indication of the quality of workmanship used in manufacture. The dosimeter shall be inspected for care in assembly, exterior finish, markings, and the presence of burrs or sharp edges. The care of manufacture is also indicated by interior features such as alignment with scale markings over the full length of the scale, absence of foreign material from the scale, and legibility of the scale and engravings or other defects which detract from the readability or accuracy of the dosimeter.

5.1 Optical System.

5.1.1 *Resolution.* The optical system of the dosimeter shall permit the fiber image to be clearly resolved relative to the scale divisions over the entire scale when the dosimeter is exposed to an illumination as low as 55 lux (5 ft candles) from a diffused source at the charging end. The focus of fiber and reticle shall permit resolution within ¼ of the smallest scale interval over the entire scale.

5.1.2 *Fiber Image.* The image of the fiber shall appear as a line parallel to the scale markings within ¼ of a minor scale division, over the entire scale, that is, with the charging potential adjusted so that the image of the fiber coincides with the bottom of any major scale division marking, the distance between the image of the fiber and the top of that same major division marking shall not be greater than ¼ the distance between two adjacent minor division markings.

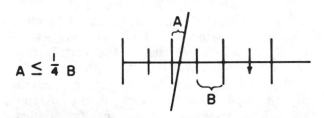

5.1.3 *Materials for Scale.* The scale shall be constructed of a material which will resist discoloration, pitting, or damage by exposure to the extremes of light, heat, or any ionizing radiation received during the tests specified herein.

5.2 Surface. The dosimeter shall be corrosion resistant in accordance with 9.3, and shall be smooth, except for identifying markings, stake marks, or detent to hold the clip.

5.3 Clip. The dosimeter shall be equipped with a clip to prevent loss. The clip shall be affixed in such a manner as to afford considerable resistance to removal and rotation. A male—female detent may be used with the male portion on the dosimeter barrel.

5.4 Serial Numbers. Each dosimeter shall be marked with a distinctive serial number.

6. Charging and Reading

6.1 Charging.

6.1.1 *Change in Reading After Charging.* Upon the completion of charging and adjusting the dosimeter to zero (0) scale reading, the pressure on the dosimeter charging switch shall be released. The dosimeter shall then be removed from the charging receptacle. The charging electrode of the dosimeter shall then be electrically shorted to the barrel of the dosimeter and the scale reading again observed.

The above procedure shall be carried out with the dosimeter and charging unit or external electrometer at $-10°C$ ($14°F$), $+22°C$ ($71.6°F$), and $50°C$ ($122°F$).

The net difference between the initial setting with the dosimeter fully depressed on the charger and the final reading shall not exceed 5 percent of full scale for dosimeters with ranges up to and including 5 R, and 2 percent of full scale for dosimeters with ranges greater than 5 R.

6.1.2 *Charging Life.* Each dosimeter shall be capable of meeting the requirements of this specification for a minimum of 2500 charging operations.

6.1.3 *Charging Voltages.* Dosimeter charging voltages shall be as specified in ANSI N42.6-1971 (formerly N3.2-1965), Interrelationship of Quartz-Fiber Electrometer Type Dosimeters and Companion Dosimeter Chargers.

6.2 Reading: Direct Reading Dosimeters Only.

6.2.1 *Test for Geotropism.* A special characteristic of direct reading dosimeters is the effect of gravity on the indicator (geotropism). This effect is observed as the fiber-image shift when the instrument is held horizontally and rotated about its optical axis.

With the dosimeter charged to midscale and held horizontally, read the dosimeter with the scale upright and horizontal, rotated 90 degrees each direction, and with the scale inverted. The change in reading shall not exceed ± 5 percent of maximum scale value. (See 9.6, ANSI N13.5-1972.)

6.2.2 *Test for Thermomechanical Equilibrium.* Charge the dosimeter to approximately midscale and read. Reduce the temperature of the dosimeter to $-10°C$ ($14°F$) and read. Raise the temperature of the dosimeter to $50°C$ ($122°F$) and read. Return the dosimeter to $+22°C$ ($71.6°F$) and read.

The readings at temperature extremes and the final reading shall not vary from the original reading by more than ± 5 percent of full scale.

7. Physical Tests

7.1 Test for Ruggedness. The dosimeter shall withstand, without damage or impairment of operation, four drops from a height of 1 m onto a hardwood surface. For test purposes, the dosimeter shall be set to midscale reading. At any time throughout this test, the reading of the instrument shall not change by more than 10 percent of full scale.

The instrument shall be dropped once on each end and twice on its side in random orientation.

At the completion of the test, no foreign materials or particles should appear on the fiber or scale images as a result of this test.

The dosimeter shall be capable of meeting all the requirements of specifications ANSI N13.5-1972 at the completion of this test.

7.2 Test for Electrical Leakage.

7.2.1 *The Dosimeter.* The discharge of the dosimeter in the absence of a radiation environment above normal background (usually 0.02 mR/h) shall be measured at $+50°C$ ($122°F$) and 90 percent relative humidity. A fully charged dosimeter shall discharge no more than 5 percent per 48 h. The test period shall equal or exceed 48 h. Initial and final readings of the dosimeter shall be made at the same temperature. This test shall be repeated at $-10°C$ ($14°F$) with any relative humidity.

7.2.2 *The External Electrometer.* The external electrometer provided for use with indirect reading pocket dosimeters is to be tested for electrical leakage at normal operating temperatures and 90 percent relative humidity. The leakage shall not exceed 2 percent of full scale for a 1 h period starting when the instrument is set to zero. If the instrument utilizes a protective cap for the charge/reading socket, the test may be performed with the cap installed.

7.2.3 *Electrical Leakage After Excessive Radiation — Dosimeters Having Ranges 5 R Full Scale or Greater.* The dosimeter at room conditions shall be exposed to X or gamma radiation totaling 2000 R delivered at a rate between 1000 and 10 000 R/h. Immediately after exposure, the dosimeter shall be fully charged and adjusted to zero scale reading. Readings shall be taken at the end of 4 h and again at 48 h, and shall not exceed 5 percent of full scale during the first 4 h and 10 percent during the 48 h period after zeroing the dosimeter.

7.2.4 *Electrical Leakage After Excessive Radiation — Dosimeters Having Ranges Less Than 5 R Full Scale.* The dosimeter at normal conditions shall be exposed to X or gamma radiation totaling 200 R delivered at a rate between 100 and 1000 R/h. Immediately after exposure, the dosimeter shall be fully charged and adjusted to zero scale reading. Readings shall be taken at the end of 4 h and again at 48 h, and shall not exceed 5 percent of full scale during the first 4 h and 10 percent during the 48 h period after zeroing the dosimeter.

8. Tests of Radiation Response Under Standard Environmental Conditions

These tests shall be carried out under "normal" environmental conditions as defined in 4.2.

8.1 Calibration, Linearity, and Precision.

8.1.1 *Calibration.* The dosimeter shall be fully charged and then exposed to radiation with the axis of the dosimeter perpendicular to the axis of the beam, to a source of radiation of the specified energy (see 9.1), ANSI N13.5-1972), to deliver an exposure equal to 80 percent of the full scale reading, at a rate at least a factor of 1000 below the maximum exposure rate certified in Section 10, item (7), of ANSI N13.5-1972. The reading obtained shall be within ± 10 percent of the true exposure delivered.

8.1.2 *Linearity.* The test shall be repeated under the same conditions as in paragraph 8.1.1 at exposures equal to approximately 20 percent and approximately 50 percent of full scale. Each reading shall be within ± 10 percent of the true exposure delivered.

8.1.3 *Precision.* The test shall be repeated under the same conditions with a minimum of ten consecutive exposures made at approximately 80 percent of full scale. The readings obtained shall not deviate from the mean value by more than 5 percent at the 95 percent confidence level.

8.2 Radiation Rate Effects. The dosimeter shall be fully charged and exposed to a source of the specified energy (see 9.1 of ANSI N13.5-1972) at the specified maximum exposure rate for which the manufacturer certifies accurate operation. For test purposes, the total exposure shall exceed 80 percent of full scale. The sensitivity of the dosimeter at the maximum exposure rate shall differ by no more than 10 percent from the sensitivity at the exposure rate employed in 8.1 above. For this test, sensitivity is defined as instrument reading divided by true exposure.

9. Tests of Radiation Response Under Extreme Environmental Conditions

9.1 Effects of Temperature, Pressure, and Humidity Extremes. Radiation exposure may occur under environmental conditions substantially different than those specified in Section 8. It is necessary, therefore, that tests be conducted for exposures occurring at the extreme environmental conditions described in Section 8 of ANSI N13.5-1972. The effect of the extreme condition is to be determined by comparison of dosimeter response for a radiation exposure which produces an 80 percent of full scale reading while at the extreme environment, to the dosimeter response for the same exposure at normal room conditions (see 4.3). Initial and final readings for each test shall be taken at the same environmental conditions.

The following tests are required with the dosimeter at equilibrium with the specified environment prior to exposing the dosimeter to radiation.

9.1.1 With pressure and humidity normal, and the temperature at −10° C (14° F).

9.1.2 With pressure and humidity normal, and the temperature at +50°C (122°F).

9.1.3 With the temperature and pressure normal, and the relative humidity at 90 percent or greater.

9.1.4 With temperature and humidity normal, and the pressure at 6.08×10^4 Pa (456 mmHg).

9.1.5 With the temperature and humidity normal, and the pressure at 12.16×10^4 Pa (912 mmHg).

9.2 Immersion. The dosimeter shall be capable of being immersed to a covering depth of 1 ft of water for a period of not less than 1 h. After such immersion, the dosimeter shall still meet all performance requirements. Protective caps used in normal operation remain in place for this test.

9.3 Corrosion Resistance. The dosimeter shall not corrode as a result of normal handling, immersion, or decontamination. The manufacturer shall specify any limitations on corrosion resistance as confirmed by specific tests performed by the manufacturer.

10. Energy Dependence and Geometrical Considerations

10.1 Energy Dependence. The dosimeters shall be fully charged and then discharged by radiations of several energies to produce readings of approximately 80 percent of full scale. Tests shall be made at extremes of the stated energy range and at several additional points between the lower extreme and 200 keV. If the lower extreme is 30 keV, tests shall be made at approximately 30, 50, 80, 120, 160, and 200 keV. In lieu of testing at the upper extreme of the energy range above 1175 keV, 1333 keV (Cobalt-60), the manufacturer may provide analytical evidence of expected dosimeter performance based on charged particle equilibrium, attenuation theory at energies of 2 and 3 MeV, materials of construction, and other major influencing factors. Exposures shall be made at a rate at least a factor of 1000 below the maximum permissible exposure rate certified for the instrument.

Exposures shall be made with the dosimeter perpendicular to the beam. The sensitivity of the dosimeter at each energy shall be normalized to the sensitivity at the specified energy (see 9.1 of ANSI N13.5-1972). The percentage difference due to energy effects shall not be greater than ± 10 percent from unity.

10.2 Isotropism Normal to the Dosimeter Axis. Exposures at radiation energies specified in 10.1 shall be made at 80 percent of full scale range. For purposes of this test, the dosimeter is oriented perpendicular to the radiation beam, and rotated 45 degrees around its longitudinal axis before each exposure. Readings obtained for each exposure shall not deviate from the mean exposure value by more than ± 5 percent of the maximum scale value.

10.3 Isotropism Parallel to the Dosimeter Axis. Exposures at radiation energies specified in 10.1 shall be made with the axis of the dosimeter at angles of 40 degrees and 140 degrees to the axis of the beam. Rotation shall be about an axis through the sensitive volume. Each reading obtained shall be not less than 70 percent of the reading obtained for the same energy in the test described in 10.1.

10.4 Other Geometrical Considerations. The manufacturer shall describe the test procedures used for instruments that are asymmetrical or which have more than one sensitive volume exclusive of sensitive volumes resulting from extracameral effects (apparent response of an instrument caused by radiation on any portion of the system other than the detector).

American National Standard
Radiation Protection Instrumentation
Test and Calibration

Secretariat for N13

Health Physics Society

Secretariat for N42

Institute of Electrical and Electronics Engineers, Inc

Approved September 13, 1977

American National National Standards Institute

Published by

The Institute of Electrical and Electronics Engineers, Inc
345 East 47th Street, New York, NY 10017

N323

American National Standard

An American National Standard implies a consensus of those substantially concerned with its scope and provisions. An American National Standard is intended as a guide to aid the manufacturer, the consumer, and the general public. The existence of an American National Standard does not in any respect preclude anyone, whether he has approved the standard or not, from manufacturing, marketing, purchasing, or using products, processes, or procedures not conforming to the standard. American National Standards are subject to periodic review and users are cautioned to obtain the latest editions.

CAUTION NOTICE: This American National Standard may be revised or withdrawn at any time. The procedures of the American National Standards Institute require that action be taken to reaffirm, revise, or withdraw this standard no later than five years from the date of publication. Purchasers of American National Standards may receive current information on all standards by calling or writing the American National Standards Institute.

Foreword

(This Foreword is not a part of American National Standard Radiation Protection Instrumentation Test and Calibration, ANSI N323-1978.)

The American National Standards Institute Joint Subcommittee N13/42 which was responsible for the development of this standard was established by authority of the Chairman of American National Standards Institute Committees N13 and N42 to represent the interests of the respective parent committees. The Joint Subcommittee comprises manufacturer and user membership in about equal numbers.

The ANSI Committee on Radiation Protection, N13, and Instrumentation, N42, which reviewed and approved this standard, had the following representatives at the time of approval:

American National Standards Institute Committee N13

M. E. Wrenn, *Chairman*   **J. Sohngen,** *Secretary*

Organization Represented   *Name of Representative*

Organization	Representative
American Chemical Society	Ira B. Whitney
American Conference of Governmental Industrial Hygienists	D. E. Van Farowe
American Health Physics Society	J. J. Cherubin
American Industrial Hygiene Association	Wilbur Speicher
American Insurance Association	Harry W. Rapp, Jr
American Mutual Insurance Alliance	William J. Uber
American Nuclear Society	James E. McLaughlin
American Public Health Association	Simon Kinsman
American Society for Testing and Materials	L. B. Gardner
American Society of Mechanical Engineers	H. J. Larson
Association of State and Territorial Health Officers	G. D. Carlyle Thompson
Atomic Industrial Forum	G. Edwin Brown, Jr
Electric Light and Power Group	Marvin Sullivan
Industrial Medical Association	Thomas Ray
Institute of Nuclear Materials Management	Ken Okolowitz
International Association of Government Labor Officials	Morris Kleinfeld
International Brotherhood of Electrical Workers	Edward J. Legan
Manufacturing Chemists Association, Inc.	P. W. McDaniel
National Bureau of Standards	Robert Loevinger
National Safety Council	Hugh F. Henry
Underwriters' Laboratories, Inc.	Leonard H. Horn
Uranium Operators' Association	L. W. Swent
US Atomic Energy Commission	Edward J. Vallario
US Department of Labor	John P. O'Neill
US Public Health Service	John Villforth
Individual Members	Merril Eisenbud
	Donald Fleckenstein
	John W. Healy
	Duncan A. Holaday
	Remus G. McAllister

American National Standards Committee N42

Louis Costrell, *Chairman*   **D. C. Cook,** *Secretary*

Organization Represented   *Name and Business Affiliation*

Organization	Name and Business Affiliation
American Chemical Society	Vacant
American Conference of Governmental Industrial Hygienists	Jesse Lieberman Department of Public Health
American Industrial Hygiene Association	W. H. Ray† US Energy Research and Development Administration
American Nuclear Society	Frank W. Manning Oak Ridge National Laboratory
American Society of Mechanical Engineers	P. E. Greenwood Newport News Shipbuilding and Drydock Co.
American Society of Safety Engineers	Vacant
Atomic Industrial Forum	Vacant

+Deceased

Organization Represented	Name and Business Affiliation
Health Physics Society	Dr. J. B. Horner Kuper Brookhaven National Laboratory
Alternate	Robert L. Butenhoff US Energy Research and Development Administration
Institute of Electrical and Electronics Engineers	Louis Costrell National Bureau of Standards
Alternate	J. Forster General Electric Company
Alternate	David C. Cook Naval Research Laboratory
Alternate	A. J. Spurgin General Atomic Company
Instrument Society of America	M. T. Slind Atlantic Richfield Hanford Company
Alternate	J. Kaveckis United Nuclear Industries, Inc
Manufacturing Chemists Association	Vacant (A. C. Clark, MCA, for information)
National Electrical Manufacturers Association	Theodore Hamburger Westinghouse Electric Corporation
Oak Ridge National Laboratory	Frank W. Manning Oak Ridge National Laboratory
Alternate	D. J. Knowles Oak Ridge National Laboratory
Scientific Apparatus Makers Association	Robert Breen The Foxboro Company
US Department of the Army, Materiel Command	Abrahan E. Cohen US Army Electronics Command
US Defense Civil Preparedness Agency	Carl R. Siebentritt, Jr US Defense Civil Preparedness Agency
US Department of Commerce National Bureau of Standards	Louis Costrell National Bureau of Standards
US Energy Research and Development Administration Division of Biomedical and Environmental Research	Hodge R. Wasson US Development Research and Development Administration
US Naval Research Laboratory	D. C. Cook U.S. Naval Research Laboratory
Individual Members	J. C. Bellian Bircon Corporation
	O. W. Bilharz General Electric Company
	John M. Gallagher, Jr Westinghouse Electric Corporation
	S. H. Hanauer US Nuclear Regulatory Commission
	Walter C. Lipinski Argonne National Laboratory
	Voss A. Moore US Nuclear Regulatory Commission
	R. F. Shea Consultant
	E. J. Vallario US Energy Research and Development Administration

This standard was prepared under the direction of the joint ANSI Subcommittee N13/N42, Radiation Protection Instrumentation, which had the following membership at the time of approval:

Edward J. Vallario, *Chairman*

E. Bemis
V. T. Chilson
A. Cohen
John Dempsey
E. E. Goodale
J. D. Henderson

H. R. Wasson

R. L. Kathren
W. R. Klein
T. P. Loftus
H. W. Patterson
H. J. L. Rechen
C. R. Siebentritt, Jr

The working group responsible for the preparation of this standard consisted of the following personnel:

J. M. Selby, *Chairman*

R. Beard
R. L. Kathren
H. V. Larson

T. P. Loftus
W. H. Ray†
A. R. Smith

†Deceased

Contents

American National Standard
Radiation Protection Instrumentation
Test and Calibration

1. Scope

This standard establishes calibration methods for portable radiation protection instruments used for detection and measurement of levels of ionizing radiation fields or levels of radioactive surface contamination. For purposes of this standard, portable radiation protection instruments are those which are carried by hand to a specific facility or location for use. Although this standard is specific to portable radiation protection instrumentation, the basic calibration principles may be applicable to radiation detection instrumentation in general.

Included within the scope of this standard are conditions, equipment, and techniques for calibration as well as the degree of precision and accuracy required. Alpha, beta, photon, and neutron radiations are considered. Passive integrating dosimetric devices such as film, thermoluminescent, and chemical dosimeters are outside the scope of this standard, but the basic principles and intent may apply. In cases where integrating capability is included along with rate measurement or detection, this standard shall apply.

Throughout these criteria, four verbs have been used to indicate the degree of rigor intended by the specific criterion. "Shall" and "will" indicate a minimum criterion that must be met, while "should" and "would" indicate a criterion that is recommended as good practice and is to be applied when practical.

2. Definitions

Technical terminology used in this standard is generally consistent with the definitions in the American National Standard Glossary of Terms in Nuclear Sciences and Technology,

N1.1-1976 [1],[1] and ICRU Report 20 [2]. The following terms are defined specifically for use within this standard.

accuracy. The degree of agreement of the observed value with the true or correct value of the quantity being measured.

calibrate. To determine (1) the response or reading of an instrument relative to a series of known radiation values over the range of the instrument or (2) the strength of a radiation source relative to a standard.

check source. A radioactive source, not necessarily calibrated, which is used to confirm the continuing satisfactory operation of an instrument.

decade. Synonymous with power of ten.

detection limit. The extreme of detection or quantification for the radiation of interest by the instrument as a whole or an individual readout scale. The *lower detection limit* is the minimum quantifiable instrument response or reading. The *upper detection limit* is the maximum quantifiable instrument response or reading.

detector. A device or component which produces an electronically measurable quantity in response to ionizing radiation.

effective center. The point within a detector that produces, for a given set of irradiation conditions, an instrument response equivalent to that which would be produced if the entire detector were located at the point.

energy dependence. A change in instrument response with respect to radiation energy for a constant exposure or exposure rate.

---

[1]Numbers in brackets refer to those of the references in Section 7 of this standard.

extracameral. Pertaining to that portion of the instrument exclusive of the detector.

geotropism. A change in instrument response with a change in instrument orientation as a result of gravitational effects.

instrument. A complete system designed to quanitify one or more particular ionizing radiation or radiations.

overload. Response of less than full scale (that is, maximum scale reading) when exposed to radiation intensities greater than the upper detection limit.

photon. A quantum of electromagnetic radiation irrespective of origin.

range. The set of values lying between the upper and lower detection limits.

readout. The device that conveys information regarding the measurement to the user.

reproducibility (precision). The degree of agreement of repeated measurements of the same property expressed quantitatively as the standard deviation computed from the results of the series of measurements.

response. The instrument reading.

sensitivity. The ratio of a change in response to the corresponding change in the field being measured.

standard (instrument or source) (1) national standard. An instrument, source, or other system or device maintained and promulgated by the U.S. National Bureau of Standards as such.

(2) derived or secondary standard. A calibrated instrument, source, or other system or device directly relatable (that is, with no intervening steps) to one or more U.S. National Standards.

(3) laboratory standard. A calibrated instrument, source, or other system or device without direct one-step relatability to the U.S. National Bureau of Standards, maintained and used primarily for calibrated and standardization.

test. A procedure whereby the instrument, component, or circuit is evaluated for satisfactory operation.

transfer instrument. Instrument or dosimeter exhibiting high precision which has been standardized against a *national* or *derived standardized* source.

uncertainty. The estimated bounds of the deviation from the mean value, generally expressed as a percent of the mean value. Ordinarily taken as the sum of (1) the random errors at the 95 percent confidence level and (2) the estimated upper limit of the systematic error.

unwanted radiation. Any ionizing radiation other than that which the instrument is designed to measure.

3. General Discussion

The operational requirements of radiation protection instrumentation are set forth in the recommendations of various commissions and committees [2],[3]. Additionally, the user may establish the need for different or more restrictive requirements. The ability to meet these requirements will depend not only on the instrument capabilities but also on periodic recalibration, preventative maintenance, and testing of the instruments.

For the purpose of this standard, new instruments are assumed to have been evaluated by the manufacturer to assure that the instruments are working properly. This evaluation, which is described in more detail by Zuerner and Kathren [4] involves a measurement of the characteristics of the instrument under design conditions. The evaluation includes determination of some or all of the following characteristics.

Nonradiological Characteristics:

(1) Physical construction, that is, safety, utility, weight, and ease of decontamination

(2) Effect of shock, sound and vibration, electric transients, RF energy, magnetic fields, high humidity, or other environmental influences

(3) Extent of switching transients, capacitance effects, geotropism, and static charge effects

(4) Power supply, including stability and battery life

Radiological Characteristics:

(1) Range, sensitivity, linearity, detection limit, and response to overload conditions*

(2) Accuracy and reproducibility*

(3) Energy dependence*

(4) Angular dependence

(5) Response to ionizing radiations other

than those intended to be measured

(6) Temperature and pressure dependence*

Certain tests from the above list (indicated by *) should be repeated routinely because aging of components, changes in available power (battery aging), and replacement of components may affect the calibration. Since the reproducibility of an instrument is critically important such a test should be performed regularly.

Periodic recalibration is distinct from a field test or simple evaluation with a check source. It includes a precalibration check followed by adjustment and calibration as described in Section 4 and perhaps a recheck of response, if any, to unwanted radiations. During the precalibration check, the instrument is tested to assure that certain operating requirements, specified by the manufacturer, are met. Thus, the instrument is determined to be in proper working order prior to calibration and adjustment.

If the instrument is to be adjusted or used for conditions other than those for which it was designed, such as use outside the designed energy range or under different environmental conditions, calibration for these conditions is necessary. Similarly, if the instrument is physically altered such that the previous calibration result could be invalidated, recalibration is required.

Components may change values with time or even fail. An instrument check must be made prior to use to ensure that (1) the instrument is operating properly, and (2) the response to a given check source is the same as it was immediately following calibration.

Radiation fields used for calibration must be thoroughly understood in terms of quality, quantity, and reproducability. Radiation sources and standard instruments that may be used in calibration are discussed in the appendix. A good review of calibration assemblies [5] should be consulted by those who calibrate instruments. Such problems as charged particle equilibrium, scattered or unwanted radiations from the source, and ambient background radiation must be taken into account quantitatively.

4. Inspection, Calibration, and Performance Test Requirements

4.1 Precalibration. The following conditions shall be established prior to exposing the in-

strument to a source for adjustment and calibration:

(1) The instrument should be free of significant radioactive contamination

(2) The meter shall be adjusted to zero or the point specified by the manufacturer using the adjustment or adjustments provided

(3) The batteries or power supply shall comply with the instrument manufacturer's specification

(4) The instrument shall be turned on and allowed to warm up for the time period specified by the manufacturer

(5) Electronic adjustments such as high voltage shall be set, as applicable, to the manufacturer's specifications

(6) Geotropism shall be known for orientation of the instrument in the three mutually perpendicular planes, and this effect shall be taken into account during calibration and performance testing

(7) The performance of any internal sampling time base in digital readout instruments should be verified as being within the manufacturer's specifications

4.2 Primary Calibration

4.2.1 General. The reproducibility (precision) of the instrument should be known prior to making calibration adjustments. This is particularly important if the instrument failed to pass the source check (see 4.6) or if repairs have been made. To check reproducibility, the instrument should be exposed to a radiation field three or more times under identical conditions. The readings obtained should normally not deviate from the mean value by more than ± 10 percent.

The response of an instrument may vary as a function of such parameters as energy, temperature, pressure, humidity, and source/detector geometry. The primary calibration should be accomplished with known values of these parameters. The calibration should be performed under the conditions specified by the manufacturer. Alternatively, any of these parameters may be fixed to the condition in which the instrument is to be used routinely, and notation made of these values. The steps that constitute the primary calibration when taken in conjunction with 4.1 are described in 4.2.2.

4.2.2 Readout Scale and Linearity Calibration and Adjustment

4.2.2.1 *Linear Readout Instruments.* Linear instruments usually have a scale selection switch. If controls are provided for each scale, adjustment of each shall be made according to the manufacturer's specifications or at the midpoint of each scale. If only one control is provided, adjustment shall be made either (1) at the point specified by the manufacturer, (2) near the midpoint of the middle scale, or (3) near the midpoint of a scale that is particularly important to the user's requirements.

After adjustment, calibration shall be checked near the ends of each scale (approximately 20 percent and 80 percent of full scale). After an adjustment or adjustments have been completed, instrument readings shall be within ± 10 percent of known radiation values at these two points. However, readings within ± 20 percent shall be acceptable if a calibration chart or graph shall be prepared and made available with the instrument.

4.2.2.2 *Logarithmic Readout Instruments.* Logarithmic readout instruments commonly have a single readout scale spanning several decades with two or more adjustments. The instrument should be adjusted for each scale according to the manufacturer's specifications or, alternatively, at points of particular importance to the user.

After adjustment, calibration shall be performed at a minimum of one point near the midpoint of each decade. After adjustments have been completed, instrument readings shall be within ± 10 percent of the known radiation values at these points. However, readings within ± 20 percent is acceptable if a calibration chart or graph is prepared and made available with the instrument.

4.2.2.3 *Digital Readout Instruments.* Digital instruments may have manual scale switching, automatic scale switching (auto ranging) or no scale switching. For instruments with either manual or automatic scale switching, the calibration shall be performed as in 4.2.2.1. For instruments without scale switching, the calibration shall be performed as in 4.2.2.2.

4.3 Calibration for Special Conditions

4.3.1 General. If the instrument is to be used under conditions (that is, radiation energy, temperature and pressure, or source/detector geometry) which vary significantly from those for which the instrument is designed, the instrument should be adjusted, calibrated, and used only for the special conditions. When an instrument is calibrated for special conditions, a special condition identification label shall be attached (in addition to any required calibration labels) to indicate its applicability *for this special use only.* However, if the instrument is also to be used within its design limits, the adjustments made during primary calibration (see 4.2) shall remain the same, and instrument readings for the special conditions shall be corrected using correction factors obtained from appropriate tables or graphs. Only one parameter should be varied at a time during calibration for the special conditions, but the interrelationships of the variables should be known.

4.3.2 Radiation Energy. Calibration shall be performed with a standard source or sources providing radiation fields similar to those in which the instrument will be used. Where instruments will be used in radiation fields of widely differing energies, the response of the instrument at several energies over the energy range shall be determined.

The response of the instrument to various energies of radiation shall be (1) plotted as a function of energy, or otherwise called out, (2) normalized to the response to a specific energy obtained during primary calibration, and (3) provided with the instrument. This type of graph is commonly called an energy dependence or spectral sensitivity curve.

4.3.3 Temperature, Pressure, and Humidity. Instruments to be used outside the manufacturer's recommended temperature range or at temperatures which differ by more than 30°C from the calibration temperature shall be calibrated over the temperature range at which they will be used. Care should be taken to ensure that instruments are not exposed to temperatures that will damage detector or electronic components.

If the manufacturer has not stated operating limits for humidity or atmospheric pressures, the instruments shall be calibrated at the approximate humidity or pressure expected to be encountered in use. Care should be taken to ensure that an instrument is not damaged by exceeding its pressure or humidity limits.

4.3.4 Detector Directional Dependence. If an instrument is to be used in a detector orientation relative to the source which is different from that used during primary calibration, correction factors should be developed.

4.4 Discrimination Against Unwanted Radiation. If adjustments or changes are made which

might alter the instrument response to unwanted ionizing and nonionizing radiations, the discrimination against unwanted radiation should be determined for all unwanted radiations that may be encountered.

4.5 Calibration Records. A record shall be maintained of all calibration, maintenance, repair, and modification data for each instrument. The record shall be dated and shall identify the individual performing the work. The record shall be filed with previous records on the same instrument in accordance with American National Standard Practice for Occupational Radiation Exposure Records Systems, N13.6-1966(R1972) [6].

Each instrument shall be labeled with the following information:

(1) Date of most recent calibration

(2) Initials or other specific identifying mark of calibrator

(3) Energy correction factors, where required

(4) Graph or table of calibration factors, where necessary, for each type of radiation for which the instrument may be used; this should relate the scale reading to the units required if units are not provided on the scale

(5) Instrument response to an identified check source (to be provided either by calibrator or user)

(6) Unusual or special use conditions or limitations

(7) Date that primary calibration is again required

(8) Special condition identification label (if applicable); see 4.3.1

4.6 Periodic Performance Test. To assure proper operation of the instrument between calibrations, the instrument shall be tested with the check source during operation and prior to each intermittent use.

Reference readings shall be obtained on each instrument when exposed to a check source in a constant and reproducible manner at the time of, or promptly after, primary calibration. If at any time the instrument response to the check source differs from the reference reading by more than ± 20 percent, the instrument shall be returned to the calibration facility for calibration or for maintenance, repair, and recalibration, as required. Reference readings should be obtained for one point on each scale or decade normally used. The check source

should accompany the instrument if it is specific to that instrument.

4.7 Calibration and Performance Test Frequency

4.7.1 Primary Calibration Frequency. All instruments shall receive the precalibration inspection described in 4.1 and the primary calibration described in 4.2 prior to first use.

Primary calibration will be required at least annually even when the performance test requirements outlined in 4.6 are met.

Where instruments are subjected to extreme operational conditions, hard usage, or corrosive environments, more frequent primary calibration should be scheduled.

Recalibration shall be scheduled after any maintenance or adjustment of any kind has been performed on the instrument. For this requirement, battery change is not normally considered maintenance.

4.7.2 Calibration Frequency for Special Conditions. Calibration for special conditions need be performed only once unless (1) the instrument is modified or physically altered, (2) the special conditions are changed, or (3) the primary calibration is altered, providing that the conditions in 4.7.1 are met.

4.7.3 Performance Test Frequency. A performance check shall be made prior to each use, during intermittent use conditions and several times a day during continuous use.

5. Calibration Equipment Required

5.1 Calibration Standards. Instruments should be calibrated either against National Standards or with Derived Standards. If National or Derived Standards are not available, Laboratory Standards, obtained in one of the following ways, should be used:

(1) Comparison of the radiation field from a user's source with the radiation field from a National or Derived Standard source in the same geometry, using a "transfer instrument" with a reproducibility of ± 2 percent. The transfer calibration shall utilize a calibration curve for the transfer instrument taken with the National or Derived source over a range that covers both the National or Derived source measurement and the user source measurement. (Such a curve reduces to a single point if the transfer calibration procedure is such

that the transfer instrument readings are identical for both measurements.)

(2) Calibration of a user's transfer instrument with a National or Derived Standard source, followed by evaluation of a user's source with the same transfer instrument. The transfer instrument shall have a reproducibility of ± 2 percent and the procedure shall utilize a calibration curve as in 5.1(1).

(3) Where no National or Derived Standard exists, as in the case of specific energies or unusual sources, by establishment of a standard source or instrument with documented empirical and theoretical output or response characteristics.

A calibration source or sources preferably should be of a radiation energy similar to that with which the instrument will be used and of a radiation exposure rate sufficient to reach full scale of any instrument to be calibrated. If the source is a radionuclide, the half-life should be long, preferably greater than several years to minimize corrections and errors. The uncertainty of source calibration shall be no greater than ± 2 percent with respect to U.S. National Standards.

5.2 Calibration Assemblies. Instrument calibration assemblies shall be mechanically precise to ensure that positioning errors of either instruments or radiation sources do not affect the radiation field values by more than ± 2 percent.

The working conditions in the calibration facility shall not cause excessive radiation exposure of personnel. Personnel exposure shall be kept as low as practicable and shall in no case under normal operating conditions exceed permissible levels permitted by agreement or law (whichever is lower).

To meet this condition, personnel shielding, remote instrument reading and positioning facilities, automatic source handling mechanisms, and other mechanical or remote operations are recommended.

A sufficient range of radiation fields shall be available to satisfy calibration requirements.

5.3 Standard Instruments. An instrument used as a Derived Standard shall have an uncertainty no greater than ± 10 percent. Calibration shall be reestablished after maintenance or repair or at intervals specified by the manufacturer but in no case at intervals greater than three years.

A periodic instrument check procedure shall be established by the user to assure continued proper operation.

5.4 Check Sources. Check sources should provide radiation of the same type or types as provided by those sources used in instrument calibration (as described in 5.1). However, check sources may provide radiation different than that used for calibration if:

(1) The source instrument geometry is well understood and easily reproduced, or

(2) The instrument response to this radiation is well understood and is not critically dependent on instrument adjustment. (For example, the use of a photon source to check instruments sensitive to beta radiation may be acceptable; the use of a photon source to check a detector utilizing a BF_3 response to neutrons is not acceptable.)

A reproducible source detector geometry shall be established and used for all performance test measurements.

6. Maintenance of Quality of Calibration

6.1 Radiation Field. Either narrow or broad beam geometry may be used to compare the response of similar instruments with that of a standardized instrument.

For calibration of X-ray machines or particle accelerators, a calibrated instrument shall be used. If a continuous monitor is available, it can be calibrated simultaneously and used in subsequent work with periodic checks on its constancy.

Alpha radiation sources shall be standardized in terms of activity or activity per unit area of the source, or both. The reference geometry, 2π or 4π, shall be stated.

Beta radiation sources shall be standardized in terms of air or soft tissue absorbed dose rate at the surface or at a specified distance from the source, or in terms of activity.

Photon-emitting radionuclide sources shall be standardized in terms of exposure rate (in roentgens per hour) at a specified distance from the source.

Neutron sources shall be standardized in terms of (1) the number of neutrons emitted per unit time and (2) the effective or average neutron energy. Concomitant photon exposure rate should be known and stated.

For photon and neutron monitoring instrument calibrations, the source-to-detector distance shall be the distance measured between the effective center of the radioactive source and the effective center of the radiation detector. Either this distance shall be greater than seven times the maximum dimension of the source or detector, whichever is larger, or suitable corrections shall be used.

The exposure rate or the flux density of the radiation field shall be known with an estimated uncertainty no greater than ± 10 percent. A continuous monitor or other device should be used to determine whether the radiation field has changed.

6.2 Calibration Facility. Free-space geometry should be achieved for photon and neutron instrument calibration. The distance to scattering objects from the source and from the detector should be at least twice the distance between the detector and the source. Where scattering contributions to instrument readings are significant they shall be included in stating the value of the radiation field for all detector positions used for calibration purposes.

The radiation background at the calibration facility shall be low, known, and stable and shall be accounted for during calibration.

Temperature, relative humidity, and atmospheric pressure shall be noted at the time of instrument calibrations. Calibrations should be performed within the temperature range 25 ± 10°C, except when the instrument is to be used outside this temperature range.

6.3 Other. If an instrument may exhibit an extracameral response, the entire instrument should be placed in the radiation field during calibration and the results compared to calibration with just the detector in the field. The fractional contribution, if any, to the instrument reading due to an extracameral response should be determined and noted on the instrument.

A reasonable delay should occur before reading to allow warmup and to accommodate switching transients and the time constant of the instrument.

7. References

[1] ANSI N1.1-1976, Glossary of Terms in Nuclear Science and Technology.

[2] ICRU Report 20, Radiation Protection Instrumentation and Its Application. International Commission on Radiation Units and Measurements, 1971.

[3] NBS Handbook 51, Radiological Monitoring Methods and Instruments. National Commission on Radiation Protection Subcommittee 7, 1952.

[4] ZUERNER, L.V., and KATHREN, R.L. Evaluation Program for Portable Radiation Monitoring Instruments. C. A. Willis and J.S. Handloser, Eds., in Health Physics Operational Monitoring, vol 2. New York: Gordon and Breach, 1972, pp 325–350.

[5] Technical Reports Series 133, Handbook on Calibration of Radiation Protection Monitoring Instruments. International Atomic Energy Agency, 1971.

[6] ANSI N13.6-1966 (R1972), Practice for Occupational Radiation Exposure Records.

Appendix

(This Appendix is not a part of American National Standard Radiation Protection Instrumentation and Calibration, ANSI N323-1978.)

Radiation Sources for Instrument Calibration

A1. Electromagnetic Radiation

The calibration of photon monitoring instruments over the energy range from a few keV to several MeV is best accomplished with X-ray producing equipment and radionuclide sources. Some radionuclides may be used in lieu of X-ray machines as sources of radiation if the radiation energy and exposure rate are suitable to particular needs. Instruments used to measure X-rays with energies to a few hundred keV may be calibrated in an X-ray beam by comparison with a Derived Standard of known response. Instrument response to photons with energies from a few keV to 300 keV can be obtained with heavily filtered bremsstrahlung spectra and fluorescence X-rays [A1]–[A15][1]. When X-rays are used for calibration, the excitation voltage (kVp) filtration, and spectral characteristics should be accurately determined.

NOTE: If calibrations are made with an X-ray or accelerator target facility from which radiation field are produced in pulsed form, care must be taken to eliminate the possibility of serious calibration error.

Instruments used to monitor higher energies may be most easily calibrated in known radiation fields produced by sources of gamma rays of approximately the same energies as those to be measured. An ideal calibration source should emit photons with an energy spectrum similar to that to be measured and have a suitably long half-life. Few sources fulfill these ideal requirements. Sources covering a wide energy range which have proven useful for this purpose are shown in Table A1.

The supplier of a calibration source should provide a certificate of either the radioactive content or the exposure rate at a specified distance from the source [A19]. If the specific exposure rate constant (Γ) is used to calculate the exposure rate at various distances from the source, the dimensions of the source must be small relative to the distance to the detector. A good guideline is to make the source-to-detector distance at least seven times the largest dimension of the source or detector, whichever is larger.

Sources used for periodic calibrations should have an output appropriate to the range of the instruments to be calibrated and should be used under conditions similar to those employed in determining the exposure rate constant (Γ). The radiation field from the source should be uncollimated (if that were the condition of calibration), source-to-detector distances should be 30 to 100 cm for small detectors, and the radiation field should follow the inverse square relationship. This latter requirement can be met by minimizing the mass of material supporting the source and instrument and placing both the source and the instrument a large distance from potential large scatterers such as the ground, floors, and walls.

Instruments for standardizing photon fields include the free-air ionization chamber [A20], [A21], cavity ion chambers [A22], [A23], and calorimeters [A24], [A25]. Thermoluminescent dosimeters have also been proposed [A26]. For most laboratories, cavity ion chambers provide the best compromise among accuracy, precision, ease of use, cost, and traceability to a National Standard.

---

[1] Numbers in brackets refer to those of the references in Section A6 of this standard.

A2. Beta Radiation

The beta radiation response of an instrument can be determined with a source of beta radiation which has been calibrated with an extrapolation chamber [A27], [A28]. To obtain uniform radiation fields, calibrations are commonly made with the detector window nearly in contact with a large area, flat, uniformly distributed source made of natural uranium, U_3O_8 or ^{90}Sr [A29], [A30]. However, sufficiently strong "point" sources could be used at distances that are large compared with dimensions of the detector window. At distances from a beta radiation point source, where fairly uniform irradiation of a detector's window prevails, considerable attenuation from air absorption will occur. The beta particles emitted from a source have a continuous spectrum of energy up to E_{max}. Table A2 includes a list of beta sources which are suitable for instrument calibration. Some of these sources may also emit photons. The instrument response to these photons together with any bremsstrahlung from surrounding materials should be taken into consideration in the calibration.

Another approach to the calibration of beta ray ion chamber instruments is to cover the detector with a thick sheet of air-equivalent material and then to calibrate the detector with gamma rays [A30]. In a field where the exposure rate is 1 R/h, the equivalent absorbed beta does rate under these conditions is about 0.98 rad/h for tissue. Electrons entering the thin window and producing the same current in the chamber would produce an absorbed dose rate in tissue only slightly higher than that produced by gamma rays. Thus, for practical purposes, the calibration for exposure rate in roentgens per hour is so nearly numerically equal to the desired calibration for absorbed dose rate from electrons in radians per hour that the difference is generally ignored.

A3. Alpha Radiation

Alpha radiation portable instrumentation is primarily used for the detection and sometimes the measurement of contamination. Standard reference alpha sources can be purchased from the National Bureau of Standards, commercially, or may be manufactured by electroplating a metallic alpha radiation emitter onto a stainless steel backing or other metal disc. Table A3 contains a list of alpha radiation sources which may be used in calibration. In some instances, alpha radiation may be present from impurities or daughters.

Table A1
Photon Emitting Radionuclides Suitable for Use in Instrument Calibration [A16]–[A18]
(in order of increasing effective energy)

Radionuclide	Effective Energy (keV)*	Half-Life		Specific Exposure Rate Constant R/(h·Ci) at 1 M†
^{241}Am	60	433	years	0.0129
^{57}Co	122	270	days	0.097
^{51}Cr	320	28	days	0.018
^{137}Cs	662	30.1	years	0.323
^{226}Ra	830‡	1600	years	0.825§ **
^{60}Co	1250	5.27	years	1.30
^{24}Na	~2000	15	hours	1.84

*Instrument response is assumed to be negligible below 10 keV. Radiations with lower energy are not included. Effective energy will depend on instrument response.

†Assume negligible self-absorption, scattering, and bremsstrahlung.

‡ ^{226}Ra emits gamma rays of many energies from 19 to 2448 keV.

§ In equilibrium with its daughter products and with 0.5 mm Pt filtration.

**Average and Effective Energy for ^{226}Ra in Equilibrium with Daughters. KATHREN, R.L., and CHURCH, L.B. *Health Physics*, vol 30, p 143, 1976.

Table A2
Beta Radiation Sources for Instrument Calibration [A17], [A18]
(in order of increasing E_{max})

Radionuclide	E_{max} MeV (percent)	E_{avg} (MeV)	Half-Life
^{3}H	0.0185 (100)	0.0057	12.35 years
^{14}C	0.156 (100)	0.049	5730 years
^{35}S	0.167 (100)	0.049	88.0 days
^{45}Ca	0.257 (100)	0.077	164 days
^{185}W	0.433 (100)	0.144	75 days
^{85}Kr	0.674 (99)	0.246	10.7 years
^{204}Tl	0.763 (98)	0.243	3.78 years
^{111}Ag	1.03 (93)	0.351	7.47 days
^{210}Bi	1.16 (100)	0.394	5.01 days
^{32}P	1.71 (100)	0.695	14.3 days
^{90}Sr$-^{90}$Y	2.27 (99)	0.566	28.5 years
^{238}U*	3.26	—	4.49×10^9 years
^{42}K	3.52 (82)	1.43	12.4 h

*100 days after separation, approximately 95 percent equilibrium; 200 days after separation, more than 99 percent equilibrium.

Table A3
Alpha Radiation Sources for Instrument Calibration [A17], [A18]
(in order of increasing energy)

Radionuclide	Alpha Energy (MeV) Abundance (Percent)	Half-Life
^{148}Gd	3.18	93 y
^{230}Th	4.617 (24) 4.684 (76)	7.7×10^4 years
^{239}Pu	5.105 (12) 5.143 (15) 5.156 (73)	2.44×10^4 years
^{210}Po	5.305 (100)	138.4 days
^{241}Am	5.442 (13) 5.484 (86)	433 years
^{238}Pu	5.456 (28) 5.499 (72)	87.8 years
^{244}Cm	5.764 (23) 5.806 (77)	17.8 years
^{252}Cf	6.076 (16) 6.119 (84)	2.65 years

A4. Neutron Radiation

Neutron monitoring instruments can be calibrated with neutrons produced by radionuclide sources [A31]–[A35], [A37]–[A42], particle accelerators [A36], [A37], or nuclear reactors [A43]–[A45].

Radionuclide neutron sources are of three types: alpha-n, gamma-n, and spontaneous fission. The alpha-n radiation sources contain an alpha emitter, such as ^{210}Po, ^{238}Pu, ^{239}Pu, or ^{241}Am in intimate contact with a low-atomic-number element or elements, such as lithium, beryllium, boron, or fluorine, and produce neutrons distributed in energy from nearly 0 MeV to the maximum allowed by reaction kinetics. These sources are physically small in in size and easily portable, can be fabricated with neutron yields up to the 10^7–10^8 neutron per second range, have relatively low-intensity accompanying photon emission, and can be calibrated by the National Bureau of Standards. ^{210}Po alpha-n radiation sources are of limited value because of the relatively short ^{210}Po half-life (138 days). Recommended alpha-n radiation sources are based on ^{238}Pu, ^{239}Pu, and ^{241}Am; however, it may be important to account for changes in neutron emission caused by in-growth of daughter alpha emitters or the initial presence of (small amounts) alpha emitters with half-lives different from the primary radionuclide, as well as decay of the primary radionuclide.

The gamma-n radiation sources consist of a gamma emitter of suitably high photon energy, such as ^{124}Sb or ^{226}Ra, placed in close proximity to a low-atomic-number element, usually deuterium or beryllium. Each reacting gamma ray, according to reaction kinetics, produces primarily monoenergetic neutrons. These sources also produce intense photon fields, which may create personnel exposure problems, as well as interference with instrument response. The low-atomic-number element is often a physically separate piece so that instrument response to photons can be assessed separately. The short half-life of ^{124}Sb (60 days) and the low neutron energy (approximately 30 keV) are important limitations for the use of this type source. ^{226}Ra gamma-n

radiation sources are more generally useful, are physically small in size, can be obtained with neutron yields up to the 10^6–10^7 neutron per second range, and can be calibrated by the National Bureau of Standards.

Spontaneous fission neutrons, emitted in one branch of the decay of ^{252}Cf, have a fission-type neutron spectrum in the energy range approximately 0.5 to 15 MeV. These sources can closely approximate an idealized point source, can be obtained in the range up to 10^8–10^9 neutrons per second, have relatively low-intensity accompanying photon emission, and can be calibrated by the National Bureau of Standards. The relatively short half-life (2.65 years) is a limitation to long-term use.

Particle accelerators produce intense neutron fields by the interaction of accelerated charged particles, such as protons, deuterons, or tritons, on low-atomic-number target materials such as deuterium, tritium, and lithium. Important characteristics of these neutron fields are variable intensity up to very high values (yields in excess of 10^{12} neutrons per second in some cases), occurrence of radiation in a brief pulse, monoenergetic neutron emission for any given beam-target-detector angular relationship, and lack of portability. Neutron output is a complex function of accelerator and target parameters and may be expected to vary with time even though measured machine parameters remain constant. Therefore, neutron output must be monitored constantly during instrument calibration work, and standard instruments or techniques must be used to establish neutron field values.

Nuclear reactions may be used as a source of neutrons for instrument calibration purposes. The energy spectral characteristics of these neutron fields can vary from an unmoderated fission spectrum, such as produced by a ^{235}U fission plate, to the heavily filtered slow neutrons from a thermal column. Neutron flux densities in excess of 10^{10} neutrons per centimeter squared-second may be obtainable for calibration purposes. Standardization of the neutron fields must be accomplished with standard instruments or techniques, and continuous monitoring of the neutron flux must be performed throughout calibration work.

Radionuclide neutron sources can be conveniently sent to the National Bureau of Standards for calibration, thereby to become Derived Standards suitable for in-house instrument calibration [A46]. On the other hand, since particle accelerators and nuclear reactors cannot be sent out for calibration, they must be standardized for instrument calibration by use of standard neutron instruments or techniques. In addition, the neutron output of these facilities must be monitored continuously during instrument calibrations. Standard neutron instruments and techniques include the precision long counter [A47]–[A49], associated particle counters (for certain accelerators) [A37], nuclear emulsions [A37], fission foils [A37], activation foils [A37], and manganese sufate bath [A35].

A5. Facilities for Radiation Instrument Calibration Laboratories

The building space, methods, and staff necessary to properly operate a calibration laboratory depend on the volume and type of work undertaken. Calibration facilities exist that employ from a fraction of one employee's time up to tens of people. The amount of building space needed depends on such diverse factors as volume of business, land and building costs, shielding, scatter, required accuracy, types of radioactive sources and instruments, and energy and strength of sources. Several types of rooms may be necessary including irradiation rooms, storage vaults, and offices. The methods for proper, efficient, and safe operation of the laboratory must be correctly established and should include such functions as receiving and handling of instruments, calibration of test irradiation systems, calibration of instruments and dosimeters, return of calibrated dosimeters and instruments, reporting of results, keeping records of laboratory activities, maintenance and development of calibration procedures. The number of laboratory staff would depend on the volume of business. In any case, the technical integrity of the laboratory should be above reproach. Typical laboratory buildings, staffing requirements, and laboratory facilities have been described in the literature and can be adapted to most needs [A30], [A50].

A6. Appendix References

[A1] CORMACK, D. V., and BURKE, D. G. Spectral Distributions of Primary and Scattered 150 kVp X-Rays. *Radiology*, vol 74, pp 743–752, 1960.

[A2] EHRLICH, M. Narrow-Band Spectra of Low Energy X-Radiation. *Radiation Research*, vol 3, p 223, 1955.

[A3] EPP, E.R., and WEISS, H. Experimental Study of the Photon Energy Spectrum of Primary Diagnostic X-Rays. *Phys Med Biol* vol 2, p 225, 1966.

[A4] EPP, E.R., and WEISS, H. Spectral Fluence of Scattered Radiation in a Water Medium Irradiated with Diagnostic X-Rays. *Radiation Research*, vol 30, p 129, 1967.

[A5] HETTINGER, G., and STARFELT, N. Bremsstrahlung Spectra from Roentgen Tubes. *Acta Radio*, vol 50, p 381, 1958.

[A6] ICRU Report 14, Radiation Dosimetry: X-Rays and Gamma Rays with Maximum Photon Energies Between 0.6 and 50 MeV. International Commission on Radiation Units and Measurements, 1969.

[A7] ICRU Report 17, Radiation Dosimetry: X-Rays Generated at Potentials of 5 to 150 kV. International Commission on Radiation Units and Measurements, 1970.

[A8] KATHREN, R.L., LARSON, H.V. and RISING, F.L. K-Fluorescence X-Rays: A Multi-Use Tool for Health Physics. *Health Physics*, vol 21, p 285, 1971.

[A9] LARSON, H.V., MYERS, I.T. and ROESCH, W. C. Wide-Beam Fluorescent X-Ray Source. *Nucleonics*, vol 13, p 100, 1955.

[A10] LAVENDER, A., THOMPSON, I. M. G. SHIPTON, R.G. and GOODWIN, J. Modification of the BNW-250 kV X-Ray Set and the Recalibration of Its Output. Report RD/B/N1263, 1969.

[A11] PLAYLE, T. S. A 250-kV X-Ray Set for Dosimeter Calibration. Report RD/B/N213, 1964.

[A12] SHAMBON, A., and MURNICK, D. Filters to Provide Nearly Monoenergetic X-Rays. Report HASL-129, 1962.

[A13] SKARSGARD, L. D., and JOHNS, H.E. Spectral Flux Density of Scattered and Primary Radiation Generated at 250 kV. *Radiation Research*, vol 14, p 231, 1961.

[A14] STORM, E., and SHLAER, S. Development of Energy-Independent Film Badges with Multi-Element Filters. *Health Physics*, vol 11, p 1127, 1965.

[A15] VILLFORTH, J. C., BINKHOFF, R. D. and HUBBELL, H. H. Jr. Comparison of Theoretical and Experimental X-Ray Specta. Oak Ridge National Labs Munich: Report ORNL-2529, 1958.

[A16] NACHTIGAL, D. *Table of Specific Gamma-Ray Constants.* Munich: Verlag Karl Thiemig, KC, 1969.

[A17] LEDERER, C.M., HOLLANDER, J.M., and PERLMAN, I. *Table of Isotopes.* New York: Wiley, 1967.

[A18] *Nuclear Data*, Sections A and B. New York: Academic Press, 1966 to 1973. (Many issues used, and not separately referenced here.)

[A19] ICRU Report 12, Certification of Standardized Radioactive Sources. International Commission on Radiation Units and Measurements, 1968.

[A20] National Bureau of Standards, Design of Free-Air Ionization Chambers (Handbook 64). Washington, D.C.: U.S. Government Printing Office, 1957.

[A21] BOAG, J.W. Ionization Chambers. In *Radiation Dosimetry*, F.H. Attix et al., Eds., vol 2. New York: Academic Press, 1966, ch. 9.

[A22] JOHNS, H.E., and CUNNINGHAM, I. R. *The Physics of Radiology.* Springfield, Ill.: Charles C Thomas, 1971, ch. 7.

[A23] BOAG, J.W. Ionization Chambers. In *Radiation Dosimetry.* G.J. Hine and G.L. Brownell, Eds., New York: Academic Press, 1958, ch. 4.

[A24] GREENING, J.R., RANDLE, K.J., and REDPATH, A.T. The Measurement of Low Energy X-Rays, *Physical Medicine*, vol 13, p 359, 1968; vol 13 p 635, 1968; vol 14, p 55, 1969.

[A25] GUNN, S. R. Radiometric Calorimetry — A Review. University of California Research Lab Report UCRL-50173, 1967.

[A26] KATHREN, R. L., and LARSON, H. V. Radiological Calibration and Standardization for Health Physics. *Health Physics*, vol 16, p 778, 1969.

[A27] DeCHOUDENS, H. Calibration of Certain Portable Detectors for Measuring Beta Doses by Comparison with an Extrapolation Chamber. Center for Nuclear Studies, Grenoble, France, Report NP-17062, 1965.

[A28] ATTIX, F. H. Electronic Equilibrium in Free Air Chambers and a Proposed New Chamber Design. Naval Research Labs, Report NRL 5646, 1961.

[A29] GOLDEN, R., and TOCHLIN, E. Characteristic Curves from Different Ionizing Radiations and Their Significance in Photographic Dosimetry. *Health Physics*, vol 2, p 199, 1959.

[A30] Technical Reports Series 133, Handbook on Calibration of Radiation Protection Monitoring Instruments. International Atomic Energy Agency, 1971.

[A31] Report 13, Neutron Fluence, Neutron Spectra, and Kerma. International Commission on Radiation Units and Measurements, 1969.

[A32] AXTON, E.J., and CROSS, P. The Establishment of an Absolutely Calibrated Neutron Source. *Journal of Nuclear Energy*, A and B, vol 15, p 22, 1961.

[A33] DANYSZ, M., and WILHELMI, Z. A Method of Determining the Efficiency of Ra-Be Neutron Sources. *Acta Phys. Pol.* vol 11, p 71, 1951.

[A34] FIELDS, P.R., and DIAMOND, H. Californium-252, A Primary Standard for Neutrons. *Proceedings of the Symposium on Neutron Dosimetry*, (Harwell, England, 1962); IAEA, Vienna, Austria, 1963.

[A35] GEIGER, K.W. Recent Improvements in the Absolute Calibration of Neutron Sources. *Metrologia*, vol 4, no 8, 1968.

[A36] FOWLER, J. L., and BROLLEY, J. E., Jr. *Review of Modern Physics*, vol 28, p 103, 1956.

[A37] MARION, J.B., and FOWLER, J.L. Fast Neutron Physics, Part I. New York: Wiley-Interscience, 1960.

[A38] NACHTIGAL, D. Average and Effective Energies, Fluence-Dose Conversion Factors and Quality Factors of the Neutron Spectra of Some (α,n) Sources. *Health Physics*, vol 13, p 213, 1967.

[A39] NBS Handbook 72, Measurement of Neutron Flux and Spectra for Physical and Biological Applications. National Committee on Radiation Protection and Measurements, 1960.

[A40] NOYCE, R.H., MOSBURG, E.R. Jr, GARFINKEL, S.B., and CASWELL, R.S. *Reactor Science Technology*, vol 17, p 313, 1963.

[A41] Reactor-Physics Constants, 2nd ed. USAEC Division of Technical Information, Argonne National Labs, Report ANL-5800, 1963.

[A42] AMALDI, E. The Production and Slowing Down of Neutrons. In *Encyclopaedia of Physics 38/2.* Berlin: Springer-Verlag, 1959.

[A43] HYDE, E. K. *The Nuclear Properties of the Heavy Elements, Volume III, Fission Phenomena.* Englewood Cliffs, N.J.: Prentice-Hall, 1964.

[A44] HUGHES, D. J. *Pile Neutron Research.* Cambridge, Mass: Addison-Wesley, 1953.

[A45] AUXIER, J.A. The Health Physics Research Reactor. *Health Physics*, vol 11, p 89, 1965.

[A46] CASWELL, R. S. Review of Measurements of Absolute Neutron Emission Rates and Spectra From Neutron Sources. Presented at the Conference on Neutron Sources and Applications, Savannah River Lab, April 1971, USAEC Report CONF-710402, p I-53.

[A47] HANSON, A.O., and McKIBBEN, J.L. A Neutron Detector Having Uniform Sensitivity from 10 keV to 3 MeV. *Physics Review*, vol 72, p 673, 1947.

[A48] DePANGHER, J. A Reproducible Precision Long Counter for Measuring Fast Neutron Flux. *Bulletin of the American Physical Society*, vol 6, p 252, 1961.

[A49] DePANGHER, J., and NICHOLS, L. L. A Precision Long Counter for Measuring Neutron Flux Density. Report BNWL-260, 1966.

[A50] CLARKE, R.W., LAVENDER, A., and THOMPSON, I.M.G. Experience Gained in Operating a Dosimeter Calibration Facility. *Health Physics*, vol 13, p 73, 1967.

Index

Index entries that appear in boldface type can be found in volume 1. Entries in lightface type are in volume 2.

A

B

Q

U

V

W

X

Z